The Genetics of African Populations in Health and Disease

The birthplace of modern humans, Africa has the highest genetic diversity in the world, yet it remains vastly understudied. With biomedical research increasingly focused on human variation, studying the large population size and number of mutations in African genomes could unravel the complexity of phenotypic traits underlying the biology of our species and hold huge potential for scientific and medical advances. An initial chapter "conceptualizes Africa," providing relevant terminology. The first part of the book covers genetic history and population structure. The next part looks at the genetic basis of common infectious diseases, such as leishmaniasis, malaria, and tuberculosis, with a final part considering common non-communicable diseases, such as diabetes, hypertension, obesity, heart disease, and cancer. Gene–environment interaction under globalization and the burden of diseases of lifestyle are included. This book is suitable for researchers and graduate students in biological anthropology, genetic anthropology, human and population genetics, and public health.

Muntaser E. Ibrahim is a professor at the Department of Molecular Biology, Institute of Endemic Diseases at the University of Khartoum, Sudan. He established the Unit of Diseases and Diversity at the university in 1998. He is a founding member of the African Society of Human Genetics.

Charles N. Rotimi is the Chief of the Metabolic, Cardiovascular and Inflammatory Disease Genomics Branch and the Director of the Center for Research on Genomics and Global Health at the National Human Genome Research Institute (NIH), MD, USA. He was founding president of the African Society of Human Genetics and spearheaded formation of the H3Africa Initiative.

Cambridge Studies in Biological and Evolutionary Anthropology

Also available in the series

53. *Technique and Application in Dental Anthropology* Joel D. Irish and Greg C. Nelson (eds.) 978 0 521 87061 0
54. *Western Diseases: An Evolutionary Perspective* Tessa M. Pollard 978 0 521 61737 6
55. *Spider Monkeys: The Biology, Behavior and Ecology of the Genus Ateles* Christina J. Campbell 978 0 521 86750 4
56. *Between Biology and Culture* Holger Schutkowski (ed.) 978 0 521 85936 3
57. *Primate Parasite Ecology: The Dynamics and Study of Host–Parasite Relationships* Michael A. Huffman and Colin A. Chapman (eds.) 978 0 521 87246 1
58. *The Evolutionary Biology of Human Body Fatness: Thrift and Control* Jonathan C. K. Wells 978 0 521 88420 4
59. *Reproduction and Adaptation: Topics in Human Reproductive Ecology* C. G. Nicholas Mascie-Taylor and Lyliane Rosetta (eds.) 978 0 521 50963 3
60. *Monkeys on the Edge: Ecology and Management of Long-Tailed Macaques and their Interface with Humans* Michael D. Gumert, Agustín Fuentes and Lisa Jones-Engel (eds.) 978 0 521 76433 9
61. *The Monkeys of Stormy Mountain: 60 Years of Primatological Research on the Japanese Macaques of Arashiyama* Jean-Baptiste Leca, Michael A. Huffman and Paul L. Vasey (eds.) 978 0 521 76185 7
62. *African Genesis: Perspectives on Hominin Evolution* Sally C. Reynolds and Andrew Gallagher (eds.) 978 1 107 01995 9
63. *Consanguinity in Context* Alan H. Bittles 978 0 521 78186 2
64. *Evolving Human Nutrition: Implications for Public Health* Stanley Ulijaszek, Neil Mann and Sarah Elton (eds.) 978 0 521 86916 4
65. *Evolutionary Biology and Conservation of Titis, Sakis and Uacaris* Liza M. Veiga, Adrian A. Barnett, Stephen F. Ferrari and Marilyn A. Norconk (eds.) 978 0 521 88158 6
66. *Anthropological Perspectives on Tooth Morphology: Genetics, Evolution, Variation* G. Richard Scott and Joel D. Irish (eds.) 978 1 107 01145 8
67. *Bioarchaeological and Forensic Perspectives on Violence: How Violent Death is Interpreted from Skeletal Remains* Debra L. Martin and Cheryl P. Anderson (eds.) 978 1 107 04544 6

68. *The Foragers of Point Hope: The Biology and Archaeology of Humans on the Edge of the Alaskan Arctic* Charles E. Hilton, Benjamin M. Auerbach and Libby W. Cowgill (eds.) 978 1 107 02250 8
69. *Bioarchaeology: Interpreting Behavior from the Human Skeleton, 2nd Ed.* Clark Spencer Larsen 978 0 521 83869 6 & 978 0 521 54748 2
70. *Fossil Primates* Susan Cachel 978 1 107 00530 3
71. *Skeletal Biology of the Ancient Rapanui (Easter Islanders)* Vincent H. Stefan and George W. Gill (eds.) 978 1 107 02366 6
72. *Demography and Evolutionary Ecology of Hadza Hunter-Gatherers* Nicholas Blurton Jones 978 1 107 06982 4
73. *The Dwarf and Mouse Lemurs of Madagascar: Biology, Behavior and Conservation Biogeography of the Cheirogaleidae* Shawn M. Lehman, Ute Radespiel and Elke Zimmermann (eds.) 978 1 107 07559 7
74. *The Missing Lemur Link: An Ancestral Step in Human Evolution* Ivan Norscia and Elisabetta Palagi 978 1 107 01608 8
75. *Studies in Forensic Biohistory: Anthropological Perspectives* Christopher M. Stojanowski and William N. Duncan (eds.) 978 1 107 07354 8
76. *Ethnoprimatology: A Practical Guide to Research at the Human–Nonhuman Primate Interface* Kerry M. Dore, Erin P. Riley and Agustín Fuentes (eds.) 978 1 107 10996 4
77. *Building Bones: Bone Formation and Development in Anthropology* Christopher J. Percival and Joan T. Richtsmeier (eds.) 978 1 107 12278 9
78. *Models of Obesity: From Ecology to Complexity in Science and Policy* Stanley J. Ulijaszek 978 1 107 11751 8
79. *The Anthropology of Modern Human Teeth: Dental Morphology and Its Variation in Recent and Fossil Homo Sapiens, 2nd Ed.* G. Richard Scott, Christy G. Turner II, Grant C. Townsend and María Martinón-Torres 978 1 107 17441 2
80. *The Backbone of Europe: Health, Diet, Work, and Violence over Two Millennia* Richard H. Steckel, Clark Spencer Larsen, Charlotte A. Roberts and Joerg Baten (eds.) 978 1 108 42195 9
81. *Hunter-Gatherer Adaptation and Resilience: A Bioarchaeological Perspective* Daniel H. Temple and Christopher M. Stojanowski (eds.) 978 1 107 18735 1
82. *Primate Research and Conservation in the Anthropocene* Alison M. Behie, Julie A. Teichroeb and N. Malone (eds.) 978 1 107 15748 4
83. *Evaluating Evidence in Biological Anthropology: The Strange and the Familiar* Cathy Willermet and Sang-Hee Lee (eds.) 978 1 108 47684 3

The Genetics of African Populations in Health and Disease

Edited by

MUNTASER E. IBRAHIM

University of Khartoum

CHARLES N. ROTIMI

National Human Genome Research Institute/NIH

CAMBRIDGE
UNIVERSITY PRESS

CAMBRIDGE
UNIVERSITY PRESS

University Printing House, Cambridge CB2 8BS, United Kingdom

One Liberty Plaza, 20th Floor, New York, NY 10006, USA

477 Williamstown Road, Port Melbourne, VIC 3207, Australia

314–321, 3rd Floor, Plot 3, Splendor Forum, Jasola District Centre, New Delhi – 110025, India

79 Anson Road, #06–04/06, Singapore 079906

Cambridge University Press is part of the University of Cambridge.

It furthers the University's mission by disseminating knowledge in the pursuit of education, learning, and research at the highest international levels of excellence.

www.cambridge.org
Information on this title: www.cambridge.org/9781107072022
DOI: 10.1017/9781139680295

First published 2020

Printed in the United Kingdom by TJ International Ltd. Padstow Cornwall

A catalogue record for this publication is available from the British Library.

ISBN 978-1-107-07202-2 Hardback

Contents

Contributors

Sonia Abdelhak
Laboratory of Biomedical Genomics and Oncogenetics, University Tunis El Manar, Tunis Belvédère, Tunisia

Sally N. Adebamowo
Department of Epidemiology and Public Health, University of Maryland, Baltimore, MD, USA

Adebowale A. Adeyemo
Center for Research on Genomics and Global Health, National Human Genome Research Institute, National Institutes of Health, Bethesda, MD, USA

Ammar E. M. Ahmed
Faculty of Medicine, University of Khartoum, Khartoum, Sudan

Khalid O. Alfarouk
Biomedical Research LLC, Temple Terrace, FL, USA

Ahlem Amouri
Department of Histology and Cytogenetics, Laboratory of Biomedical Genomics and Oncogenetics, University Tunis El Manar, Tunis Belvédère, Tunisia

Jennifer L. Baker
Center for Research on Genomics and Global Health, National Human Genome Research Institute, National Institutes of Health, Bethesda, MD, USA

Abdelhamid Barakat
Human Molecular Genetic Laboratory, Institut Pasteur du Maroc, Casablanca, Morocco

Endashaw Bekele
Human Genetic Variation Center, Addis Ababa University and Ministry of Health, Addis Ababa, Ethiopia

Amy R. Bentley
Center for Research on Genomics and Global Health, National Human Genome Research Institute, National Institutes of Health, Bethesda, MD, USA

Jenefer M. Blackwell
Telethon Kids Institute, The University of Western Australia, Australia

Mariem Bozguiya
Department of Dermatology, Ben Ghazi Hospital, Ben Ghazi, Libya

Lotfi Chouchane
Laboratory of Genetic Medicine and Immunology, Weill Cornell Medicine – Qatar
Division of Translational Medicine, Sidra Medical and Research Center, Doha, Qatar

Collet Dandara
Division of Human Genetics, Department of Pathology & Institute of Infectious Disease and Molecular Medicine, University of Cape Town, Cape Town, South Africa

Ahmed M. Elhassan
Pathology Institute of Endemic Diseases, University of Khartoum, Medical Campus, Khartoum, Sudan

Ghada El-Kamah
Clinical Genetics, National Research Centre, Cairo, Egypt

Liena E. O. Elsayed
Department of Biochemistry, University of Khartoum, Khartoum, Sudan

Kamal Hamed
Faculty of Medicine, University of Khartoum, Khartoum, Sudan

Ahmed Houmeida
Unité de Recherche sur les Biomarqueurs dans la Population Mauritanienne, Nouakchott, Mauritania

Muntaser E. Ibrahim
Department of Molecular Biology, Institute of Endemic Diseases, University of Khartoum, Khartoum, Sudan

Rym Kefi
Laboratory of Biomedical Genomics and Oncogenetics, University Tunis El Manar, Tunis Belvédère, Tunisia

Shomarka Omar Y. Keita
Department of Anthropology, Smithsonian Institution, Washington, DC, USA
National Human Genome Center, Howard University, Washington, DC, USA
Department of Anthropology, American University, Washington, DC, USA

Alice Matimba
Advanced Courses and Scientific Conferences, Wellcome Genome Campus, Hinxton, UK

Olfa Messaoud
Laboratory of Biomedical Genomics and Oncogenetics, University Tunis El Manar, Tunis Belvédère, Tunisia

Hiba S. Mohamed
Department of Biology, Taibah University, Medina, Kingdom of Saudi Arabia
Institute of Endemic Disease, University of Khartoum, Khartoum, Sudan

Sulma I. Mohammed
Department of Comparative Pathobiology, Purdue University,
West Lafayette, IN, USA

Mourad Mokni
Service de Dermatologie et Unité de Recherche sur les Troubles Héréditaires de Kératinisation, Hôpital La Rabta, University Tunis El Manar, Tunis, Tunisia

Melanie Newport
Brighton and Sussex Medical School, Brighton, UK

Mohamed Othman
McGill University, Montréal, Canada

Lilia Romdhane
Department of Life Sciences, Laboratory of Biomedical Genomics and Oncogenetics, University Tunis El Manar, Tunis Belvédère, Tunisia

Charles N. Rotimi
Center for Research on Genomics and Global Health, National Human Genome Research Institute, National Institutes of Health, Bethesda, MD, USA

Konduru S. Sastry
Laboratory of Genetic Medicine and Immunology, Weill Cornell Medicine - Qatar
Division of Translational Medicine, Sidra Medical and Research Center, Doha, Qatar

Daniel Shriner
Center for Research on Genomics and Global Health, National Human Genome Research Institute, National Institutes of Health, Bethesda, MD, USA

Giovanni Stevanin
Institut du Cerveau et de la Moelle épinière and Ecole Pratique des Hautes Etudes, Sorbonne and PSL Research Universities, Paris, France
Pitie-Salpetriere Hospital, Paris, France

Neji Tebib
Department of Pediatrics, La Rabta Hospital, University Tunis El Manar, Tunis Belvédère, Tunisia

Fasil Tekola-Ayele
Center for Research on Genomics and Global Health, National Human Genome Research Institute, National Institutes of Health, Bethesda, MD, USA

Afaf Tiar
Laboratory of Biomedical Genomics and Oncogenetics, University Tunis El Manar, Tunis Belvédère, Tunisia

Ambroise Wonkam
Division of Human Genetics, University of Cape Town, Cape Town, South Africa

Mohamed Zghal
Service de Dermatologie Hôpital La Rabta, University Tunis El Manar, Tunis, Tunisia

1 Reflections on Conceptualizing Africa for Biological Studies with a Historical Component

A Small Essay

Shomarka Omar Y. Keita

1.1 Prolegomenon

Pliny the Elder said, to paraphrase, that there was always "something new out of Africa." The place that was Africa for Pliny and the Romans was in supra-Saharan and northern Saharan Africa, specifically a region that encompasses the current state of Tunisia, the location of the ancient city-state of Carthage. By Pliny's time, Carthage and the surrounding territories were a colonized region, specifically a Roman (and Romanized) province (called Africa). During the Islamic period, the name Africa passed into Arabic as *Ifriqiya*, and the same region was known by this name for a millennium until it began to be called Tunis. The Ottomans continued to use this name. During French colonial occupation this "protectorate" was called Tunisia and the name was kept after independence. In many instances, Pliny's Africa, as Africa, has all but been forgotten in the conceptualization and naming of Africa, an irony of sorts – given that it was an Amazigh king from this region, Massinissa, who first said "Africa for the Africans," in the heady days of various interactions within and between the region and Rome.

Today, Africa designates an entire continent, although sometimes its supra-Saharan and northern Saharan regions, ironically including Tunisia, are often "forgotten" to be in Africa in various kinds of studies and disciplines. Or it may be seen as Africa minor or in the case of Egypt being called erroneously an "isthmus"(!) (Schuenemann et al. 2017). Africa is often equated with "sub-Saharan," which is the equivalent of equating Asia only with far eastern Asia and ignoring Arabia, India, and the Central Steppe regions, or equating Europe and Europeans only with Northern Europe (as being the only authentic Europe), in both cases excluding the rest of these regions. (Supra-Saharan and Saharan are used to have symmetry with the often used "sub-Saharan," with infra-Saharan being a better term because it is free of ideological baggage related to the colonial enterprise.) Other descriptive terms have no equivalent analogs; for example, South Africa is not the counterpart of North Africa and East Africa and West Africa are not equivalents. There is also the designation "Black Africa," often used as a synonym for sub-Saharan, which is problematic and has no parallels with other continents. Tropical Africa, as another term, is interesting in this regard because it includes a part of Egypt and parts of the Sahara desert of multiple Maghreban states, and renders the notion of "tropical" as a denotation only of forested regions as wrong. (Tropical Africa lies between the two tropics and offers the possibility of terms like infra-tropical and supra-tropical.) It is possible to construct various divisions of interest based on various criteria and relevance to particular studies. There is no one way to be African: No claim is made that Africa and Africans are uniform or homogeneous across space and time; it follows that there are and have been many African diasporas generated by voluntary and involuntary migration. There are also internal African diasporas.

The "forgetting" of Africa's northern regions as Africa, although not as common in some scholarship as it once was, has something to do with a set of notions born in the era of European-sponsored human trafficking and enforced settler colonialism. These include ideas about "race" and origins, notions of isolated populations, beliefs about static ethnicities and unchanging peoples, typological thinking, racism, the sociology of knowledge, and nothing to do with geography and interconnected human beings as subjects with value and agency. It is also related to the creation of ranked social identities inscribed onto biology based on racial schema (see e.g., Smedley 2007). This perspective crystallized during the antebellum and official colonial periods and often continues to influence various kinds of scholarship and political and social policy in some countries. The observation that "Africa" in practice is often restricted to its "tropical" regions has led to the question "Where in Africa does Africa start?" (see MacEachern 2007). As noted, what is "tropical" by the map may not be the same as preconceived concepts associated with the term. The issue of what is "African" and what is not becomes interesting in discussions of human biological, cultural, and linguistic origins and diversity, and would influence the collection and interpretation of data, including genetic data. As noted, the issue of African (and descendant) identity exists in some political administrative and social identity contexts (as in the US Census or even requirements for certain research, where some policies define supra-Saharan and some northern Saharan Africans out of Africa in grand racialist tradition). These identity issues will not be discussed here in any depth, but they cannot be avoided because they influence how science is done in terms of sampling and interpretation.

There is a larger set of philosophical issues about the notion of any of the continental/communal "identities": when do they start, what is to be their basis, what must their time depth be, are there notions of ancestral purity embedded in certain perspectives, and who gets to decide these questions? Populations are known to have changed religion and/or language, and acquired ancestry from other groups; sometimes identities have changed in the process. There is also the issue of the relationship of more specific bioethnic ancestry to social and/or cultural identity at any moment in history. This is most easily illustrated by using recent populations. For example, having some Greek ancestry (being of Greek descent) does not mean that someone would be or is necessarily *Greek-identified*: this would depend on social and political contexts, and what gets emphasized or is even known about ancestry. There are numerous other examples of this. As is well known, many if not most Latin Americans whose families date to the nineteenth century and before have some Native American and/or African ancestry but are not specifically part of *communities* that are labeled (by the state) or self-identify as "*indigenas*" or are *"Afro"-identified* – the non-European ancestries are dispersed in the general population, held in varying degrees by different individuals and families. Likewise, many people in the *communities* identified as Indian or "African descendant" have varying degrees of European ancestry, usually due to the actions of males during the conquest and enslavement periods. Many of those now identified sociopolitically as "African Americans" (in this case meaning identified as Afro-North Americans of middle passage descent, or legacy African Americans) have European ancestry, but are not socially identified as belonging to a European ethnic group ("European descent") – even if one parent is "solely" of European descent (in the recent sense). There are groups that are known to have and may self-acknowledge multiple ancestries, such as the Melungeons, Brass Ankles, and Nanticoke Moors. These examples from the "New World" can be taken to illustrate interactions that have taken place elsewhere and over millennia. The connotated issues of classification have implications for science, medicine, and history. Ethno-ancestral identities do not project infinitely back in time.

This non-exhaustive essay, with limited references, addresses some issues related to evolution, the race concept, and social organization in the conceptualization of Africa and Africans, and the nature and sources of African biological diversity against cultural and geographical backdrops. There is an awareness here of the lack of significant participation of African scientists and philosophers in much of the human biology and other research done in the human sciences on Africans: from paleontology to social anthropology to genetics/genomics. An effort will be made to present some different perspectives that could be brought to bear on existing data and new studies. No explicit evaluation of research programs such as the 1000 Genomes Project is undertaken here. The bias here is toward an evolutionary/biogeographical perspective about African human biology informed by history and culture, and the history of ideas. The very notion of "population history" becomes of interest in this regard. The concept of an *emergence* of complex biocultural entities is taken as a given in contrast to more traditional notions that imply units of purity. The perspective taken here is that the locale of biocultural history makes the subject of Africa – that any migrations into Africa from outside at any time in which migrants have been largely assimilated has led to populations whose biology has been reworked by an African biocultural environment. (Excluded are recent migrants from the colonial period or endogamous immigrant communities during the last 500 years.) An accurate biogeography is needed, one based on a constant dialoguing with data and theory. Populations are processes and ensembles, assemblages, not akin to monoliths consisting of uniform particles of a single element that emerged one time. It is important to remember, all other things being equal, that over time a biologically defined group could change cultural practices and/or language (by adopting new customs or languages), and that a culturally defined group can change biology (via intermarriage), both only limited by the sociopolitical milieu of the interactions. There are many variations of this interplay of culture, ancestry, and identity, which has implications for studies. The philosophy of science has something to offer here as well: how populations are delimited is often a product of "tradition" or subjective parameters, and *not mind-independent entities* (Gannett 2003). This has relevance to a range of studies.

1.2 "Blackness" and Notions of Africa

The stereotypical notion of *African* in recent times, in phenotypical terms, is usually linked to the trait of dark skin color, specifically that of being "black" (maybe beginning at dark brown). In fact, "black African" – the problematic common phrase of use – is often if not always made synonymous with "African" in certain literature. "Black" and "African" are often (incorrectly) used interchangeably, with *African* having a particular cultural content thought of as being indigenous in connection to a conception that makes those so identified part of an entity that was conceived to be isolated from the rest of the world. *African*, in this case, would seem to be used to connote living peoples with a particular phenotype, whose bodies collectively are the repositories of a particular genome. This notion of "blackness" derives from the received race concept and its taxonomic schema modulated through a tradition of stereotypes. The irony here, of course, is that the region that encompasses so-called "black Africa" is the same place that contributed heavily to the emergence of our species, which was likely assembled with ancestries from supra-Saharan/Saharan, eastern, and southern infra-Saharan regions.

"Black" has been and is used in other Western discourses as a referent to various non-African peoples, specifically certain of those in the Pacific region and Asia. It has its own subjectivity: How dark must skin be to be called "black"? It is different in different places. Australian Aborigines, Dravidian Indians, New Guineans, and Melanesians, who are also dark or "black" in color, were all called "black" in the context of colonialism. "Black" has sometimes been used to mean simply non-European and non-Far East Asian in the British context; Professor Leila Ahmed, a light-skinned Egyptian, noted being called black when she was a student in England in the 1960s (Ahmed 2000), a practice now discontinued due to a "refinement" of social taxonomies in England (or maybe the objections of some groups who used to be called black). It is well to note that in ancient northern Africa, the Africa of Roman times, there was a variation in the phenotype of people (Snowden 1970; Thompson 1989); this was also true of the northern Nile Valley.

In the early Islamic period, there are some interesting ironies related to notions of "blackness." Ubadah ibn as-Samit, the appointed envoy of Amr, commander of the "Islamic" or "Arab" forces that entered Egypt, was initially rejected by the Roman prelate apparently *because* he was "black" (Butler 1902: 257). The "Arabs" of that era told the prelate that they held "blacks" and "whites" in equal respect and that Ubadah was one of the best among them. Ubadah informs the prelate:

> There are a thousand blacks, as black as myself, among our companions. I and they would be ready each to meet and fight a hundred enemies together. We live only to fight for God, and to follow His will. We care nought for wealth, so long as we have where withal to stay our hunger and to clothe our bodies. This world is nought to us, the next world is all. (Butler 1902: 257)

This passage indicates that Ubadah feels himself to be a part of this community, in this era Islamic scholars consider to be Islam's Golden Age, and one of equality, an equality that would later change with various developments, including some level of bias against the darkest skin colors in some regions (Evans 1980), as well as other biases. (Of course, "black" and "white" are relative terms, and socially modulated.) Kamil (1970) documents numerous "blacks" who were a part of the early Islamic community, and even regarded as heroes, some apparently buried by the Prophet Mohammed with his own hands. And there were the "black" pre-Islamic notables, like Antar, and later Islamic figures like al-Jahiz who have down through history been called "black" or some equivalent. These cannot be overlooked. Statistically the numbers of those who became notable/remembered is suggestive that there was an even larger population. How many of these Arabian "blacks" and their families moved into supra-Saharan and Nile Valley Africa is unknown. It is important to say that they were not slaves.

What were the geographical origins of Ubadah and his other "black" companions and/or their ancestors? Was there an indigenous "black" population in early Arabia but Asian in origin – like the Dravidians – or a population that straddled eastern Africa and Arabia (quite possible), or early Holocene "prehistoric" migrants "that participated in the building of the primitive [sic] Arabian population" (as per Abu-Amero et al. 2009)? Were they the descendants of eastern African converts who returned to Arabia with the early *umma* (community) that had been sent to Ethiopia by the Prophet for safety in the earliest days of Islam? Or were they descendants of other known eastern Africans (slaves or migrants) resident in Asia – or solely historic slave descendants – a favorite trope of some researchers, or some composite of these?

The example of Ubadah and others in early Arabia negates a notion of "blackness" equaling Africa in the way that this is often simplistically done. Yet this view lingers across disciplines and in popular discourse. Of course, it would be of interest to understand how a notion of ethnicity fits into this early history. It is also of interest to understand how and when anti-"black" prejudice emerged in the Arabo-Islamic world, breaking from the early egalitarian tradition, given that some "Arabs" sometimes called themselves "black" at least in comparison to more northern peoples (Evans 1980, after Vollers 1910, footnote, page 24). Such prejudice was not apparently developed in the same way or degree that occurred in the West, and had no effect on some Arab groups as they migrated into tropical areas of Africa, with the men marrying local women or taking them as concubines. In other cases, whole groups merged with indigenous populations. It is beyond the scope of this chapter to explore this issue further, but it is clear that received notions of race and racial schema are highly problematic. It is important to say that all indigenous/endogenous African populations are not homogeneous in appearance.

The Islamic period is often presented as having absolutely changed the northern regions, somehow making that history not a part of an African world, or totally changing its people. Some view the migration that occurred as a simple massive settler colonialism of Arabs – or those composing the early Islamic migrants. The reality is far more complex than this.

1.3 Raciotypological Thinking

The stereotypical idea about Africans and Africa as conceptualized in racial thinking in the received racial schema has had an influence on the interpretation of morphological and molecular data. This has come from a model of raciotypological thinking applied to human populations in general (Keita and Kittles 1997). Races were conceptualized as foundational or fundamental units and conceived as being not only represented but defined by an idealized narrow set of physical characteristics, otherwise called a type. A type, as a concept, could also be based on gene frequencies or specific lineages. Variation from the type was interpreted to be evidence of admixture or an aberration. In this way of thinking, names imply very discrete groups or individuals which are different in kind, and seen as being only distantly connected, and sometimes in the nineteenth century as being of different species. Even when the larger racial schema are not used, there seems to be a tendency to think in terms of discrete populations consisting of uniform individuals, which were sometimes described as types (such as Nordic or Alpine). In real breeding populations, however, there is variation at the molecular and other levels (although perhaps restricted phenotypically to the eye), and types are abstractions, although extreme variants may be real if not actually representative. Traits or trait complexes were conceptualized as having necessarily a single origin. In fact, individuals and populations may have multiple genealogical connections (to other populations), even if they are not very genetically similar overall; some of these may be ancient, reflecting a deep time relationship due to shared ancestry or ancient gene flow, others may be more recent due to recent gene flow. Assumptions that generalize to regions from one or two populations (or individuals) – as in some discussions of gene flow – are merely versions of racial thinking. The use of samples from single populations like the Yoruba to represent "the African" or even infra-Saharan African or tropical African is not defensible on theoretical grounds because it ignores the heterogeneity of Africa. It is congruent with notions of the "Negro" "type" only as African. That this idea persists in various work reflects something about the sociology of knowledge, not a critical evolutionary science.

The results of analyses of groups intended to portray population history or historical processes are often expressed in dendrograms or plots, which are acceptable as long as their limitations are understood and disclosed. The results so expressed are intrinsically attractive, but are incomplete without a discussion of process and interaction, the sources of similarity (including micro-convergence or parallelism), and the problems of assuming that all data are equally good for making comparisons. Misleading results could occur if various kinds of data are not critically handled. Care has to be taken when describing or interpreting patterns of culture, biology, and language. Attaching genes to current speakers of language families in an obligatory manner under the assumption that in earlier history the speakers of the ancestral language family were the sole ancestors of those currently speaking the descendants of the language is problematic and wrong. Attempting to classify languages based on the genetics of their current speakers is in principle wrong. The splits within a language phylum are not causative of biological population splits based on mutations or natural selection and vice versa, although they may sometimes track with each other in some specific circumstances if geographical space is involved – but social customs and endogamy can be isolating as well. Early population history cannot be assumed to be a story of entities with obligatorily linked genes, languages, and cultures. Even strictly in terms of biology, population history is not simply gene history or vice versa (Weiss and Long 2009). This is an area where critical scholarship is very important.

There is also the issue of the context of interpretation of data: When should a given discipline lead the narrative? Does one look at genetics first to interpret a cultural or historical pattern, or does one look at chronologies or patterns determined by archaeology, history, or language, and use genetics within? Scale and time depth are important to this question, as well as the kind of narrative that is of interest. It cannot be said enough that gene history is not population history, and population history is not to be equated with social, political, or cultural history. Furthermore, how and why genes "flow" across space and populations cannot be answered by the genes themselves. The factors that spur migration are gene-independent. The social interactions that permit gene flow are not a property of the genes themselves. How we use genetics in developing a narrative of a specific people or kingdom is an important open question, and not a question at all for some scholars. For example, can genetics/genomics teach us anything about Funj *history* as opposed to their ancestral "identity" in a narrow, particular kind of way? And would the genetics give us the appropriate time depth to speak of the Funj as Funj specifically, as opposed to elements in their ancestry? This question about the ranking of data deserves ongoing critical examination. A naïve empiricism will not work. The errors of affirming the consequent and misplaced concreteness are to be avoided.

Another notion that influences how Africa is or has been conceptualized has had to do with stereotypes or perceptions of social organization. Africa, specifically "tropical Africa" in keeping with the stereotypical use, has been and is conceptualized as a place with numerous isolated peoples and/or "tribes" whose individuals have a putative common ancestry with a primordial origin. The idea of closed and isolated groups, in essence, bounded biocultural entities, has often been implicitly employed in much work. This will be further discussed later.

These ideas form a backdrop, an implicit mental template, to the way Africa is thought about, even when not explicitly stated. For example, published work on the "genetic history" of Africans often does not generally include the whole continent, or examine the interconnections of all of its regions, even though it is generally agreed that modern humans arose in Africa. It is not always clear what the concept of "genetic history" encompasses.

1.4 Africa: Evolution and Peopling

It has been postulated for some time that modern humans emerged in Africa (Stringer and Andrews 1988), perhaps 200,000–300,000 years ago by recent estimates (confidence intervals are large). By "modern" is meant a suite of characteristics (that permit an identification) in genetics, anatomy, and behavior (McBrearty and Brooks 2000) that relate living people to these distant ancestors. This emergence in Africa is believed by proponents of this theory to have been followed by migration within and out of Africa. The ensemble of biology and culture called "modern human" seems to have emerged in Africa from the interactions of populations from several regions, including the western supra-Saharan area. It is also possible that modern humans learned some forms of symbolic behavior from Neanderthals which speaks to a cognitive flexibility. Various scenarios and routes for the exodus from Africa and the locations for the emergence have been proposed. Interpretations of some genetic data suggest interactions leading to gene flow from other hominid populations, often called "archaic" and understood as being of different species, although it might be preferable to see all of these other populations as being subspecies. However, it may well be that some of the genes being labeled archaic are simply asymmetrical retentions from the ancestor that led to both modern humans and the other species or subspecies (Denisovans, Neanderthals, and possibly others in Africa), but this is not thought to be likely by most investigators. Modern humans may not represent a speciation event which would have largely "reset" aspects of the genome. Whatever the position taken on this matter of taxonomy, the narrative would seem to indicate the emergence of a core population or network of populations in Africa whose biology is predominant in all living humans. These early humans are postulated to have left Africa in a permanent or successful colonization of non-African regions between 40,000 and 70,000 years ago. (There is also some evidence of sporadic earlier outmigration as well.) This means that, between the time of emergence and migration, 100,000 years of biological (and cultural) history elapsed, making the early modern human story an African one; the full implications of this have not been explored in terms of biology or anything else. Serial migration and founder effects can be said to characterize subsequent human biological history, with new mutations being under adaptive and perhaps social selection augmenting differentiation along with gene flow and drift among local populations, and back-migration to Africa. A simple tree with separate non-interconnecting branches will not adequately display this history and production of variation. Rather, a range of ways to depict these can be envisioned.

During this long period, it is reasonable to postulate that modern humans had sufficient time to become differentiated at multiple levels. This differentiation is inferable from many datasets, although mutation rates are not known with certainty and confidence intervals are large. In this period of at least 100,000 years, it is possible that many lineages as represented by the uniparental markers emerged and became extinct, along with other markers, and even morphological traits. It must be said that this differentiation, be it in molecular or morphological traits, cannot be called raciation (Keita and Kittles 1997). Furthermore, the range of phenotypic variation is not completely known. Genes for skin color, such as light skin color, have been found to have multiple – including African – origins (Campbell, pers. comm.), and more of this kind of discovery should be expected. However, what is consistent with this long evolutionary history is that before modern humans left Africa they were highly variable, and likely achieved potential for even greater genomic plasticity. Gene flow from other subspecies or species of *Homo*, along with drift, serial founder effects, and adaptive selection in new environments would have contributed to variability, even though those who left Africa were only a subset of people existing at the time.

Fossils from this time period are not numerous. There is variability. There is evidence for modern anatomy or something near it, and modern behavior in northwest Africa (Bouzaggar et al. 2007), the Nile Valley (Van Peer et al. 2010), central, eastern, and southern Africa, as is well known. The Middle Stone Age of Africa is the time and context for the emergence of documentable modern humans from an African ancestor. Conceptually this geographical spread is important to understand because it negates the idea that early modern human developments were confined to southern and eastern Africa.

The oldest described mitochondrial and Y chromosome haplogroups found in living human populations emerged in Africa, but exactly where cannot be said: Current locations are not necessarily a guide to the locale of ancestral emergence due to migration. The lineages largely confined to Africa are also the earliest documented "modern human" lineages, now selectively called "African" since there are other lineages outside of Africa – although these too are derivative from African sources. Although there is DNA evidence of interbreeding with modern people, neither Neanderthal or Denisovan mitochondrial or Y chromosome lineages have been found in living people, and it seems that the Y haplogroup found in one population of living people (the Mbo) is apparently older than the age traditionally given to fossil remains designated as fully modern. This is not surprising (and it does not mean the Mbo are living fossils). Modern humans emerged from "pre-modern" populations in Africa, and by definition there is a connection no matter how invisible it may be. The recent findings that Neanderthals demonstrated symbolic behavior is not surprising, given the observations about other species. Whether or not "archaic" *Homo* like Neanderthals are actually another species or merely a subspecies (race) of *Homo* may be a matter of taxonomic preference as opposed to a matter of "nature," but obviously more research may be informative on this matter. It will be of interest to find ways to systematically explore the genome to look for evidence of disease genes or detrimental gene–gene interactions of non-"modern" origins. Regardless, Africa remains the place of the origin of the "source code" of living humans.

1.5 Human Variation in Africa

The molecular variation of humans in Africa has been said to be the greatest in the world (Armour et al. 1996; Elhassan et al. 2014; Jakobsson et al. 2008; Jorde et al.

1995; Schlebusch et al. 2012). This has been one element in the argument that modern humans emerged in Africa and is presented as supporting the fossil and/or symbolic evidence from the Horn, Maghreb, Congo, Nile Valley, and southern Africa.

It has been found that diversity in "sub-Saharan" Africa is great at the morphological, metric, and molecular levels. Descriptions, even in older literature, can be informative after wading through older or prejudicial language and interpretations. We can use the writers of older literature as informants for descriptions, not interpretations, even though descriptions may also be problematic. There was great diversity that was observed in the past. The explanation for this diversity is another matter. In terms of morphology, this may be somewhat surprising for those who may have a stereotyped idea of what constitutes the "real" African phenotype. The late Jean Hiernaux (1975: 53), with a critical eye, noted that in one region of east to central Africa there is a "wide range of population means or frequencies" for "many anthropological characters":

> In some of them, the whole of the world range is covered in the sub-continent. Here live the shortest and the tallest human populations, the one with the highest and the one with lowest nose, the one with the thickest and the one with the thinnest lips in the world. In this area, the range of average nose widths covers 92 percent of the world range: only a narrow range of extremely low means are absent from the African record. Means for head diameters cover about 80 percent of the world range; 60 percent . . . the cephalic index.

He noted this long before the surge in molecular data. Hiernaux (1975: 53–54) indicates that there is also great diversity in dermatoglyphics, hair form, skin color, aspects of cranial form, and serological traits in tropical Africa. When the supra-Saharan region is considered, this diversity increases. The observations made by Hiernaux about tropical African morphological variation were not unknown to some Westerners before the late twentieth century. After a description of the extreme morphological variant or the ensemble of traits called in some literature the true Negro or "Negro type," one writer from the nineteenth century noted that:

> No single tribe, however, possesses all these deformities (sic). The colour of the skin passes through every gradation from ebony black . . . to . . . copper coloured . . . As to the skull in many tribes . . . the jaws are not prominent, and the lips not swollen . . . In some tribes the nose is pointed, straight, or hooked; even Grecian profiles are spoken of, and travelers say with surprise that they cannot perceive anything of the so-called negro type among the negroes. (Peschel 1888: 463)

Elsewhere this variability is observed, again with some quaint language:

> What is called the "negro type" – that is the low type of the coast of Guinea – is comparatively the exception – perhaps as much so as the low Irish type in the Keltic race . . . there are jet black negroes with woolly hair and noble Aryan features . . .; or black people with frizzled and even straight hair and regular features . . .; or blacks with flowing hair in ringlets . . . or light brown people with woolly hair and European forms and face . . . or light brown with negro features. (Brace 1863: 308–309)

We can ignore the use of the term "low" as a subjective notion of uncertain meaning to current readers, but with likely negative connotations at the time of its writing.

Others also present similar descriptions, which are sometimes accompanied by aesthetic valuations, and also the idea that only one physical type in terms of craniofacial traits is truly African, thus the emergence of the root of an (incorrect) idea that sometimes still haunts ontological conceptions about Africa and Africans.

The amount of diversity at the molecular level has been established and confirmed in tropical Africa in many studies. The morphological and molecular data contradict the idea that "sub-Saharan" denotes an entity consisting of uniform individuals, a type; this notion can be summed up as "the stereotyped idea that most of us have about what a 'true' African looks like" (Hiernaux 1975: 54). Perhaps the best example is the famous anatomist Petrus Camper's criticism of European painters' depictions of the Magus traditionally said to be of African origin (see Gould 1987). Camper, the inventor of the facial angle, said that Europeans had merely painted Europeans dark or black and gotten the details of the underlying bony structure wrong, i.e., the traits which later came to be declared the signature of the black or African were not present. For some later anthropologists and anatomists, these traits would be seen in combination as a marker of the "Negro" type later called "True Negro" (Seligman 1930, 1957), a concept which never had a surviving European or Asian analog in "science" even as a foil. There is no "True White" or "True Yellow" type or group that was ever consistently used in discussions – or even devised based on a reading of various older works. In other writings, "true" was equated with "pure." Camper was actually wrong in his criticism of artists in saying that they were inaccurately depicting the Magus as an "African" because he apparently assumed that all Africans looked a certain way (Korst 2008). However, he was correct in criticizing them for merely painting Europeans dark or black, because while similar in physiognomy, the eastern Africans are distinct on inspection and by craniofacial mensuration. The artists probably had not been exposed to the range of variability of Africans, even those in the tropics. However, as noted, other Europeans were aware of physiognomical variability in those having darkly pigmented skins.

The morphological variability in Africa has also been described in the twentieth century by researchers other than Hiernaux, usually in works that look at more local populations in various regions. The interest here is in the descriptions and not the taxonomies of the past or taxonomy in general, although it is interesting and sometimes useful; the issue of the interpretation of the origin of this variation is another matter. (Usually, this is rooted in racial or categorical thinking with simple admixture between foundational groups being the explanation for variation.) For example, Coon et al. (1950: 127) presented a description of a group they called "Hamites," which they restrict to parts of eastern Africa: "Some of the peoples of East Africa seems to be skeletally Mediterranean, but dark skinned." Other writers, following Seligman (1930, 1957), used the term "Hamite" to cover a range of peoples and phenotypes found primarily in eastern and northern Africa, below, in, and above the Sahara; in terms of morphology, Montagu (1960: 421) (following Seligman) stated:

> Among the Hamites skin color varies all the way from the white to black; head hair varies from straight to frizzly; the face is never prognathous; the lips vary from thick to thin, but are almost never everted as in the Negro; there is, in general, a paucity of body hair. The head is mesocephalic.

It is interesting that Montagu cited Seligman here because Montagu, an opponent of racism for most of his life, quotes in the same textbook the core cultural contention of Seligman's biased Hamitic hypothesis, namely that the Hamites (conceptualized as a distinct community of Asian immigrants) brought all noteworthy culture to Africa. It is puzzling that Montagu did not critique this, did not see the contradiction of this idea to his own beliefs, as well as interrogate its validity on numerous points. The Hamitic hypothesis resonated negatively in some colonized countries, even playing a role in the Rwandan genocide due to teachings about the groups there called Hamite or Hamitic-influenced. Ideas have consequences. In Rwanda there had not emerged a widely taught human science based on the work of critical endogenous scholars who could have countered teachings that divided the population based on a notion of superiority. The social context of science is not value-free.

In northern and to some degree eastern Africa there is/was diversity that one label denoting a category of uniformity will not cover if that is the purpose of the name. This very diverse grouping – all called "Hamites" – included but is not restricted to ancient Egyptians, Somali, Fulbe, and the diverse Berber speakers (Amazigh) encountered by the Phoenicians. What seemed to tie these peoples together for certain researchers was an elongated face with a narrow nose and face and orthognathous profile, irrespective of color or hair form. It was also assumed that these peoples spoke closely related languages, which is only true for some of them. The coastal northern region of Africa, like eastern Africa, has a history of known interactions between specific peoples. Berber speakers are often noted for sometimes having very pale skin and light eyes, but they can actually have a range of colors, hair types, and facial features, and this was more likely true in the past. Based on ancient writings, the populations there were diverse, including dark-skinned people, although today this has been perhaps obscured in recent memory both by the descendants of recent infra-Saharan and Saharan "blacks" who may be seen as strangers and "Others"; for example, the "Ethiopians" of the ancient Maghreb do not have an unbroken connection to a distinctly identifiable community that lives today – and in any case this would only be expected if those ancient "Ethiopians" and others had not been a part of the general community that merged over time, i.e., if there was discrimination or a particular self-isolating cultural tradition. The assimilation of European captives and others over the last 1000 years would have also had an impact.

The Nile Valley inhabitants, specifically the Egyptian fellahin and Copts, are described as being of medium height, brownish skin, with straight nasal bridges, a range of eye colors, lips of medium thickness, and medium stature, although there is much variation, with some Copts being notably lighter in skin color. It would be interesting to know what the attitudes of the ancients toward each other vis-à-vis this diversity were, versus later attitudes. Neither the Greeks and Romans nor the ancient Egyptians record commentary or ideology that can be called color prejudice, although there was conflict and no doubt ethnocentrism. The description of Hamites presented by Montagu, based on Seligman (1930, 1957), is a signal to any critical analysis that there is a problem. It is of historical interest that in Seligman's view the Hamites were "Europeans" from Asia who imparted all measure of interesting culture to Africa – except curiously farming to tropical Africa – and brought the "Hamitic" language family. (Seligman did acknowledge that others thought their origin was in Africa, but his ideas still have an unrealized influence in many ways, such as the idea that Afrasian languages are not African in origin.) The idea that "Hamitic" languages (now part of a group called Afrasian, Afrasan, or Afroasiatic) originated in Asia is not supported based on general linguistic approaches using the principles of greatest diversity and least moves, along with the location of the member of the family which seems to retain the most ancient features of the ancestral group. There is also other evidence: For example, the reconstruction of proto-Afrasan indicates that its speakers were hunters and gatherers, and zoological inventory includes ancestral donkeys – indigenous to northeastern Africa. It is also known that some of the groups phenotypically classified as Hamitic by Seligman do not speak Afrasian languages (e.g., Tibu, Fulbe, Tutsi).

What is the source of the biological variation? The old classic answer involved various migration theories that invoked admixture between different "races" and/or "types," and in the example of Hamitic hypothesis was frankly biased against those conceived as "indigenous" Africans; however, some researchers did invoke adaptation to explain at least some or even most of this diversity, at least in the tropics (e.g., Coon et al. 1950; Hiernaux 1975). It is probable that adaptation, gene flow, genetic drift, founder effect, and perhaps social custom chance have affected variation, but all in the context of African environments. It is known that with a small amount of one-way gene flow per generation into a population, with or without selection favoring the genes in question, that frequencies can change appreciably in several hundred years. It has been shown that the modern Egyptian population was affected by migration during the Islamic period (Pagani et al. 2015). It was also affected during earlier periods where there is evidence for marriage with Near Easterners and others. However, any ancient migrants from this era and before have been assimilated into the local populations subject to their social and physical environments such that the genes now reflect a local Nilotic population history. More interesting than the dissection of the genetics/genomics of the population or its individuals is the place of emergence of the synthesis and the circumstances of interaction. Other populations in the eastern half of Africa show evidence of gene flow, but only in some cases can this be traced to known ethnic movements from specific places. It is likely that genes flowed through networks of several populations and perhaps were affected by drift or adaptation.

There has not been a full discussion or perhaps even acknowlegment of the overlap in autosomal variation between the modern population(s) that left Africa and those that remained, i.e., the source population (analogous to the NRY marker M168). Autosomal variation in modern humans in Africa would have been regionally variable. Not all regions were likely represented in the emigration from Africa. The modern humans who left Africa were likely from a restricted region, e.g., from somewhere in the northeastern quadrant of Africa (from the Mediterranean to the Horn). That population would not have been autosomally the same as one in southern Africa, or Mendelian populations elsewhere. It can be asked why some of the autosomal variation that is called "Asian" or "Eurasian" when found in Africa is not actually African in origin. This should logically be the case given evolutionary history, and likely is so. This observation will change the characterization of aspects of the Saharan–supra-Saharan genome.

As noted, "Africa" is often restricted to "sub-Saharan" Africa, but there is no scientific reason to do so. There is fossil evidence for early or near-modern humans in Saharan and supra-Saharan Africa (Taramsa and Nazlet Khater in Egypt, Dar as Soltan, Morocco). The molecular data show that all of the populations/bioethnic groups of the region still have the evidence of indigenous African peoples that predate the known historical contacts with Near Easterners and Europeans (e.g., Phoenicians, Greeks, Romans, Hyksos, Persians, and the later diversity of Near Easterners during the Islamic period). One trope that needs to be interrogated intensely is that of the trans-Saharan slave trade – a favorite explanation for certain findings that is overdone. The example of Ubadah and his comrades has been mentioned. The early Holocene Sahara during its wet phase was resettled by diverse populations who interacted, and on desiccation their descendants migrated to circum-Saharan regions. This "wet" Sahara can be conceptualized as the site of a metapopulation in which populations fused, fragmented with, and experienced some extinctions as desertification proceeded. Migrants would have gone in various directions to circum-Saharan regions carrying a number of genetic variants. It is of some interest that the Benin haplotype of the sickle cell gene is found in western tropical Africa, the Maghreb, and the Nile Valley – specifically Egypt; a study of predynastic ancient Egyptian remains revealed the sickling gene – which was not further characterized, but the clinical description of the remains is possibly consistent with the Benin haplotype.

The Y chromosome biallelic lineages are perhaps the most interesting indicators of the connection of the regions of Africa. The majority of males in most regions of Africa bear E haplogroup lineages, defined by the M96 marker. The E haplogroup is either a majority or plurality in all large regions. They show that the regions of Africa are interconnected at a time depth *after* modern humans left Africa by E haplogroup Y chromosome lineages. The PN2 or P2 is "ancestral" to both the V38/M2 found in most of infra-Saharan tropical Africa and M215/M35 defined lineages, the latter found from eastern Africa up to Egypt over to Morocco. The E haplogroup family tree shows that the P2 descendant lineages are deep within the family tree, not at its base. It is interesting that the basal split in the family occurred in tropical Africa. The P2 "descendants" are the most numerous, although there are other terminal subclades within E. It bears repeating that the lineages that seem to be most common in supra-Saharan and infra-Saharan Africa have a common origin in having the P2 ancestor; this contradicts the foundation of the received racial model of Africa that implies no direct connection other than that of a shared ancient humanity, or that created by gene flow via voluntary or forced migration. It is not known when the E group reached its current distribution. Its defining marker is old, but some derived lineages are young, having likely arisen within the mid-Holocene, specifically P2/M35/M81 (M183) (Sole-Morata et al. 2017) which is found in very high frequencies in some Amazigh (Berber) communities and in others in the Maghreb. The young age of M81 raises interesting questions about the interactions of populations that would have led to its spread and about origins in regions that are no longer populated – what is that history and who were the subjects? This is the history that is of greatest interest: What were the social circumstances of the interactions over time that led to the widespread distribution of this lineage among Amazigh? M78, the brother clade to M81, is older and also found in the Maghreb, Horn, and Nile Valley, with migrants taking it out of Africa perhaps by sea to the Aegean during the late Paleolithic/Epipaleolithic.

The other African haplogroups in origin, namely A and B, are also found in eastern Africa, notably in Sudan and Ethiopia and scattered among others such as the Batwa (Central Africa) and Khoisan speakers in Namibia, southwest Africa. A and B are sporadically found in other regions of Africa, including the northern Nile Valley (Egypt), with an old A lineage having been found in northwest supra-Saharan Africa. In the border regions of Africa, haplogroup J is also found (Egypt, Ethiopia, Somalia), and it can be disputed when it arrived, although the era of grain agriculture has been suggested; given the migrations into northern Africa during the Phoenician and Islamic periods, a J contribution surely occurred at this time.

Haplogroup R1-V88, normally associated with Europe and the Near East, is found in high frequency in the central south Sahara/Sahel among Chadic speakers; this lineage likely came to Africa in a "back to Africa" migration. Its age is older than the Chadic language family and it is not a marker of that family's emergence or movement, which is an example of the layering of population effects. There is an irony associated with the R1 presence in these Chadic speakers that bears pointing out. There was resistance by some scholars to the Chadic family being a member of the Afroasiatic based on the external phenotype of its speakers (a stereotypical "African" variant). The fact that this particular population of Afroasiatic speakers – although the samples are limited – usually has a higher frequency of a "European/Asian" lineage than Berber speakers/more northern coastal peoples is the kind of "new" thing that Pliny was speaking about. It is indeed ironic that those with what can be called the stereotyped African facial morphology and very dark skin are precisely those with the European haplogroup, while others who phenotypically more resemble "Europeans" are often predominantly of the E haplogroup.

The R haplogroup is also found in notable frequencies in some other populations of Saharans, namely Nilosaharan-speaking Tibu (Toubou/Tubu), and also in a Sudanese sample of Fulani. None of this means that any of these specific *groups as formed biocultural entities* came to Africa from Asia, let alone Europe. The lineage has been transmitted through several populations. The most likely scenario is that when the bearers of R1 came into Africa they got spread into a region in or adjacent to the Sahara, and may have been spread around during the time of its greatest inhabitability, the period of the so-called "Green Sahara." This is when the Sahara could be seen as having a metapopulation consisting of groups belonging to various language families whose speakers did not live in a sealed world of exclusion, but migrated between groups, as was the case in precolonial Kenya (Ogot 2005). A random (not representative) sorting out of male and female lineages in a broadly panmictic environment likely explains differences in frequencies of male and female lineages due to various social and chance circumstances. In some cases it is likely that pastoralists absorbed each other or allowed intermarriage in more recent times, something known from the Beja in Egypt and Sudan who absorbed Arab males (Fogel 1995), thus adding a different Y chromosome lineage. There is the genesis of the Swahili peoples, whose ancestors developed a coastal and maritime culture that came to absorb many from the Indian Ocean world, and even to look outward away from the African hinterland. Then there is Madagascar, whose main language is Malagasy, whose genealogical background is about equally half eastern African and Asian (specifically Indonesian). The various scenarios resulting from bioethnic interactions of varying sorts deserves serious attention by African scholars who are perhaps more acutely aware of it. More examples will be given of interesting historical scenarios ripe for exploratory discovery science that considers the physical and social environments. It is also important to note that in all geographical regions there is a diversity of male lineages and this has a history that can be recovered.

In summary, the E haplogroup connects the regions of Africa, obviously to varying degrees. This observation, extracted from the data of various Y chromosome studies, negates the Saharan boundary. This was not the goal of the researchers who generated the basic data, but it is something that the generalist scholar interested in contextualizing information in the history of ideas would see and make important. The predominant lineages from the E haplogroup are P2 "descendants" who account for the lion's share of male descent in the whole of Africa. The E haplogroup most likely emerged in tropical infra-Saharan eastern Africa, as did all of its subclades. The best total evidence is consistent with the E-M96 marker emerging in eastern Africa in the Late Stone Age. Borrowing from the language of evolutionary process used in reference to speciation, but projected onto the infraspecific level, it can be said that eastern Africa was the possible region of sympatric or peripatric differentiation of the E haplogroup, and notably of the P2 subclades. The differentiation was not allopatric. The M96 ancestor did not spawn different clades on different sides of the Sahara as a basal split. The cultural associations – if any – of this have yet to be worked out in terms of possible meanings in archaeology and deep-time linguistics. When lineages of known non-African origin are excluded, such as those found in border regions of Africa, but also in the interior of the Sahara, as with Chadic speakers in the Cameroons or Chad, the frequency of the lineages of African origin increases. These data show the genealogical connections of substantial numbers of males from all regions of Africa, irrespective of phenotype.

It is worth repeating what effects gene flow can have, and remembering that genes do *not* construct identity. Mating practices could also theoretically greatly influence gene frequencies and magnify the effect of the number of founders; for example, polygamy could magnify a particular Y chromosome lineage in the population, assuming the birth ratio was normal. So foreign males taking numerous local wives, all things being equal, would via their male offspring alter and magnify the frequency of "foreign" lineages over time in their new geographical residence, beyond the numbers of founder males. Ironically, at a smaller scale, if they maintained ethnic boundaries they would, through polygamy, increase the frequency of foreign mtDNA lineages within their own group. This has implications for sampling strategies, population nomenclature, and generalizing findings to whole regions from possibly unrepresentative samples. Interpretations that estimate the number of founders based on the percentage of lineages could also be skewed if mating practices are not considered.

1.6 Defining versus Describing Africa/Africans

There is a problem of how things are defined when perhaps the issue is how they should be described and with what new language. How are the African world and peoples to be conceptualized in studies involving biology for purposes of understanding genetics in health and disease – as well as certain historical (and evolutionary) questions? What are the hidden biases or grossly wrong assumptions? In some sense, the African–non-African dichotomy which is often extended even to African soil represents a sort of static schema that does not incorporate a fully evolutionary or historical perspective. In fact, based on lineage data, the world divides into a subset of Africans and the rest-of-the-world, which includes Africans, but this observation is only heuristic.

The biological diversity of Africa has been mentioned, but it is well to note that its linguistic diversity is also great on different levels – not just language families, but also other aspects of the elements of language.

This reality leads to asking about defining "African." Should it include recent settlers like the Europeans and Asian Indians in South Africa as former colonies? What about the Arabic-speaking populations of Nilotic and supra-Saharan Africa who have been previously mentioned? Supra-Saharan Africa would seem to be an example of acculturation with language shift in the Maghreb, with more migration from the Near East having taken place to the northern Nile Valley. Egyptian government spokesmen in the 1950s "claimed that the 'supposed racial difference between Africa south of the Sahara and Arab Africa in the north' was a tool of imperialism to create a rift and division" and also stated that there had been a mixing of peoples up and down the Nile Valley; they also stated that while modern Egypt had a "share" in the East and West that it was still an African entity (Ismael 1971: 103). Even if the reasons at the time for stating this were purely political in the context of Egypt being surrounded by enemies, and uncertain distant kinsmen who spoke Arabic, it is an accurate assessment of Egypt's Africanity.

It has to be repeated for emphasis and clarity that there is a history of ideas about Africa and Africans that focuses on tropical Africa as being the "real" Africa, as has been discussed. Underlying or associated with this perspective is the traditional racial construct and schema, rooted in ideas that a certain number of racial types emerged at some point in the past, migrated, and mixed to produce variation.

The idea of Africa has often been confined to tropical or sub-Saharan (and sometimes Saharan) Africa, the implication being that populations in northern Africa were only geographically African, but had no grounding in Africa. Conceptual clarity and descriptive precision are what is needed in new discussions of Africa. Various peoples have bioculturally assimilated non-Africans, but also sometimes changed languages themselves. Of course, during the Greco-Roman and Islamic periods there was substantial migration from non-African regions – and biocultural assimilation between groups. It is important to remember that there were indigenous people there, specifically Amazigh and the ancient Egyptians, both of which had assimilated others after their ethnic emergence and had complex biological histories prior to the crystallization of their identities. Neither of these peoples "disappeared," but via interactions they were changed. We still have Amazigh, Nubians, Egyptians, Beja, and others who may have assimilated or been assimilated. They often adopted a new language, religion, and customs, but are aware of their origins. We have Arab "tribes" who assimilated indigenous Africans in the Sahara and other regions. Another instance is that of Arabs who were assimilated by Amazigh. There are Arabized Africans – the ancient Nilotic peoples from Alexandria to Khartoum and parts of the Maghreb. Many populations have multiple ancestries and there are multiple descendants. Identity is another matter. It is possible to be a descendant of X group but not socially identified as X at present. Social rules have a role to play in determining social identity, which may not have much to do with "blood quanta."

This is not a genetic question at this point per se. It is about how we wish to conceptualize biocultural history. It bears repeating that all things being equal, a small amount of one-way gene flow of even 1–2 percent per generation with all offspring remaining and marrying into the receiving population and normal survival can have a profound effect on gene frequencies and, therefore, the biological profile over a not-so-long period (Cavalli-Sforza 1991; Harrison 1984).

The regions of northern Africa all have primary genealogical connections to other, more southerly regions, to one degree or another. The regions of northwestern and Nile Valley Africa all have fossils of modern humans that pre-date the likely expansion out of Africa. They both currently speak or use, or have a history of speaking, languages of families that arose in Africa. While they have biological profiles that indicate non-African admixture, the process that led to this took place in Africa and was subsequently shaped by African environments. These populations pose an interesting question, for some a "problem," although it need not be seen that way – that of untangling what is unique to them that arose in Africa. We can even extend this by asking a rhetorical question: If there is a concept of Eurasia, why is there no concept of Africasia – which in various biological situations would have as much or more validity, and would likely be useful in numerous discussions. As stated earlier, the bounding of populations is not usually "mind-independent," i.e., "natural."

Given the interest in indigenous populations here, it can be argued that "African" should, for the purposes of this discussion, be restricted to groups who have substantial genealogical connection to populations whose biocultural identity (including language) emerged in and was shaped in Africa, even if these have been modified over time. The locale of biological and cultural history is the focus. We can modify a position from a focus on the indigenous, a sometimes loaded term with some ties to colonialist discourse, and speak of the "endogenous" instead. Endogenous in this instance is adopted from Hountondji's (1999) usage to describe the state in which communities have assimilated various cultural practices and made them their own. These new practices are not seen as foreign; they have become a part of the fabric of daily existence. Endogenous, when deployed to culture-bearing biological populations, would mean groups that have bioculturally assimilated others into their sociocultural and physical "spaces" – and ancestry – over time, such that the specifics of the encounters with new people have been forgotten in some sense, or simply accepted as a part of who they are at the time. This perspective is useful from an etic perspective, and acknowledges that there may be tension between it and some emic perspectives.

There is no one way to be African.

1.7 Biological Variation and Social Organization

Social organization has been shown to correlate with levels of genetic heterozygosity, one measure of variability (Beals and Kelso 1975). One explanation would be that more complex societies – with the apex assumed to be states or state-like societies – have tended to incorporate more diverse populations, as well as naturally increased their population sizes, over time, such that even if the society has a self-concept of being one nation or one people, it has accumulated genetic diversity due to its history. (A critique of the narrow state-driven concept of social complexity is beyond the scope of this presentation.) More complex societies also tend to be larger, and population size has an intrinsic positive correlation with genetic diversity. All of this is relative, since recent studies on some groups who are conceptualized as having been isolated evince more molecular diversity based on their age, or at least the age of some component of the ancestry (Schlebusch et al. 2012). It is important to make this distinction since no living population should be viewed as a kind of fossil or relict. Diversity can be measured in different ways.

The issue of social organization is also nevertheless important to consider because it is important to avoid the idea that in sampling a particular group that one is sampling, as noted earlier, an entity like the notion of tribe as critiqued by Ogot (2005). Africa in all of its regions was home to various kinds of social organization (see the *UNESCO History of Africa* and the *Cambridge History of Africa*). State-level societies emerged in all quadrants of the continent before the colonial period, although all peoples did not create them. The more generally known of these are in or near the Nile Valley part of northern Africa, such as ancient Egypt, Nubia, and Axum. In West Africa, there were the Sahel belt states and kingdoms of old Ghana, Mali, Songhai, Kanem-Bornu, and the Segu-Tukulor state, as well as smaller entities like Bondu, Futa Jallon, and Futa Toro. In the forest belt of West Africa, there were the states of Ashanti, Oyo, Benin, and evidence of greater centralization (in having kings) in the past in Iboland at the site of Igbo-Okwu. In west Central Africa, the kingdoms of the Congo, the Luba, and others were also the products of local development. Zimbabwe, Zulu, and Swazi kingdoms/states have been well described. In northern Africa the ancient Amazigh societies emerged on the coasts as well as the Sahara, with a range of social organizations through time, with evidence of early interactions with other peoples; these include the Gaetulians, Mauretanians, Numidians, and Tingitanians who fought the Romans (and Carthaginians who were transformed from Phoenician settler colonists into a largely admixed entity who claimed their region of Africa as home). The ancient Egyptian state acted and interacted with peoples from the south, west, north, and east during its long history, having emerged on the banks of the Nile from a complex interaction. There were Nubian polities from the early times. The Garamantian kingdom emerged in the Sahara before the "Common Era." Later, in the Islamic period, there were a series of states in northern Africa which had various connections to Near Eastern centers but were never without some indigenous roots, such as the Fatimids and Almoravids.

Various kingdoms rose and fell in this period. Egypt in this era was a major center; northern Africa experienced a political history that included many peoples coming into its space. In the Sahara or its fringes, there were other state societies such as the emirates in northern Nigeria and Darfur. In eastern Africa, there were various kingdoms and communities that exhibited complexity in other ways. The Swahili city-states such as Kilwa, Mombasa, and Sofala are the best known of the eastern African entities south of Ethiopia, but there were other societies of early iron smelters in the interior. In the interlacustrine regions of central Africa there were several kingdoms such as Toro, and the kingdoms found in Burundi and Rwanda. Some of the pre-European entities, such as the Hausa emirates, still exist as moieties at some level within the postcolonial newly named states. In what is now Sudan there was the Funj sultanate – in some sense perhaps a successor state to the Nubian Christian kingdoms, such as Alodia. Sampling in these areas to ascertain ethnogenetic heritage may or may not reveal something about the histories of those who created and participated in different phases of the kingdoms.

Another kind of social complexity that does not necessarily involve the process of the creation of states deserves some attention. Precolonial Kenya was home to societies that were more interconnected than indicated by conceptions often associated with societies assumed to be small, closed, and homogeneous. However, some societies/peoples that are not in themselves states or kingdoms are heterogeneous in origin, and new as named entities, even if composed of old ancestries. One example is the Taveta who emerged in the eighteenth and nineteenth centuries when various groups came together and created a new people, society, and social organization. Ogot made the following general observation:

> By the end of the nineteenth century, the African communities in the future Kenya were already all contaminated [sic] by each other in a complex interdependent human world. There were no watertight ethnic categories. Numerous clans, lineages, and sections of clans expanded and contracted, gaining and losing members. The migration of segments or absorption by other ethnic groups produced considerable complexity. New communities and new languages were often the result. Hence the colonial idea of a "tribe" as an isolated and closed group is certainly a myth and ultimately, a racist concept. (Ogot 2005: 268)

This passage makes some very serious points and amplifies concerns relevant to critical work. If this description is accurate, it has implications for sampling and for doing human biological research differently. It may indicate a new way to do certain types of research. It could be called a kind of wealth because of its relative uniqueness.

Returning to history, Ogot (2005: 268–273) points out that "30–40 percent of the Baluyia clans were originally Kalenjin"; that "several Luyia [Luhya] clans are of Maasai origin," and that "[T]he evolution of the Abaluyia [Luhya] was still in progress when the east African protectorate was declared in 1895." Ethnic entities emerged that had multiple origins but adopted one language, like the Abatarichi and Taveta. There were relationships between Samburu and Rendile pastoralists despite differences in language family and subsistence style, with divisions of the Cushitic speaking Rendile assimilating into Samburu clans, including joining their age sets and marrying Samburu women. This was not an isolated case:

> This picture can be multiplied across the territory that was to become Kenya in 1920 . . . There were no pure ethnic groups. Each group was a dynamic and living unit whose continuity depended less on its purity or single origin than on its ability to accommodate and assimilate diverse elements. Most of the myths, legends, epics, and rituals once come across in stories of migration and settlement are meant to facilitate the process of integrating people whose origins are divers [sic]. (Ogot 2005: 273)

Ogot (2005: 273) gives a more telling example of complex ethnogenesis hidden by a name. The Taveta people of the Kenya coast may be taken as an excellent example of nation-/community-building. Their history reveals that refugee groups comprising the Pare, Shambaa, Kamba, Taita, Chaga, and Arusha fleeing from the famines and conflicts in their respective home areas settled in the Taveta forests in the seventeenth century. By the nineteenth century, this heterogeneous group had developed a distinctive common culture and evolved land-holding clans and central institutions which unified the migrants into a single people. There was also widespread intermarriage between the clans and the adoption of the Pare language by all these clans, which greatly assisted in welding them into a cohesive nationality. The roles of culture, common central institutions, language, and intermarriage in nation-building are clearly demonstrated in the case of the Taveta. When colonial literature later portrayed the Taveta as a "tribe" in the sense of an exclusive "primitive" and static society, it was in effect inventing a primitive community whose presumed needs and hopelessness could be used to justify colonialism.

In the case of the Taveta, we have a group/community/breeding population whose origins are diverse. Taveta as an entity is the genealogical intersection of many groups at the levels of biology and culture. The Taveta are one people out of many. The recentness of their *emergence* does not negate the reality of their existence. They can be an example to African peoples and others who make much out of differences, especially African peoples who have assimilated negative ideologies and redeployed them in ways that hinder development.

Applying this information to work that is population-centered requires a change in thinking that assumes that groups are continuations of primordial groups that had singular origins and that have remained endogamous. While there are examples of some ethno-religious groups that have retained a high degree of endogamy, this is not true in many places in Africa where Western geneticists have been at work. The results of a study used to construct population history could be wrong in its conclusions if only genetics is used. An example would be an analysis that concluded that the Taveta diverged from another group a certain number of thousand years ago; this would clearly be wrong since there were no Taveta before 1800. Assuming that the methodology is correct, the conclusion, assuming it to be computationally valid, would reflect the idiosyncrasies of the sample. One might say that the "average" genome of those identified as Taveta diverged from the comparative groups – and while possibly statistically valid this would be meaningless, i.e., unreliable in a substantive or historical sense. The result might say something about some of the ancestries of the Taveta, and determining this would require another kind of effort. We must always keep in mind that two or more populations could be genetically similar but culturally or linguistically different and that the reverse is true – that culturally or linguistically similar groups could be genetically different. Much depends on how we measure differences and the goals of the work. In general, the writing of population history as seen in the genetics literature on Africa has not undergone the kind of scrutiny and vetting that philosophers of history write about in terms of the kinds of bias that may exist in historical writing (McCullagh 2000, 2004). Population history is, after all, a kind of history, and as such should be required to not only present accurate genetic information but also the issues of accuracy and plausibility that concern historians.

The facts of the social organization and history described by Ogot are not compatible with the view that the populations connected to them can be treated as isolates, or the "primitive isolate" or the "anthropological populations" of an earlier anthropology. It is clear that, depending on the purpose of a given study, any conclusions of a historical nature have to be informed by some historical and social understanding of the populations and indeed individuals from them, since identities may be multiple. All human populations and individuals are related, but to varying degrees and in varying ways. Human variation, depending on the levels of interest, the data used, and questions asked, can be described in terms of networks, clines, phylogenies, and clusters.

1.8 Summary

Conceptualizing variation in Africa deserves to be informed by an understanding of, and sensitivity to, the history of ideas that has shaped how Africa is/was viewed and their ongoing legacy. The term Africa was first applied to parts of its northern coastal regions and their hinterlands, where peoples with a diversity of external phenotype were observed. Biological Africanity is not confined to one physiognomy, genome, phenotype on any level, or genotype. Biohistorical Africans are highly diverse and complex at anatomical and molecular levels and this is consistent with evolutionary theory. This diversity has to be a fundamental starting point and no effort should be made to parse it into units of uniformity or "purity." The various evolutionary forces should be actively recognized as contributing to variability. Africa did receive ancient and more recent migrants who were largely bioculturally assimilated to varying degrees, with some African peoples adopting new languages before the period of European domination. Raciotypological thinking has to be abandoned in approaching Africa. Some data are interesting in this regard. The quadrants of Africa are connected by males of the E haplogroup, which is the most common male lineage in Africa. The population bearing the E haplogroup apparently emerged in eastern tropical Africa after modern humans left Africa and spread to all major regions at varying time depths. This is also true for A, B, and R haplogroups in the sense that they do not tell you about the phenotype or culture of an individual who may have them. These genealogical connections cut across geography, language families, breeding populations, ethnicity, and physical appearance. How genetics is integrated into the study of African population history will benefit from considering time depth, social structure, and practices and data from other fields. Genetics will not necessarily have a primary role in generating narratives at macro or micro levels in the period after the emergence of current language families. When genetics should be allowed to dictate the framework, if ever, of a narrative versus archaeology, history, or language is a methodological issue to still be resolved as it makes a difference and may change the nature of the project. Social organization in Africa exhibits different forms of complexity; this has implications for research and interpretation and for understanding different kinds of history beyond states, kingdoms, and migrations. Scholars with the skills and insights of Ogot can contribute significantly to the reconceptualization of Africa and approaches to synthetic work. Indeed, as new schools of European historians have emerged, one would expect this of Africa.

This chapter has been a sketch of some issues. It is only a beginning.

Acknowledgments

I wish to thank Professor M. Ibrahim for inviting me to participate in this project and for his patience. Thanks also go to Mr. J. Gaillard and Ms. A. Webb for some technical assistance. Both Dr. A. J. Boyce and Professor G. A. Harrison were important influences due to their work in zoological taxonomy methods, and migration/gene flow, as was Dr. Rom Harre, who taught philosophy of science and ethics in science. Dr. Koffi Maglo was an early sounding board for ideas behind aspects of this work. This chapter is dedicated to my mother, my late father, my wife, and my children.

References

Abu-Amero K, Hellani A, Gonzalez AM, et al. (2009). Saudi Arabian Y-chromosome diversity and its relationship with nearby regions. *BMC Genetics*. DOI: 10.1186/1471-2156/10-59.
Ahmed L (2000). *A Border Passage: From Cairo to America, a Woman's Journey*. Penguin.
Armour JA, Anttinen T, May CA, et al. (1996). Minisatellite diversity supports a recent African origin for modern humans. *Nat Genet* **13**(2): 154–160.
Beals K and Kelso AJ (1975). Genetic variation and cultural evolution. *Amer Anthrop* **77**: 566–579.
Bouzaggar A, Barton N, Vanhaeren M, et al. (2007). 82,000-year-old shell beads from North Africa and implications for the origins of modern human behavior. *PNAS* **104**(4): 9964–9969.
Brace CL (1863). *The Races of the Old World: A Manual of Ethnology*. John Murray.
Butler A (1902). *The Arab Conquest of Egypt and the Last Thirty Years of the Roman Dominion*. Clarendon Press.
Cavalli-Sforza LL (1991). Genes, peoples and languages. *Sci Am* **265**(5): 104–110.
Coon C, Garn SM, and Birdsell JB (1950). *Races: A Study in the Problems of Race Formation in Man*. Charles C Thomas.
Elhassan N, Gebremeskel EI, Elnour MA, et al. (2014). The episode of genetic drift defining the migration of humans out of Africa is derived from a large East African population size. *PLoS One* **9**(5): e97674.
Evans WM (1980). From the land of Canaan to the land of Guinea: the strange odyssey of the "Sons of Ham." *Amer Hist Rev* **85**(1): 15–43. (See footnote on page 24.)
Fogel F (1995). Of Nubies, Nubians: scientific traditions and local identity. *Anthropol Egypt* **24**: 75–86.
Gannett L (2003). Making populations: bounding genes in space and time. *Philos Sci* **70**(5): 989–1001.
Gould SJ (1987). Petrus Camper's angle: the grandfather of scientific racism has gotten a bum rap. *Nat Hist Mag* **96**(7): 12–18.
Harrison GA (1984) Migration and population affinities. In Boyce AJ, ed., *Migration and Mobility, Symposia of the Society for the Study of Human Biology*, vol 13. London: Taylor and Francis, pp. 57–68.
Hiernaux J (1975). *The People of Africa*. Charles Scribner's Sons.
Hountondji P (1999). *Introduction: Endogenous Knowledge*. Codesria.
Ismael T (1971). *The U.A.R. in Africa: Egypt's Policy under Nasser*. Northwestern University Press.
Jakobsson M, Scholz SW, Scheet P, et al. (2008). Genotype, haplotype and copy-number variation in worldwide human populations. *Nature* **451**(7181): 998–1003.
Jorde LB, Bamshad MJ, Watkins WS, et al. (1995). Origins and affinities of modern humans: a comparison of mitochondrial and nuclear genetic data. *Am J Hum Genet* **57**(3): 523–38.
Kamil AA (1970). *Islam and the Race Question*. UNESCO.
Keita S and Kittles R (1997). The persistence of racial thinking and the myth of racial divergence. *Amer Anthrop*. DOI: 10.1525/aa.1997.99.3.534.
Korst JK (2008). *Het rusteloze bestaan van dokter Petrus Camper (1722–1789)*. Springer.
MacEachern S (2007). Where in Africa does Africa start? Identity, genetics and African studies from the Sahara to Darfur. *Jour Soc Arch* **7**(3): 393–412.
McBrearty S and Brooks A (2000). The revolution that wasn't: a new interpretation of the origin of modern human behavior. *J Hum Evol* **39**: 453–563.
McCullagh CB (2000). Bias in historical description, interpretation, and explanation. *History and Theory* **39**(1): 39–66.
McCullagh CB (2004). What do historians argue about? *History and Theory* **43**(1): 18–38.
Montagu MFA (1960). *An Introduction to Physical Anthropology*, 3rd edn. Charles Thomas.

Ogot BA (2005). *History as Destiny and History as Knowledge: Being Reflections on the Problems of Historicity and Historiography.* Anyange Press Limited.

Pagani L, Schiffels S, Gurdasani D, et al. (2015). Tracing the route of modern humans out of Africa by using 225 human genome sequences from Ethiopians and Egyptians. *Am J Hum Genet* **96**(6): 986–91.

Peschel O (1888). *The Races of Man and their Geographical Distribution.* D. Appleton.

Schlebusch CM, Skoglund P, Sjödin P, et al. (2012). Genomic variation in seven Khoe-San groups reveals adaptation and complex African history. *Science* **338**: 374–376.

Schuenemann VJ, Peltzer A, Welte B, et al. (2017). Ancient Egyptian mummy genomes suggest an increase of sub-Saharan ancestry in post-Roman periods. *Nat Comm Res* **8**: 15694. DOI:10.1038.

Seligman CG (1930). *Races of Africa.* Thornton Butterworth. Republished by Oxford University Press 1957 and 1966.

Seligman CG (1957). *Races of Africa*, 3rd edn. Oxford University Press.

Smedley A (2007). *Race in North America, Evolution of a Worldview.* Westview.

Snowden F (1970). *Blacks in Antiquity: Ethiopians in the Greco-Roman Experience.* Harvard University Press.

Sole-Morata N, Garcia-Fernandez C, Urasin V, et al. (2017). Whole Y-chromosome sequences reveal an extremely recent origin of the most common North African paternal lineage E-183 (M81). *Nat Sci Reports* **7**: 15941. DOI: 10.1038/s41598-017-16271.

Stringer CB and Andrews P (1988). Genetic and fossil evidence for the origin of modern humans. *Science* **239**(4845): 1263–1268.

Thompson L (1989). *Romans and Blacks.* University of Oklahoma Press.

Van Peer P, Vermeersch P, and Paulissen E (2010). *Chert Quarrying, Lithic Technology, and a Modern Human Burial at the Palaeolithic Site of Taramsa 1, Upper Egypt.* Leuven.

Weiss K and Long J (2009). Non-Darwinian estimation: my ancestors, my genes' ancestors. *Genome Res* **19**: 703–710.

2 History and Genetics in Africa

Multidisciplinary Efforts

Shomarka Omar Y. Keita and Muntaser E. Ibrahim

2.1 Introduction

Population studies that integrate genetics and aspects of culture in order to examine historical topics, or that use history or archaeology to help explain geographical patterns of genetic variation, have become more common in the literature. This body of work includes, but is not restricted to, the exploration of the relationship between patterns of genetic and linguistic differentiation, genetic variation in societies defined as politically simple or complex (even if they fall within the boundaries of states), and the possible association of genetic variation and the spread of cultural innovations such as the practice of plant and animal domestication. In all of these examples, geography and/or migration are major background themes because genes and aspects of culture are evaluated in terms of movement across space and/or time. Such work at its best in theory requires the effective use of historical linguistics, archaeology, ethnology, and written and oral historical texts in the context of population genetics data and theory. The term "population history" is not necessarily used consistently. The subject – "population" – may vary in its referent. Genetics and other fields would likely be better understood as having different roles to play in any writing of history from any region in the world. This leads to the question of what kind of data/evidence should lead the construction of a historical narrative, and when: This interesting question may be obvious in some cases, but not so obvious in others. Ideas about identity and whether culture, with or without language, may be more important than ancestry or phenotype in a narrative are points of contestation. In some models of research, genetic profiles could be interpreted in terms of ideas about the geographical origins of genes found in a population, which may give us a "history" of interactions of gene bearers, but it must be kept in mind that gene history is not the same as population or culture history: The mediators (i.e., "gene bearers") of particular genes may not be the population, or from the region, of the ultimate origin of the genes. Time and geography are important. The genetic profile at one time or place for a society does not necessarily tell you about its population's total cultural or biological "origins" – a notion that is not always used consistently or clearly: Different fields may bring different assumptions to the analysis of the data. It also does not tell you about its descendants. Although statistical modeling might be helpful, it is clear that it has real limits: Statistics, Bayesian or otherwise, could not have predicted the change in population in North and South America, or the subsequent history that led to numerous "admixed" communities. (We have to remember that "admixture" is not a new phenomenon.) It is important to say that "genetic history" is not the same as social or political history, and to ask how or if genetics would help understand various historical events versus the other data. This chapter will explore some concepts related to these issues.

Studies limited solely to linguistics, ethnology, or genetics may successfully restrict explanations to the parameters of their own research traditions – and meet the expected canons of those disciplines. However, interdisciplinary studies have to satisfy the canons of multiple disciplines and attend to the issue of tension between the fields, at least in theory. Genes, language, and culture are not intrinsically related or linked, and their association is casual, not causal; at the level of population they may or may not travel together and may change relationships over time. They pertain to different levels of order in complex systems, with the information content of biological systems in some sense being more reducible for the purposes of historical enquiry. Integrating disparate layers of uncertainty and predictability between information from biological and social systems awaits the establishment of appropriate and perhaps often novel tools. Although language and culture are to be analytically separated depending on the definition of culture, we acknowledge that sometimes they get treated together, and that this may be appropriate in some situations. However, the reduction of genes to language families and cultures with the implication of causality or a necessary linkage is to be avoided as it can lead to assumptions about the ability of genetics to, for example, determine whether a given language or language family belongs in a particular phylum. Linguistic evidence is used to make linguistic classifications; biological evidence is used to make biological classifications; archaeological classifications are based on archaeological data; music is correctly classified based on musical elements, not the social identity of those playing the music – even if there is discord around "who" "should" be playing what kind of music. The association between genetic patterns and linguistic assignments to families is not a directly causal one rather than an outcome of "isolation" mechanisms that operates in creation of reproductive isolation. Theoretically people with similar languages and cultures are more prone and likely to exchange mates with each other which does not mean that observed associations between genes and language/culture are causal (e.g., one cannot construct a taxonomy of languages based on craniofacial traits).

All of this raises another point: The majority of people now speaking a given language, or practicing a particular religion – in some sense a cultural phenomenon – or who are making particular artifacts may not be genetically/genealogically the closest to or the same as the founding population or the one initially associated with a particular phenomenon. Any population may have multiple ancestries; how it is identified is based on other rules. An example is people in the United States and Latin America who may have ancestry from various continents. A person may be of *Irish descent*, but not *Irish identified* due to social practices: This example describes many legacy Afro-North Americans (of middle passage descent), with "legacy" denoting African ancestors who were enslaved or indentured persons on US ground, but the term can be extended to those in Latin America and other places. In Latin America there are many people with some African ancestry via the middle passage, but who are not identified as "Afro-Latinos" – a term that seems to be usually reserved for those with "obvious" tropical African ancestry living in certain communities that may be culturally somewhat distinct with others like them (along with those in this era who choose to self-identify in this way as individuals). It can be hypothesized that most of the African ancestry in Latin America is dispersed in the general populations and not located in the Afro-identified enclaves. There are many people of middle passage descent in various regions if the phrase is treated as a denotation of some African ancestry.

Studies that use multiple sources to look at populations engender numerous difficulties in themselves due to the absence of good (or perhaps good enough) data quality from multiple disciplines. However, they are made even more complicated because of the difficulty of keeping abreast of the developments in many disciplines, not only points of contention and various views of a topic, but major re-interpretations of evidence and the nuanced opinions found within specialized fields. This problem is shared by researchers and readers alike, and it could be argued that all such published work should give some review of previous ideas and research, and the associated sociology of knowledge when appropriate.

Common agreed-upon criteria to evaluate multidisciplinary studies still await development. Until that time, assessments must use the canons of the involved disciplines. Using standard philosophical perspectives will also help: Broadly speaking, explanation in science is based on natural laws or principles, but in history on the role of contingent unique events, not amenable to "natural laws" in the conventional sense. The terms nomothetic and idiographic, respectively, are often used to describe the former and latter modes of explanation. Population "histories" have a double standard to meet, namely adequacy in terms of biological theory, and the credibility and fairness required by historians in order to acknowledge and avoid bias in descriptions, interpretations, and explanations (McCullagh 2000, 2004) – or admit their existence, and to produce interpretations that are not just "credible" but also "fair," "balanced," and "not misleading." Apparently, most historians do care and argue about these issues, although in the context of post-modernist thought and practice the idea has arisen that some general interpretations are often idiosyncratic. Most certainly no scientist would espouse such a perspective in doing what might be called the work of classic science; however, when such work, specifically genetics, is used as part of historical enquiry, a new element enters into an enterprise that has now become synthetic. The credibility of a causal explanation should be supported by objective evidence that supports one narrative more than another. In some cases this may not be possible, and this should be admitted as well.

In the most general terms, historical works can be evaluated by examining how well they explicitly present and discuss alternative explanations, avoid generalizing from incomplete information, and show proper consideration for all available information or differing views of the same evidence. Broadly speaking, historical writing, according to McCullagh (2000), can be biased in four ways: (1) the misinterpretation of evidence which could include an incomplete review or misreading of data; (2) the presentation of unfair or unbalanced accounts which may be true at some level, but otherwise omit significant information; (3) a general description of the past that implies facts that are known to be false; and (4) a selectivity in the presentation of causal explanations, where not all of the important causes are given to the reader, hence leaving a false impression. These shortcomings can be made by mistake, in which case they are simply unjustifiable or wrong. For McCullagh (2000) they become *biased* when they occur because the scholar wants the particular outcome produced by these failures. To this can be added outcomes that are expected or subconsciously produced due to embedded cultural beliefs. These may not always be recognizable. It is possible, however, to produce credible accounts that are not comprehensive, as long as this is explained, thus allowing a critical evaluation. Interdisciplinary work may be more prone to these kinds of errors.

Geneticists doing work on recent populations that is conceptualized and presented as historical work have some burden due to all of the issues mentioned above. The pattern of genes across geographical and/or ethnocultural spaces cannot be used as the sole evidence for constructing a narrative of "population" history, or "genetic history," although they identify ancestral components on one level based on the names in their databases. One theoretical issue, not often confronted, is the ranking of data in terms of what primarily determines the framework of a narrative (Keita et al. 2010). Should genetic data ever be used to construct the narrative, i.e., construct its framework, with other data being fit into patterns of genetic findings? Or should the genetic data be subsumed to historical frameworks determined by texts, archaeology, linguistics, etc.? Is "reciprocal illumination" possible without falling into circular reasoning, and what are the criteria of its limits? When do genetic data become "primary"? It is clear that some concepts, facts, or "factoids" could lead to erroneous narratives, and some of these are rooted in bad concepts or conceptual schema. Historical explanations and general interpretations of the past are expected to be fair and not misleading (McCullagh 2000), even if there is no claim that they are comprehensive (McCullagh 2004). Here we suggest that better narratives are likely to result when teams work in what might be called historical genetics. Multiple disciplines are more likely to produce reports that are conscious of bias in concepts and interpretations.

Studies of African-based or related topics provide interesting cases with which to explore the utility of good collaboration in multidisciplinary work. One anecdote serves to illustrate this. Gould (1987) reported that Petrus Camper, the famous eighteenth-century Dutch anatomist and inventor of the facial angle, criticized contemporary artists for their renditions of the African Magus in nativity paintings. Camper noted that the depictions were simply Europeans painted black or dark brown, and that their faces were morphologically incorrect in terms of the underlying bone structure. He observed that the physiognomy, in terms of nose shape and facial profile, were incorrect. Gould (1987) seems to agree with this assessment. However, Camper overstated his case. He apparently based his criticism on his knowledge at this time of only the physiognomy of a subset of certain enslaved Africans from the forest belt of West Africa, who primarily possessed the so-called "Negro" anatomical conformation, which was abstracted into a "type" resulting sometimes in the error called misplaced concreteness in discussions about "authentic" Africans. Even in Camper's time it was known by some Europeans that there was also variation in the bony anatomical "complex" associated with dark skin and frizzy hair. There are dark-skinned Africans with physiognomies that are similar to those of "Europeans" in having narrow noses, orthognathism, etc., but the trait complexes in question should best be described with anatomical terms, and not in group terms from a raciotypological paradigm that implies that individuals with a given morphological complex all necessarily have a common "origin" – an idea that is antievolutionary at its core. Had Camper known at the time of that writing about the range of variation even in West Africa, he perhaps would have made other criticisms, since there are other aspects of bony facial variation that will differentiate populations who broadly share these traits. (Camper apparently learned about the variation in tropical Africa later [Meijer, pers. comm.].) This anatomical complex composed of a narrow nose, orthognathism, and narrow face, no matter where found, was once interpreted as being necessarily of "Caucasian" or "Caucasoid" origin via migration or gene flow in a model of "racial" origins (Keita and Kittles 1997). This assumes a necessary group or geographical origin for these traits versus their being products of adaptive microevolution or drift. In the tropical African context, Hiernaux (1975) suggested the term "Elongated African" for this physiognomy and variants, a phrase that emphasizes a relatively narrower nose and face that evolved in a hot-dry climate but has some variation. The authors would prefer that the phrase "elongated African trend" be used; its counterpart would be "broad African trend," and it is admitted that this is phenotypically an incomplete suggestion. The use of the word "trend" conveys the existence of a continuum which should not be broken into discrete types, but can be mathematically described with descriptive multivariate approaches. This is also admitted to only refer descriptively to craniofacial shape in some sense consistent with morphometric findings. It says nothing about other traits. Some migration and admixture are not denied as a part of African biohistory, but if most of the gene flow has been so long ago as to have been reworked by African selection pressures and social circumstances, then it constitutes a part of a genuine African biogeographical and biocultural history (Hiernaux 1975). Overlap in a range of biological traits between biogeographical Africans and non-Africans should be expected based on evolutionary theory and the concept of serial founder effect, as well as recurrent gene flow in multiple directions.

The primary purpose here is not criticism, but an exploration of cases, because in general multidisciplinary cooperation has not been maximized, and African participation has been minimal. While older literature is featured for examples, this is not a review. Older ways of thinking have a way of re-emerging and persisting in our knowledge templates and cognitive maps. While some critique is implicit in this discussion, it is only in the context of an examination of data or ideas in order to increase awareness of the problems of such work, hypothesis generation, and not for criticism's own sake. Topical "cases" are presented in an effort to illustrate the value of having a multidisciplinary team. The analyses are intended to illustrate the range of possible explanations by integrating various lines of evidence. It is important to be vigilant against simplistically answering one discipline's question with another's data.

2.2 Genetics and Bantu Speakers

Bantu languages, collectively a group within the Niger–Kordofanian or Niger–Congo (NC) family, are spoken west to east from the Cameroons to Kenya and then south to the Cape of Good Hope. Bantu is a linguistic term; its speakers are not a single people, although one sees the word listed as though it were describing a specific ethnic group, breeding population, or even "racial" group. Sometimes it seems to be used as a "euphemism" for "Negro/Negroid," which is unacceptable. The term "Bantu expansion" refers to the spread of this linguistic grouping. The direction of movement is generally accepted as having been from east of Nigeria in a southern and eastern direction to the subequatorial regions, but how much was due to the mass movement of people versus language diffusion from community to community is a point of discussion. The PN2/V38/M2 biallelic lineage in part maps to the distribution of the family, as does haplotype IV of the TaqI 49a,f RFLP system, which in Africa and adjacent regions apparently marks the same clade (see al-Zahery et al. 2003; Underhill, pers. comm.). The spread of this family is frequently identified with the distribution of these variants in nearly a causal fashion. In other words, the biallelic allele V38/M2 is said to be a *marker* of the Bantu expansion, which some earlier writers, who mistakenly called languages in this region Bantu, even thought had gone into West Africa (see, e.g., Guthrie 1962).

However, RFLP haplotype IV(V38/M2) is found in very high frequencies in Africa west of the Cameroons from Nigeria to the Atlantic, reaching a frequency of ~80 percent in a sample from Senegal. Just as interesting is its reported frequency in one study of Egypt (27 percent) and Nubia (39 percent) (Lucotte and Mercier 2003). There are currently no Bantu speakers in these regions and no evidence that they expanded into the region as settlers, although some ancestral Egyptians (inhabitants of the Sahara) may have had contact with early Niger–Congo and Nilo-Saharan speakers based on likely loan words (Takacs 1999). Hence the "Bantu expansion," a problematic concept, especially as often conceived, in any case cannot be used to explain M2 presence found in some studies of Nile Valley populations (and there is no need to invoke the over-worn trope of the slave trade, which cannot explain everything). The Bantu expansion should not be conceived as having been a mass movement of a single people, analogous to an *mfecane*, or the migration of the Banu Hilal.

Archaeology and historical linguistics may *possibly* help explore credible explanations, but only at the time depths where migrations of early speakers might have been a real historical phenomenon. The V38/M2/haplotype IV marker is found at great frequencies in Niger–Congo speakers in general. It is possible that V38/M2 existed in the speakers of the early ancestral family – proto-Niger–Congo – and got distributed into the speakers of all of its branches as the population biolinguistically differentiated through space and time. This explanation does not work for Egypt and Nubia at a macro level since languages spoken there belong to other families, and if a close correlation between languages and populations is assumed, which can be problematic, the caveat is that there is evidence for Nilo-Saharan and Niger–Congo loan words in ancient Egyptian (Takacs 1999). However, archaeological data indicate a late Pleistocene recolonization of the eastern Sahara after a probable population hiatus between 50,000 to 15,000 years ago (Wendorf and Schild 2001). The peoples involved can be expected to have been highly biologically and culturally diverse. The early Holocene wet (sometimes called "green") Sahara, with its inhabitable places, can be envisioned at various times as having been the site of a metapopulation, with there having been contacts between various groups. This marker may have entered the Nile Valley with mid-Holocene population Saharan migrations into the Nile Valley (Hassan 1988), which contributed to the peopling of the valley. The "primary pastoral community," as defined by Wengrow et al. (2014), had extensive north–south Sahara–Nile Valley geographical roots and distribution, and was apparently a trans-communal entity whose common cultural base would have likely facilitated gene flow between populations.

Another possibility is that Nilo-Saharan – to which Nubian belongs – and Niger–Congo form a larger language phylum called Kongo–Saharan (Gregersen 1972) or Niger–Saharan (Blench 1995), whose earliest speakers shared a biohistorical heritage that included the V38/M2/RFLP IV marker. This could also explain the notable frequencies in Egyptian Nubians and NC speakers as a whole, and not just Bantu speakers.

However, there is a caveat: From a strictly biological point of view it is important to note that M2 likely arose before any of the language families emerged, based on standard estimates of the ages for language families, and most certainly their branches. It is wrong to treat them in effect as having a basal or causal relationship. This would also be true of Afroasiatic. It is worthwhile making the observation that the P2 marker or clade, sometimes called E1b1, which is "ancestral" to both the M2 and 215/M35, is "father" to the male clades whose bearers are speakers of the three different language macrofamilies in Africa, with one of them being Afroasiatic (also called Afrasian or Afrasan). This is intriguing because the ancestral language of P2 "father" – at the population level – is not known. Was it a language that went extinct, with a subsequent sorting of languages into different families? Was there some sort of ancestral areal linguistic entity that does not fit current normative constructs? Was there an early proto-African language family that led to the current language families? Most Western linguists would say "no" to this last question at the time of this writing, at least to a simple notion of a pan-African ancestor, although this is worth exploring. However, for those historical linguists who believe that larger groupings of ancestral language families can be reconstructed, a parsimonious answer would be that the P2 population spoke some fragment of "proto-world" that remained in Africa after modern humans left. This hypothetical proto-world fragment may have had some substructuring that explains some of the recovered similarities between different African families and those outside of Africa. It is generally hypothesized that all language phyla, like all human populations, are related, just to different degrees. Space does not permit a full discussion of this matter. The simplistic racialized gene–language maps or dendrograms that were once drawn as a validation of a schema of disconnected "race" populations fall apart when the Y chromosome lineages are examined against the geography of language families (and their origins).

A multidisciplinary approach clearly can help to avoid over-generalizations with regard to Bantu speakers. Misinterpretations can skew interpretations. This in turn could lead to poor study designs in future work. Another issue is the use of Bantu as a euphemism for "Negro" (Robertson and Bradley 2000) from the old unscientific racial schema, which seems to be how some geneticists and morphologists are using the term. While this issue is beyond the scope of this chapter, it deserves mentioning given the emphasis placed on the sequencing of a "Bantu" genome, which strictly speaking would have to mean looking at genes that were thought to have arisen at the time the Bantu linguistic branch emerged in those speakers and not those now found in the bodies of current Bantu speakers – which may not be the same as the original speakers or their ancestors, before Bantu even arose as a language unit. The descendants could be different due to language shift and/or substantial microevolution (which includes gene flow).

2.3 Genes, Food Production, and Afrasan

The spread of animal and plant domestication is now frequently argued to be associated with the dispersal of language families, via *demic diffusion*, the actual migration of expanding populations over space (Ammerman and Cavalli-Sforza 1984), instead of cultural diffusion. Such movement would have implications for spatial patterns of genetic variation. This claim for a coterminous spread of genes, language (families), and/or food production is most successfully made by reconciling data and interpretations from genetics, historical linguistics, and archaeology. When sufficient evidence is available, a multidisciplinary approach requires testing all models.

In terms of Africa, one interesting case regards the Afrasan (Afroasiatic) language phylum or family. The spread of domesticated wheat, barley, goats, and sheep into northern Africa from southwest Asia (the Near East) and its relation to the Afrasan language family is a good example. A minority argue that Afroasiatic is Asian in origin (which requires opposing standard linguistics evidence), the majority that it is African. Either position on origins influences any subsequent interpretations. Unfortunately, both models are not usually presented in published work, so an imbalanced picture is presented. Most genetic studies and/or interpretations have proceeded from models in which either ancestral Afrasan or a later branch, e.g., Berber (Arredi et al. 2004), are posited to have come from the Near East with the spread of food production, without the consideration of alternatives, which is problematic when not simply wrong (see Ehret et al. [2004] for a corrective to this). Generally, population migration versus cultural diffusion is given the primary responsibility as the mechanism of change. Here is explored the ramifications of taking into account the views of mainstream linguists, as well of archaeologists, in the examination of the genetic data. The issue is the construction of the most credible narrative that is balanced and fair.

The study of individual Afrasan languages, and then later the family itself, formally called Hamito-Semitic, or Semito-Hamitic, has a long history. The great interest, at least in part, is because this family gave rise to the Semitic languages found in the texts used in Judaism, Christianity, and Islam. Ancient Egyptian is the other member that received substantial attention. Afroasiatic has another distinction related to the study of African peoples, because it is connected to a theory of biological and cultural origins called the "Hamitic hypothesis" promulgated by C. G. Seligman (see Sanders 1969). This thesis posited that Southwest Asian peoples called "Hamites," distant relations of Semites, came into Africa bringing a new language family, pastoralism, "superior" culture, and unique (and "superior") biology marked by narrower noses and faces, straighter hair, and lighter skin color, although curiously not the skills of plant agriculture. Nearly all instances of culture in Africa deemed to be noteworthy from a Western perspective were attributed ultimately to contact with Hamites. Thus many African societies by this myth were rendered "non-African" in a model of culture that was clearly thought to be genetically determined.

The Hamitic hypothesis was overturned by Greenberg and others and abandoned by Africanist historians. The majority of historical linguists who study the family read the evidence as supporting that Afrasan is a family of African origin and history, with one branch, Semitic, having had an ancestor that was carried to Asia (the Near East). They generally agree, based on evidence, that the proto-language was spoken by hunters and gatherers, not farmers or pastoralists (i.e., food producers).

In a recent study by Gebremeskel and Ibrahim (2014), the Y-E haplogroup was found to map with this family, which is not to say that genes determine languages or language family classification. The principle of greatest diversity indicates Africa to be the geographic cradle, and the concept of "least moves" that the Horn of Africa or the southeastern Sahara is the more specific locale of the ancestor. It is generally true that the place of greatest diversity of an "evolving" population or language, all other things being equal, is its likely place of origin. Care has to be taken to understand that in theory if a broad sample of a population migrated, that the level of diversity or something close might travel with it, although this is seldom the case. Of equal interest is that the branch of Afrasan, Omotic, believed to be the least derived in its relationship to the ancestral proto-family is found only in the Horn of Africa, specifically Ethiopia; if this branch is not considered to be in the family, it becomes the nearest relative to the group and still anchors the proto-family in Africa. Studies involving genetics, Afrasan, and food production should always examine hypotheses based on the different models of geographical origins, and the internal structuring of a language phylum. This does not mean that genetics can be used to determine internal structuring or membership in the phylum.

What happens when genetic and linguistic data are considered in terms of the African origins model? The recent spread of Arabic has to be ignored and the presence of Semitic languages in Ethiopia based on the standard linguistic interpretations. The Y chromosome data are used because they are the most complete for the greatest number of members of the family. The core distribution of the family runs from eastern Africa, north to Egypt, and west along the Mediterranean littoral. The subfamilies spoken or once spoken here are Cushitic (Kenya, Tanzania, Somali, Ethiopia, and Sudan), ancient Egyptian in the Egyptian Nile Valley, and Berber spoken in Siwa Oasis of Egypt and west to Morocco, and south into Mali and Mauretania. Collating the data from various studies indicates that the TaqI 49a,f haplotype V is the predominant variant in this region overall; it has notably lower frequencies in the Near East. This is exactly what would be expected for an African model of origins – assuming that the language family was initially and primarily spread by the migration of humans, and not by cultural diffusion. Furthermore, genetic studies that were not concerned with language confirm this. For example, Underhill et al. (2000) state that the V38/M35 lineage was carried into the Near East from Africa in the "Mesolithic"; this V38/M35 in Africa corresponds to the TaqI 49a,f V. It is far more likely that pre-proto-Semitic speakers went into the Near East at this time, where Semitic languages have long since dominated, but the predominant haplotypes in most Semitic-speaking populations are variants VII and VIII in the TaqI 49a,f system (al-Zahery et al. 2003), which are equivalent to M89 associated lineages. The most parsimonious explanation is that ancestral Semitic was adopted by those bearing haplotypes VII and VIII. (It is not argued here that there has not been some two-way migration between Africa and Asia, but in this instance the predominant migration was from Africa; the underived V38/M35 lineage is found primarily in East Africa, and most of the variation in derived and related clades are in Africa.) Wheat and barley agriculture in Africa is first attested in northern Egypt around 5200 BC, dating to some 2000 years after its emergence in Asia (the Near East) based on the standard archaeological evidence; however, it appears as another item in an indigenous foraging strategy, not as a total change in subsistence pattern (Wetterstrom 1993). Newer work, not cited here, confirms this perspective. This finding is not consistent with a mass migration hypothesis of settler colonists involving farmers from Asia (the Near East). By this time Afrasan had already begun to differentiate or was already differentiated at some level. Among the earliest agriculturalists in Asia (the Near East) were speakers of common Semitic based on vocabulary reconstructions (Diakonoff 1998). If there had been a mass migration into Africa of rapidly multiplying farmers, those who came to be known as the ancient "Egyptians" would have been Semitic speakers. Curiously, not even the terms for sheep, goat, wheat, or barley in ancient Egyptian are Semitic loan words; this has interesting implications in light of recent evidence for seed collecting and management in the "green" Sahara around 10,000 years ago. This example reveals the complexity of attempting to synthesize various kinds of data. There can be no a-priori expectations.

Afrasian and its speakers offer other fascinating findings that invite a critical historical enquiry. Chadic speakers living primarily in the central Sahara and northern Nigeria have a very high frequency of an M89 lineage of an R1 subclade. These Chadic speakers phenotypically are generally very dark-skinned with frizzy hair and the facial conformation often thought of as representative of "the African," a point that will be revisited later in relation to the classification of Chadic within Afrasan. Curiously, the frequency of this lineage is far greater in some Saharo-tropical population samples (over 90 percent) than it is in the Near East – its putative place of origin. This is likely due to genetic drift. The frequencies of R1 are not nearly as high in Nile Valley Egypt and other parts of supra-Saharan Africa. What are the historical explanations for this lineage and its distribution, or what could they be? Some writers have postulated that the presence of the lineage corresponds to bearers of the Chadic family of Afrasan spreading into the central Sahara based on the presence of the R-V88 marker (Cruciani et al. 2010); for some this would be reminiscent of the *Hamitic hypothesis*, but should not be confused with that project because of the relatively low frequency of this clade in other branches of Afrasan. Interpretations of Y chromosome data including short tandem repeat (STR) variability from Equatorial Guinea suggest that the V88 marker may have arisen on an R1 background south of the Chad basin and spread north instead of the reverse (Gonzalez et al. 2012). It may very well be that the language family spread independently of the lineage, with the lineage being in the region before the Chadic language family. The presence of the R haplogroup should probably be simply conceptualized as a Paleolithic "back migration" without any necessary cultural implications; in this scenario it becomes a part of the biological ancestry of regional peoples subject to local evolutionary forces. Its distribution in the circum-Saharan areas may reflect later movements of Saharan peoples during mid-Holocene droughts. Another explanation is that some areas of the central Sahara and Nigeria were settled in part by Assyrian refugees in the last millennium BC (see Lange 2011); this interpretation of the archaeological and cultural evidence has not gained wide acceptance among historians.

It is of some historical interest and still very revealing that although some linguists in the early twentieth century thought that Chadic was Afrasan, others greatly resisted its classification into the family (Newman 1980). This was not based on linguistic data but rather on the fact that the Chadic speakers were "black" or "blacker" than other members of the family (Ruhlen 1991). This illustrates the problem, leaving alone the bias for a moment, of having expectations that all connections or relationships would be between entities perceived as being uniform in all respects. This issue is made even more ironic when it is understood that some groups of sampled Chadic speakers have a higher percentage of a "Eurasian" Y lineage than do others who were categorically seen as closer kin to Europeans.

The Afrasan-speaking regions of Africa are anciently connected to speakers of other families via their common P2 "father." This deep genealogical biogeographical relationship precedes that of more connections gained through migration causing gene flow. The E haplogroup places their male affiliation in Africa, no matter the time depth for any given branch. A critical narrative to explain the mtDNA profiles has not yet been developed, and may not "match" expectations based on the Y chromosome findings.

2.4 Genes, Ethno-Population Origins, and Identity

The subject of "origins" is complex because of the term's various meanings. The "origin" of a *population* has multiple dimensions. These include language, biology, and culture. As noted, these are not causally related or linked, but sometimes may "travel" together at least over the short term; assumptions cannot automatically be made about this. Another issue is related to the question of origins in terms of time. How far back into the past can a specific ethnicity/ethnic group as known in the ethnographic present be traced? It is important to remember that the external biological traits of a socioculturally defined group can change based on intermarriage if social rules permit and cultural values are not exclusionary based on external phenotype; gene flow connotes changes in ancestry (unseen genetic variation can obviously also change). Likewise, the cultural traits of a group defined biologically (at any level) can change by the adoption of new languages and customs. Of course the definitions can also change over time and space. All of these increase the work for the population biologist interested in the dimensions of history or reconstruction of genealogy. Research on the origins and current genetic profiles of groups with particular ethnic designations can illustrate the benefits of better multidisciplinary cooperation. A case in point is that of the Fula, who are perhaps best known in West African history as Islamic reformers in the nineteenth century.

The Fula, variously called Fulani, Fulbe, Pula, and Peul, have sometimes been seen as an "anomaly" in tropical West Africa due to the physiognomy and skin color of the most distinctive subgroups, which on average have been noted to usually be different than the West African stereotype. Many have narrow noses and faces and "copper"-colored skin. Fulani communities can be found in the Sahel belt from Senegal to Sudan, with southern extensions into the Cameroons. In a recent report, the "origin" of the Fulani is said to be "unknown," with "tradition" relating them to "Hyksos and Nubians," and their language is said to be related to Berber, an Afroasiatic family member (Rosa et al. 2004). However, current research places the group's "homeland" in West Africa (reported in Keita et al. 2010; McIntosh and Scheinfeldt 2012), and the Fulani language in the Niger–Congo family (Ruhlen 1991). The Hyksos invaded Egypt from the Near East in about 1800 BC and likely were Semitic speakers. They were defeated and eventually returned to Asia. There is no evidence that they reached western Africa or penetrated into the populations of the Sahara or Sahel, and there is nothing about the Fulani that would seem to be primary evidence that indicates a specific Hyksos origin. The Nubians, who live in southern Egypt and Sudan, speak languages in the Nilo-Saharan family, and while there are populations that speak languages in this family far to the west of the Nile Valley, there is nothing specifically Nubian that would make for convincing evidence that the Fulani have a specific Nubian origin. In short, neither Hyksos nor Nubians can be shown to have likely been the ancestral group at the root of the historical origins of the Fulani. In terms of historiography, these accounts are not justified.

The Balanta are another example, a population having some mtDNA signatures that are "unexpected" in the Senegambia. Specifically, they have haplogroups that are called "Eurasian." This group from the Senegambia region is said to have a tradition that they were related to "Sudanese" "from whom they could have separated 2000 years ago with the first spread of kushites [sic] migrations" (Rosa et al. 2004). Another hypothesis is offered that derives them from "a Bantu branch" which separated in the Pleistocene near the Nile. Many West African peoples have traditions of origins from the east; some of these may reflect the influence of Islamic groups, or migrations related to climatic or other events. The word "Sudan" is Arabic and would suggest a period of interactions later than the ancient "kushites," who were not Arabs or Arabic speakers. Sudan is also a generic term, the last part of the phrase "Bilad as Sudan" (Land of the "Blacks"), and in African historiography "western Sudan" refers to the Sahel belt of West Africa. Current research suggests that the "kushites" of the Nile Valley are believed to have spoken a language from the Nilo-Saharan family, although another opinion is that the language was Afrasan. Furthermore, the Balanta do not speak a Bantu language, and this family is not thought to date to the Pleistocene, but rather be of late mid-Holocene at the earliest. The Niger–Congo (or Niger–Kordofanian) family does have a branch in the Sudan – Kordofanian – but it is not linguistically close to the Balanta. Current research does not speak of "camitic" invasions, and it is not certain to which "invasions" Rosa et al. (2004) refer. "Camitic" is a synonym for Hamitic, a term no longer in common usage in modern African studies because the theories associated with it have been proven wrong. There is little reason to think that the Balanta as a *people* have come from the Sudan, but migration of an element of their population from the east is very likely. This does not deny migration, since examples of long-distance migration in Africa exist, such as Arabs coming into Africa, Fulani migrating across the Sahel, and West Africans making the Hajj and not returning. The traditions of long-distance migration to explain origins may have some reality, but not in the literal terms that they are expressed; it has been established that they generally cannot be accepted literally as historical documents, especially if they use terms that are chronologically dissonant, and the events are very remote in time. Specialists in the historiographic use of oral traditions, a debated field for some time depths, have developed methods to interrogate them.

The ideas reported about the Balanta and Fulani are based on older literature and set up investigators to give "unfair" accounts. A non-West African origin for these peoples (*as whole peoples*) is not supportable based on the current research of specialist historians (but genetics do in the case of the Fulani). The distributions of these peoples, written and oral texts from the regions, linguistics, and archaeology invite a priori a holistic approach based on more recent research. It is very important to say that the authors do not say that they necessarily accept the reported views, but unfortunately they do not present other alternatives, which makes it difficult for the readers. It is not possible to know which ideas are of historical interest only, as opposed to being presented as current viable hypotheses. The mtDNA profiles of the Balanta and others may reflect ancient interactions or medieval ones. It is worthwhile recalling that according to al-Omari in the Masalik al Abasar that Mansa Musa, king of Mali, brought back to Mali in the fourteenth century various females purchased in the markets of Cairo, some of whom were of Asian ("Turk") origin. There were many kinds of interactions in the Sahel regions, and although the Balanta were not part of the imperial Malian system, their eastern origin stories may place some of their ancestors in the Sahara, a place of the interactions of many peoples.

Another example of the origins issue conflates material culture similarities with identities, and non-standard linguistic analyses (see Arnaz-Villena et al. 1999). Genetic similarities in the circum-Mediterranean region have been postulated to be related to population movements that also have linguistic correlates of questionable validity. It is argued that many languages of the Mediterranean region are related in a family or grouping called Usko-Mediterranean (UM), which includes Etruscan, Berber, Egyptian, and Basque. Early Nile Valley populations from the Badarian culture were suggested to have migrated to Spain based on pottery similarities, which on critical examination are not supported. Other movements of ancient Saharans to Europe were postulated, and while these cannot be dismissed, better evidence would be welcomed to support this idea. The UM grouping is not a part of current language classifications, which does not mean it should be discounted out of hand, but its unorthodoxy should be acknowledged, and its validity demonstrated by the totality of current linguistic methodology. It is not inconceivable that there were early movements back and forth between Mediterranean Europe and Africa, but the theory of deep cultural unity implied by the UM hypothesis is not well supported. In the case of the Badarians, they were a part of deep cultural tradition in the Saharo-Nile region (Wengrow et al. 2014).

This case illustrates how collaboration would have strengthened or altered the hypotheses. Standard archaeological and linguistic interpretations would have been contributed by specialists.

2.5 Genetics and the Northern Nile Valley: Egypt

An example of the utility of historical resources to explain spatial genetic variation can be found in studies of Egypt, a country not only at the crossroads of three continents but with a well-studied history based on texts, archaeology, and climatology. Egypt flanks the Nile Valley and can be conceptualized as a linear oasis in the eastern Sahara desert. It has had numerous settlements dotted along its length since the mid-Holocene.

Patterns of Y chromosome and mtDNA genetic variation in the Egyptian Nile Valley corridor both show opposing clines for different haplotypes (see Lucotte and Mercier [2003] for the Y chromosome study and Krings et al. [1999] for the mtDNA analysis). One Y chromosome study using the TaqI 49a,f probe found geographical substructuring in Egypt of the most prevalent variants, which are haplotypes V, IV, and XI. Haplotype V (called "northern") decreases in frequency from north (northern Egypt) to south (Egyptian Nubia); IV and XI (called "southern") increase in the opposite direction. Upper (southern) Egypt had intermediate frequencies that were not overall statistically different from those in Nubia, located immediately south of Upper Egypt (Lucotte and Mercier 2003). In another study, mtDNA haplotypes designated "northern" and "southern" were also found to be distributed differentially (Krings et al. 1999). The geographical labeling of variants clearly implies a model of explanation that posits two fairly distinct populations that have created a zone of intergradation in the Nile Valley by bidirectional gene flow.

The observed distribution of these haplotypes has been suggested to perhaps be related to three ancient military campaigns: the invasion and/or occupation of Nubia by Egypt on two occasions (Middle Kingdom ~2000–1600 BC, and New Kingdom ~1550–1300 BC), the last for 500 years, and in turn of Egypt by Nubia (750 BC) for less than 100 years. Later migrations from Europe and the Near East during the Greco-Roman Islamic and Ottoman periods may have made a significant contribution. The spatial distribution has been suggested to be related to movements caused by military conflicts in the two studies. No alternative explanations were offered. While the military explanations are somewhat attractive to explain the pattern of the most common variants, they are likely to be inadequate (Keita 2005). At these dates in the Nile Valley population sizes were large and unless there was a specific policy of population replacement or marrying local women, soldiers are not likely to have so large a genetic impact as indicated by the Y chromosome data. The findings do encourage the exploration of non-genetic data to understand social practices, which can affect gene frequencies. It is known in the late period that Alexander encouraged his troops to marry local (Egyptian) women. There is no historical evidence suggesting that foreign troops gave their daughters or sisters in marriage in any period to those being occupied in any consistent fashion, thus making it less likely that the military incursions explain the mtDNA patterns. Historiographically, the claim based on mtDNA implies evidence that does not exist, and there has been an incomplete enumeration of reasonable possible causes for the patterns. Together these make the explanation an example of misplaced concreteness. There is evidence for strong trade relationships with the Near East during the pre-dynastic period, and Egyptian settlements in the Near East as early as the proto-dynastic period. These may have established a basis for unions to cement trade relationships. A steady stream of such marriages over three millennia would most certainly have had an impact.

A holistic approach that considers varying time depths will augment or supplant this explanation. One issue concerns the original "peopling" of the Nile Valley, and how this happened. This means examining data that can plumb different time depths. The evidence can come from archaeology, various texts, in addition to the biological data. Is the north–south description of variants valid for all markers? Is it possible that the model positing the interaction of idealized discrete groups is false, and that a model that postulates more populations more as processes is more accurate? Archaeological data support the post-glacial re-peopling of the eastern Sahara after the long Late Pleistocene hiatus (Wendorf and Schild 1980, 2001), probably from multiple directions. These Saharans were mobile and would have likely interacted; it is hard to envision that there was no gene flow between populations. Evidence indicates Holocene migration to the Nile Valley from the Sahara, thus contributing heavily to the population resident there (Hassan 1988; Wendorf and Schild 1980), and a cultural basis for interactions with Sudan. During and after the rise of the Egyptian state, archaeological and textual evidence supports ongoing interactions between Nubian and Egyptian populations before the three military invasions. In summary, there are pre-Middle Kingdom events that should be invoked to explain the distribution of the most prevalent alleles.

Pagani et al. (2015) have shown that there was substantial migration/gene flow into Egypt during the first millennium of the Islamic period, which indicates a change from the original population. Movement from southwest Asia coupled with the social practice of polygamy would have increased non-P2 lineages in the male descendant population and southwest Asian lineages among females. Migrations from other sources would play a role. Social selection may have further amplified these effects. Drift may have contributed. There would have been gene flow from southwest Asia and Europe into Egypt during the preceding Christian period, as well due to European and southwest Asian influence, and during the Pharaonic period also based on the Amarna Letters, which indicate marriages to southwest Asian women. What must not be forgotten is that there was an indigenous population in the Nile Valley that emerged from processes in northeastern Africa.

2.6 The Case of the Nilotes/Nilotics: Considerations for Population History

The Nilotics/Nilotes, as represented in older literature, are a group of culturally diverse populations that reside in southern Sudan, Sudan, southern Ethiopia, Kenya, and countries of the Great Lakes in central Africa, who speak what are usually called Nilo-Saharan languages. The term Nilotic can (and is sometimes used to) refer to any peoples associated with the Nile, including modern and ancient peoples from Egypt and Nubia, but the term here will refer to the older anthropological usage that may be problematic. Many of the more traditional Nilotes/Nilotics of today practice predominantly a subsistence economy based on cattle pastoralism. This pastoralism may be ultimately related to what Wengrow et al. (2014) have called the "primary pastoral community," which was also ancestral to ancient Nubian and Egyptian cultures. Other names could be chosen for this, given its wide-ranging impact, such as "ancestral African pastoral tradition" or "Saharo-Nilotic ancestral pastoral tradition." It likely dates to the sixth millennium BC, although its clearest and geographically greatest distribution is demonstrable in the fifth millennium BC. Currently, an original "homeland" and past expansions of "Nilotes" have largely been based on older work which has not had the full benefit of a synthesis of linguistics, archaeology, and genetics. The construction of a narrative of the history of much of this group – especially the northern groups of the earliest times – is still a work in progress and will require a multidisciplinary approach. Ehret (1971: 3), in the introduction to *Southern Nilotic History*, a historical work on the Southern or Highland Nilotes, such as the Kalenjin in Kenya, states the nature of this kind of historical work:

> A people's history does not begin with their memories. It begins rather with their emergence as a separate group. If their emergence and early development lie beyond the reach of the historian's familiar source, human memory as it appears in documents and oral traditions, the historian cannot therefore say that such events are darkness and not history . . . It would be a different kind of history. It would not name individual leaders, or describe particular wars, or encompass its events with precise dates. But it would nevertheless be a history of population movement, cultural change and interchange, and technological development. The history of the Southern Nilotic peoples of East Africa in the eras preceding the 1600s must be this kind of history.

In general, he makes use of historical linguistics and archaeology in his work. Interest in the Nilotics/Nilotes increased following the publication of the classic *The Nuer* by Evans-Pritchard (1940), which brought them to greater notoriety in the Western scholarly community. The accuracy of these accounts as depicting Nilotic history and society is contentious, as might be expected for these older works. Older biases in theory and data collection must be systematically examined in order to make progress. Burton (1988) sees Nilotic studies as largely dependent on contingent cross-sectional observations, and stresses the need for further historical and scientific data to consolidate or refute the conclusions/narratives based on often anecdotal accounts. It is possible that the data from genomics/genetics can provide evidence that can be used in the construction of a more complete narrative history of the Nilotic peoples (as usually defined).

Together with some other eastern Africans, the "classic" Nilotes of southern Sudan carry in substantial frequencies some of the most ancient haplotypes in both male and female lineages. These haplotypes relate them spatially and temporally to early chapters of modern human evolution as suggested by recent work done in Sudan (Hassan et al. 2008; Underhill et al. 2000). The recovered Y chromosome lineages – A (M13), B (M60), and E (M78) – are most likely to represent local deep historical descent (pre-dating the Nilotes of record and the Nilo-Saharan language family) rather than being an outcome of a more recent gene flow. It is noteworthy that haplogroup A is predominantly confined to eastern Africa, and the San of southern Africa (Underhill et al. 2000). Their mtDNA similarly displays ancient and ancestral haplotypes, but nevertheless is more heterogeneous than their Y chromosome counterparts. It is likely that the ancestors of those commonly called Nilotes used to extend over a wide geographical range based on observed distributions. The distribution of the unique, nearly population-specific, Y chromosome markers along with elevated population frequencies may be assumed to represent a relic of the historical movement and expansion of the carriers of these haplotypes combined with evolutionary forces, e.g., genetic drift and perhaps some hitch-hiking effects. However, the distribution of these markers should not be used solely as a record of historical expansions, apart from sociocultural and historical contexts. It is important to reiterate that these ancient markers are older than current ethnic, geographical, and linguistic units. Certain populations harbor more of the most ancient lineages, but we must be wary of slipping into thinking that these human repositories are *living fossils*, and that their ethnocultural or linguistic identities are equal in time depth to those lineages.

The predominance of the Y chromosome group A in southern Africa (Underhill et al. 2000) has sometimes been used as a marker of one of the earliest human migrations from eastern Africa, where early modern human evolution is believed to have taken place. The San represent one of the few groups that carry this haplogroup in considerable frequencies, but with a different set of associated mutations than the Nilotes, who also have a high frequency of A. The San bear little morphological similarity to the Nilotes (as used here). This may indicate that the predominance of haplogroup A among the San is: (1) a product of genetic drift; (2) a relic of an early population divergence from the human ancestral population; and (3) that no single physical trait or feature could be ascribed to the ancestral group of humans, which in any case is not known. Neither San nor Nilote morphology likely characterize early moderns. The presence of ancient markers like A in northwestern Africa, along with certain fossils, has led to some hypothesizing of origin in that region for modern *Homo sapiens*. In fact, the phenotype of early modern humans is not known beyond their skeletal anatomy, and this is not completely known because there are not many fossils. The dark skin tone and tall stature of the present Nilotes might be an adaptation to the continuous habitation of the Sahara and the Sahel area, where rock drawings feature figures resembling potential ancestors of present-day Nilotics. The relatively light skin of the San may also be extremely ancient or relatively "modern" on the scales of tens of thousands of years. Both rock drawings and complex morphological traits are difficult to date, although it is known that the very light skin of some human groups is a relatively new development, likely dating largely to within the last 10,000 years. These biological traits could also vary over time – with possible reversals. Caution is necessary in developing the relevant hypotheses – or tentative explanations. This is also true for linguistic features like clicks – it is difficult to be certain of when they arose. The migration of ancestral Nilotes and no doubt others from the Nile basin to the Sahara and back has been attributed to East African and Saharan climate pulses (Maslin et al. 2014).

Diop (1991) hypothesized east to west migration based on his interpretation of various data for specific peoples, first presenting his suggestions in the 1950s. His ideas, which center on a specific Egypto-Nubian/Nile Valley origin (*sensu stricto*) for some West African populations and a relationship with their languages, have not been widely accepted, but varying aspects of his work have currency among some scholars. Broadly, he sees legacies of Nile Valley culture in parts of West Africa, an idea that is similar to that of some colonial era Western writers working in diffusionist traditions. A distinction has to be made between primarily the diffusion of ideas alone versus the mass migration of peoples. Ideas and genes can travel between adjacent populations, in theory making their way over great distances over time. In general the hypothesis that there was migration from east to west in the Sahel or Sahara is interesting. The details of any such hypothesis are what have to be elucidated. If there are cultural legacies, how much would this be due to point-to-point migration versus a more complex transmission through geographically intermediate populations? Is the "legacy" related to what was once called the "Aqualithic," a mid-Holocene cultural horizon that shows similar material culture traits from the Nile to the Niger during the first half of the Holocene during the so-called "Green Sahara?" Or is it related to more recent migratory events in the first millennium BC related to population upheavals in the eastern Sahel? Or to earlier desiccation events in the less arid Sahara? There were likely many migration events that can be marked in different ways. Given that modern humans most likely originated in eastern Africa (although some suggest supra-Saharan western Africa), then east to west migration is a given, and it is possible that modern humans arose in Africa due to the interactions of multiple early African populations, not emergence from just one. This would be the peopling of western Africa and likely had different layers and phases. The early Niger–Congo language or proto-language family spread from east to west. The question of when the M2 Y lineage arose, which is now common and predominant in West Africa, would be relevant to questions of migrations, as would be efforts to ascertain the biological profile of the earliest Songhay. Of course, populations can be swamped, die out, or be displaced in ways that obscure the complete history of a region. Today's Songhay, specifically those that live in Timbuktu, are also descendants in part of the Moroccan soldiers who began to marry local women after the conquest of Songhay in 1591. Songhay-related speakers live along the Niger. Just as interesting is the Koranje language, of likely Nilo-Saharan affiliation, spoken in Algeria. As to direct migration from the Nile Valley at any given time, the data would have to be gathered and examined carefully based on modern standards. It may also be that the Chadic, or rather its speakers, had a bidirectional impact, given Chadic's affiliation with ancient Egyptian, or that there were overlapping areal effects.

Were there several "waves" of migration after the establishment of modern humans in the region, one washing over the other? What was the mode of migration? Were there local movements of people from one community to another that brought changes in languages and/or transmitted other aspects of culture? Was there a *mfecane*-like movement at different times by different peoples? Unlikely. These are a few of the questions that could be explored. There is likely evidence for migrations at different times. For example, possible Christian churches have been found in Chad, perhaps of Nubian origin. The Songhay group of languages, spoken along the Niger River, are generally considered to be Nilo-Saharan. This would make them the geographical outlier in the Nilo-Saharan family, being the furthermost western group. It is not known with certainty when their speakers first arrived, but this is an interesting question. These proto-Songhay may have brought a pool of ideas born of a particular Saharan physical and cultural environment, and these may have spread to other groups. The task for archaeologists, linguists, biologists, and cultural anthropologists is to contribute data for the construction of a narrative that meets the previously mentioned standards.

Biological and cultural assimilation can take many paths, and sex bias is often a feature. For example, skewed and biased mating practices are often epitomized in ideological or cultural beliefs, such as in the marriage culture among Indian castes, and manifested in a downward mobility of the upper castes' Y chromosome (Bamshad et al. 1998). This has been seen in other situations in which a strong dominance hierarchy exists that is supported by sociopolitical power. The larger possibilities are also of interest. Groups could change language and culture without much biological assimilation, or a given group could assimilate others via marriage or concubinage and hence change its biogenealogy without changing culture and language. There are many intermediate scenarios. One can envision an example of such change in situations in which a pastoral family unit or community loses its herds and joins farmers as a means of survival. Something similar could happen – has happened – in situations in which farmers have become successful pastoralists, thus changing their identity because their farming community does not have a pastoralist community and identity is based on subsistence strategy. Polygamist pastoralist societies are usually male-dominated, and they may adopt wives from local farming communities rather than allowing access by these communities into their female gene pool. In theory, and depending on the larger socioenvironmental context, this might eventuate in language loss or assimilation over time due to languages being transmitted via the mothers – a more likely scenario if the mother's language is widely spoken and the family lives in the mother's community. This obviously would not happen all the time: The foreign woman who is brought into a large, stable community would eventually learn its language. It would be less likely if marriage involves moving into regions where the father's language and ethnic group are dominant. This could be modeled using population genetics and network theory in deteriorating environments such as the mid-Holocene Sahara, a situation of a metapopulation, based on recent archaeological work. If metapopulation theory with extinctions and coalescence is invoked, one can see the emergence of language isolates, great biological heterogeneity, "unusual" combinations of lineages, and other population phenomena which do not conform to expectations. This should be expected to characterize Africa in many regions.

Recent analysis of the Y chromosome E haplogroup provides supporting evidence of migrations. The origin and distribution of the E haplogroup – and its branches – does indicate migration, but again the question is the time depth and archaeological correlates of migration. A brief digression is in order. The E haplogroup and its Yap "parent" have been a point of contention, with both being given an African origin by some researchers – the original position of most workers, but now with some postulating an Asian (Near Eastern) origin. There is no convincing archaeological signal in the Middle Stone Age or Late Stone Age for a movement from Asia into Africa. It is just as parsimonious to derive Yap (DE) from M168 in Africa, with both leaving at the same time, as it is to create a scenario that has multiple migrations coming back into Africa. The DE in Asia would become D and E in Africa.

Genetic markers on both Y chromosome (specifically the M78) and mtDNA of the Nilotes indeed shows the presence of haplotype sharing with the pastoralists of other linguistic families (Hassan et al. 2008; Gebremeskel and Ibrahim 2014). These may be dated to a time subsequent to the domestication of cattle during the early Holocene, namely the root marker M78; a date suggested by Trombetta et al. (2015) for the mutation to arise is in a time window between 20.3 ka (95 percent CI: 16.2–25.4 ka) and 14.8 ka (95 percent CI: 11.6–18.5 ka), which is not very different from the 12 ka date suggested by Elhassan et al. (2014) for the human expansions associated with animal and plant domestications. However, in the western half of the "old world," plant – wheat and barley – domestication is thought to have first occurred in the Levant, specifically among Semitic speakers as found by the late Igor Diakonoff (1998). Some of those ancestral to the early Neolithic populations likely had a more direct eastern African ancestry as determined by an analysis of ancient DNA data of some Natufian remains (Shriner, pers. comm.). There is no evidence that there was demic diffusion of wheat and barley agriculture plus ovacaprine domestication into the Nile Valley or Sudan. Demic diffusion means the migration of a cultural trait due to the actual migration of those who practice the specific culture, which would have led to the early Egyptians and some Saharans being Semitic speakers. However, this is not the case; ancient Egyptian is not Semitic and neither are the original languages of the Sahara. Berber is not Semitic. The earliest Holocene dates suggested for cattle domestication in the Sahara by the late Fred Wendorf and his colleagues no longer have wide acceptance, but there is little to dispute conceptually in the idea of indigenous cattle domestication in Africa. Pastoralists likely carried some of these markers to various regions because the locale of cattle domestication and the emergence of M78 coincide; this possible claim is made with great caution.

Interestingly, few of the distinctive markers common among Bantu speakers (see below) have been detected in non-Niger–Congo-speaking pastoralists, with the possible exception of the Luo. If the original "homeland" of all cattle herders is the eastern part of the Sahara, then is it possible to make an estimate of dates of their southward movements to the Lakes? Was it a single movement or successive movements, and to what extent did the environment and political history motivate these waves? Of course, it is important to avoid conflating culture, biology, and language. While these sometimes travel together in whatever combination, they are not obligatorily connected or associated.

Analysis of genetic loci with varying evolutionary rates – e.g., microsatellite, mtDNA, Y chromosome, and autosomes – might provide reliable estimates of population origins, divergence times, and migration, particularly if these estimates can be dated or linked with confidence to climatic, archaeological, linguistic, and demographic data. This is in keeping with the concern expressed here about how to rank different kinds of data. A possible example is the sickle gene, in which a case of admixture between the Nilotic Luo and Bantu speakers has been inferred based on the presence of the sickle gene among the Luo in Kenya. Current Nilo-Saharan speakers in general are not known to carry the sickle mutation and by extension to have not been populations in which the mutation either developed or became elevated to a high frequency.

This observation has been used to date the hypothetical migration of the sickle gene across the Sahel to eastern Africa (Bereir et al. 2007); of course, ultimate geographical origins are not known.

Such approaches may also be applied to ancient DNA recovered from archaeological and burial sites. The sickle mutation has been found in a late pre-dynastic Egyptian mummy (Marin et al. 1999), which may indicate something about the network of populations in northeastern Africa and the Sahara during the fifth and sixth millennia BC. Analysis of DNA samples from burial sites in Northern Sudan (pre-Kerma) support a predominance of Y haplogroup A during the "Mesolithic," but sample sizes are not large and such results must be viewed and used with great caution. This changes over time.

The history of the Nilotes is an example of where genetic data may provide clues and insights on specific historical topics like migration when the historical and archaeological records are insufficient, either because of historical time depth or reasons pertaining to cultural replacement, and hence neither linguistic nor archaeological evidence can give more "complete" information. However, genetics cannot give all of the answers or relevant data either. The goal is to produce an account that is fair, balanced, and not misleading, and that is also free of biases, both explicit and implicit. All of the evidence must be carefully integrated; there is exciting work for young scholars, especially from Africa. Those from communities in the Nile basin, from the river's origin to Egypt, where it discharges into the Mediterranean, may be especially interested in this undertaking.

2.7 Brief Discussion and Conclusions

These cases and examples collectively illustrate the value of more aggressive collaboration with a range of specialists from other fields. The examples could be easily multiplied. Multidisciplinary teams can in theory help improve the accuracy of research design and interpretations. In work in which patterns of genetic variation are used as the primary data to drive interpretations, specialists from other fields can help raise hypotheses about such singularly based interpretations. In work that derives a framework from non-genetic data, geneticists can help devise models that will account for known facts. The reciprocal illumination afforded by such give and take will increase the value of this work so that future studies have a firmer foundation. African scholars from the Mediterranean coastal countries to the Cape of Good Hope can be envisioned to have a major role in new research efforts.

Acknowledgments

This work is a collaborative effort of MI and SOYK, and includes parts of a revised version of a paper by SOYK that was published in *The Backbone*, an online journal of the Cobb Laboratory, Howard University, reproduced with permission: Keita SOY (2015) History and genetics in Africa: the need for better cooperation between the teams. *The Backbone* 1(1): 1–13.

References

Ammerman AJ and Cavalli-Sforza LL (1984). *The Neolithic Transition and the Genetics of Populations in Europe*. Princeton University Press.

Arnaz-Villena A, Martinez-Laso J, and Alonzo-Garica J (1999). Iberia: population genetics, anthropology, and linguistics. *Hum Biol* **71**: 725–743.

Arredi B, Poloni ES, Paracchini S, et al. (2004). A predominantly Neolithic origin for Y-chromosomal DNA variation in North Africa. *Am J Hum Genet* **75**(2): 338–345.

Bamshad MJ, Scott Watkins W, Dixon ME, et al. (1998). Female gene flow stratifies Hindu castes. *Nature* **395**: 651–652. DOI: 10.1038/27103.

Bereir RE, Hassan HY, Salih NA, et al. (2007). Co-introgression of Y-chromosome haplogroups and the sickle cell gene across Africa's Sahel. *Eur J Hum Genet* **15**(11): 1183–1185.

Blench R (1995). Is Niger-Congo simply a branch of Nilo-Saharan? In *Proceedings: Fifth Nilo-Saharan Linguistics Colloquium*, Nice, 1992, 83–130. Eds Nicolai, R and Rottland, F. Rudiger Koppe.

Burton JW (1988). Nilotic studies: some past problems and prospects. *Anthropos* 83(4–6): 453–468.

Cruciani F, Trombetta B, Selitto D, et al. (2010). Human Y chromosome haplogroup R-V88: a paternal genetic record of early mid-Holocene trans-Saharan connections and the spread of Chadic languages. *Eur J Hum Gen* 18: 800–807.

Diakonoff I (1998). The earliest Semitic society: linguistic evidence. *J Semit Stud* 43: 209–219.

Diop CA (1991). *Civilization or Barbarism: An Authentic Anthropology*, 3rd edn. Hill Books.

Ehret C (1971). *Southern Nilotic History: Linguistic Approaches to the Study of the Past.* Northwestern University Press.

Ehret C, Keita S, and Newman P (2004). The origins of Afroasiatic. *Science* 306(5702): 1680.

Elhassan N, Gebremeskel EI, Elnour MA, et al. (2014). The episode of genetic drift defining the migration of humans out of Africa is derived from a large East African population size. *PLoS One* 9(5): e97674. DOI: 10.1371/journal.pone.0097674.

Evans-Pritchard EE (1940). *The Nuer: A Description of the Modes of Livelihood and Political Institutions of a Nilotic People.* Clarendon Press.

Gebremeskel EI and Ibrahim ME (2014). Y-chromosome E haplogroups: their distribution and implication to the origin of Afro-Asiatic languages and pastoralism. *Eur J Hum Genet* 22: 1387–1392.

Gonzalez M, Gomes V, Lopez-Parra AM, et al. (2012). The genetic landscape of Equatorial Guinea and the origin and migration routes of the Y chromosome haplogroup R-V88. *Eur J Hum Genet.* DOI: 10.1038/ejhg.2012.167.

Gould S (1987). Petrus Camper's angle. *Natural History.* August: 12–18.

Gregersen EA (1972). Kongo-Saharan. *J Afr Ling* 11(1): 69–89.

Guthrie M (1962). Bantu origins: a tentative new hypothesis. *J Afr Lang* 1: 9–21.

Hassan F (1988). The predynastic of Egypt. *J World Prehis* 2(2): 150–180.

Hassan HY, Underhill PA, Cavalli-Sforza LL, Ibrahim ME (2008). Y-chromosome variation among Sudanese: restricted gene flow, concordance with language, geography, and history. *Am J Phys Anthropol* 137: 316–323.

Hiernaux J (1975). *The People of Africa.* Charles Scribeners Sons.

Keita SOY (2005). History in the interpretation of the pattern of p49a,f *Taq*I RFLP Y-chromosome variation in Egypt: a consideration of multiple lines of evidence. *Am J Hum Biol* 17: 559–567.

Keita SOY and Kittles RA (1997). The persistence of racial thinking and the myth of racial divergence. *Amer Anthropol.* DOI: 10.1525/aa.1997.99.3.534.

Keita SOY, Jackson F, Borgelin L, Maglo K (2010). Letter to the editor: commentary on the Fulani – history, genetics and linguistics, an adjunct to Hassan et al. 2008. *AJPA* 141: 665–667.

Krings M, Salem AH, Bauer K, et al. (1999). MtDNA analysis of Nile River Valley populations: a genetic corridor or a barrier to migration? *Am J Hum Genet* 64: 1166–1176.

Lange D (2011). The founding of Kanem by Assyrian refugees ca 600 BCE: documentary, linguistic, and archaeological evidence. Boston University Working Papers in African Studies 265.

Lucotte G and Mercier G (2003). Y-chromosome haplotypes in Egypt. *Am J Phys Anthropol* 121: 63–66.

Marin A, Cerutti N, Massa ER (1999). Use of the amplification refractory mutation system (ARMS) in the study of HbS in predynastic mummies. *Boll Soc Ital Biol Sper* 75(5–6): 27–30.

Maslin MA, Chris M, Brierley CM, et al. (2014). East African climate pulses and early human evolution. *Quat Sci Rev* 101(1): 1–17.

McCullagh CB (2000). Bias in historical description, interpretation, and explanation. *History and Theory* 39: 39 66.

McCullagh CB (2004). What do historians argue about? *Hist Theor* 43: 18–38.

McIntosh S and Scheinfeldt LB (2012). It's getting better all the time: comparative perspectives from Oceania and West Africa on genetic analysis and archaeology. *Afr Arch Rev* 29: 131–170.

Newman P (1980). *The Classification of Chadic within Afroasiatic.* Universitaire Pers.

Pagani P, Schiffels SG, Gurdasani D, et al. (2015). Tracing the route of modern humans out of Africa by using 225 human genome sequences from Ethiopians and Egyptians. *Amer J Hum Genet* **96** (6): 986–991.
Robertson J and Bradley RA (2000). A new paradigm: the African Early Iron Age without Bantu migrations. *Hist Afr* **27**: 287–323.
Rosa A, Brehm A, Kivisild T, Metspalu E, and Villems R (2004). MtDNA of West Africa Guineans: towards a better understanding of the Senegambia region. *Ann Hum Gen* **68**: 340–352.
Ruhlen M (1991). *A Guide to the World's Languages: Volume 1, Classification.* Stanford University Press.
Sanders E (1969). The Hamitic hypothesis: its origin and function in time perspective. *J Afr Hist* **10**: 521–532.
Takacs G (1999). *Some Problems of Egyptian's Position within Afro-Asiatic and among African Languages. Etymological Dictionary of Egyptian, Vol 1, A Phonological Introduction.* Leiden, pp. 35–48.
Trombetta B, Atanasio E, Massaia A, et al. (2015). Phylogeographic refinement and large scale genotyping of human Y chromosome haplogroup E provide new insights into the dispersal of early pastoralists in the African continent. *Genome Biol Evol* **7**: 1940–1950.
Underhill PA, Shen P, Lin AA, et al. (2000). Y chromosome sequence variation and the history of human populations. *Nat Genet* **26**: 358–361.
Wendorf F and Schild R (1980). *Prehistory of the Eastern Sahara.* Academic Press.
Wendorf F and Schild R (2001). *Holocene Settlement of the Egyptian Sahara. Vol I. The Archaeology of Nabta Playa.* Kluwer Academic.
Wengrow D, Dee M, Foster S, Stevenson A, and Ramsey CB (2014). Cultural convergence in the Neolithic of the Nile Valley: a prehistoric perspective on Egypt's place in Africa. *Antiquity* **88** (339): 95–111.
Wetterstrom W (1993). Foraging and farming in Egypt: the transition from hunting and gathering to horticulture in the Nile Valley. In Shaw T, Sinclair P, Andah B, and Okpoko A, eds., *The Archaeology of Africa: Food, Metals and Towns.* Routledge, pp. 165–226.
al-Zahery N, Semino O, Benuzzi G, et al. (2003). Y-chromosome and mtDNA polymorphisms in Iraq, a crossroad of the early human dispersal and of post-Neolithic migrations. *Mol Phyl Evol* **28**(3): 458–472.

3 Disease, Selection, and Evolution in the African Landscape

Muntaser E. Ibrahim and Endashaw Bekele

Africa's geographic location may have dictated its predestination as a climate refuge and the scene of consecutive chapters of biological adaptive radiation throughout time. The continent is still home to some of the most spectacular forms of biodiversity on the globe, and is believed to be the stage for the heroic, perhaps rather tragic, early chapter of human evolution. These distinctive features invite a renewed investigation and interpretation of this epic saga.

Humans as relative newcomers in the evolutionary time scale have been experiencing disproportionately rapid biological and cultural changes in adapting to the immense challenges posed by the environment and the consequences of their unique speciation process. Our evolutionary fate has hence been shaped and remains so by adaptation to natural selection by diseases, disparate climatic terrains, and new cultures. Both individual adaptation (adaptability) and population adaptation (adaptedness) are under genetic control. The adaptation of populations and individuals that produce offspring in the population extends over many generations and hence is clearly an evolutionary phenomenon. These intricate relationships have been ramified in a spectrum of co-evolutionary patterns manifested in the breadth and width of genetic variation, particularly in Africa where our species originated.

Understanding the biological basis of adaptation and the underlying vast reservoir of variation is not only essential to satisfy intellectual and scientific curiosity, but is imperative nowadays for the implementation of new approaches in individualized and population medicine, a necessity for effective and parsimonious health delivery and intervention. Furthermore, Africa is a place where cultural diversity and lifestyle may have a measurable bearing on health even today. For example, it is the only continent where heart disease and other lifestyle diseases are not the leading causes of mortality in adults. The alarming rise in the incidence of heart diseases, hypertension, and diabetes in the developing world (including Africa) reported by the World Health Organization (Morabia and Abel 2006), together with an impending burden of infectious diseases, poses a unique health challenge to the continent. Africa may be following a pattern seen in the industrialized world, where the sweeping advance of lifestyle diseases may soon overshadow the continent's notorious historical burden of infectious disease. Notably, socioeconomic development, increased adoption of the lifestyles of industrialized nations, reduction in physical activity, and large increases in consumerist behavior – including diet – are already taking their toll on the health of Africans.

With the scientific community increasingly turning to genome-wide approaches that require both large biological datasets and well-defined sets of phenotypic information, Africa is right at the center of many quests – for example, in the "nature versus nurture" conundrum in attempting to unravel the genetic basis of health and disease, particularly diseases of complex inheritance. Few studies, however, actually involve Africans due to the prohibitive cost and lack of well-defined cohorts of target diseases. There is a dire need to carry out such studies in Africa, largely justified by the fact that the majority of biological and phenotypic traits in biomedicine are based on phenotypic measurements and indexes of non-Africans. Data on African Americans represent only a historical subset of a broader African diversity. Genetically, Africans remain the most understudied population and Africa the most underrepresented continent in genetic databases. This is in spite of the continent's endowment with the greatest share of genetic diversity and potentially the widest repertoire of phenotypic traits underpinned by such diversity, whether clinical, anthropological, or biochemical.

A number of factors, including population variation, environmental diversity, and economic disparity, have combined to make Africa an endemic focus for various infectious diseases, including some of the most notorious parasitic diseases of all time, such as malaria, leishmaniasis, and trypanosomiasis. In addition, peculiarities in African genetics include the "focality" of diseases and traits, which is probably a feature emanating from the pronounced population structure – an outcome of geographic and cultural barriers and history. Millennia of pathogen–host coevolution have resulted in major resistance–susceptibility loci, which are manifested in unique population and geographic clustering best exemplified in the distribution of common malaria-protective polymorphisms such as sickle cell anemia (SCA), with its diverse haplotype structure attesting to such "focality."

A lengthy history of existence for humans in Africa, extending more than anywhere else and resulting in differentiation of major traits, entails an interesting conjecture: The majority of variation underlying such traits is embedded within the African gene pool, particularly that of East Africa. In this chapter, we discuss the importance of understanding the influence of genetic population structure in health and disease, and the role of natural selection and drift in shaping that structure and the phenotypic features of populations.

3.1 The Peculiarity of Genetic Structure In and Out of Africa

Numerous lines of evidence suggest that *Homo sapiens* evolved as a distinct species in Africa around 60,000–100,000 years before present (YBP) and began major migrations within and out of Africa 40,000 YBP. By 20,000 YBP, our species had effectively colonized the entire Old World, and by 12,000 YBP *Homo sapiens* already had a global distribution.

Such obvious success may have emanated from unique biological features leveraged during early human evolution and retained within the current genomes in the substructured populations of Africa. The exceptional number of unique variants in

Africans, especially East Africans, stands in contrast with a dramatic reduction in genetic diversity in populations living in other continents (Chiaroni et al. 2009; Deshpande et al. 2009). The population of East Africans turned out to be not only at least twice that of other populations, including Africans, but the population has also retained the greatest number of ancestral alleles of sex-specific markers (Elhassan et al. 2014) and autosomes (Keinan et al. 2009). The human original founding group estimated size is 1000–10,000 individuals – a decent amount of diversity for a "kick-off." This number is believed to have fluctuated over time, often suffering severe bottlenecks (Tenesa et al. 2007). Populations with the highest effective size (N_e) harbor a larger number of variants at the stem of the human evolutionary tree, as shown in Figure 3.1. Furthermore, populations of low effective population size were shown to have lower constraints on their allele fixation rate (Eyre-Walker 2002), thus affecting in turn their N_e and the genotype–phenotype degree of correlation in the long term.

A peculiar African population genetic structure is also reflected in the multiple facets of the evolutionary history of Africans, including the low and different linkage disequilibrium patterns (Reich et al. 2001), which may in turn be due to the peculiarity of African histories of migration and admixture that break the linkage disequilibrium between loci over time, and the relatively large effective size of its founder population (Elhassan et al. 2014), or both. Tishkoff et al. (2009) have defined the broad genetic structure of African populations, which is strikingly concordant with Africa's linguistic families and geography. These peculiarities have their implications for association studies and the mapping of simple Mendelian and complex genetic traits, as has been demonstrated not only in the clustering of physical traits and diseases, but also by findings from the *MalariaGen* consortium, where the impact of the ethnic population structure has been shown to affect in a fundamental manner the outcome in such association analyses (Jallow et al. 2009). However, and in spite of such structure, indirect evidence supports the existence of extensive mobility and gene flows. In East Africa, for instance, a shared origin of the Nyangatom with other populations (such as the Toposa, Turkana, and Jiye) in the Nilo-Saharan Karimojong group in northern Uganda has been postulated on the basis of ethnologic and linguistic evidence. Studies on African populations need to consider such complexities when choosing a population to be sampled. Sample sets must be based on fine-scale mapping – a bottom-up approach – and should take into account cultural, linguistic, and lifestyle ethnography, physical boundaries, urban versus rural environments, as well as internal and external diasporas (Poloni et al. 2009). Among the benefits and utilities of defining the structure of a population is the single nucleotide polymorphism (SNP) tagging approach. It has been shown that it is possible with a limited set of population markers (e.g., on the Y chromosome) to guide both linkage (Miller et al. 2007) and association (Eid et al. 2010) analyses efficiently.

Prior to agriculture and animal domestication, humans are believed to have lived in patchy communities through periods when the force of drift was acting more effectively. This led, through the relative isolation of human communities, to the formation of some of the common phenotypes and traits seen in humans today,

(a)

(b)

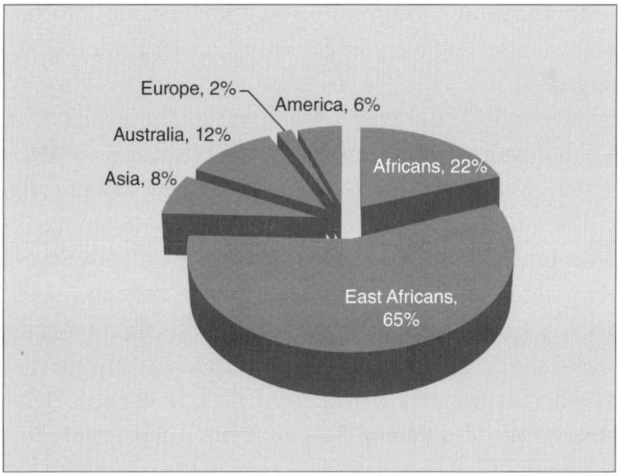

Figure 3.1 (a) A sketch depicting the human evolutionary tree, where due to the large effective size of African populations, most basal polymorphisms exist at the base of the tree. The branches portray the out-of-Africa scenario and the accompanying genetic drift. As a consequence, more straightforward genetic–phenotypic relationships are expected in the tree branches. (b) The pie chart attests to the above, with the effective east African population accounting for nearly two-thirds of the effective variations.

mostly prior to the migration out of Africa. De Magalhaes and Matsuda (2012) presented evidence of high divergence in genes previously suggested as targets of selection related to skin pigmentation, immune response, male reproductive system, and dietary adaptations. Genome-wide scans have also revealed novel candidates contributing to population-specific traits. These include genes related to neuronal development and behavior that may have been influenced by cultural processes (de Magalhaes and Matsuda 2012).

Several polymorphisms are bound to either have emerged after the migration out of Africa or subsequently amplified as a result of peculiar adaptation to specific environmental challenges under natural selection. The *TLR1* locus, for example, is extremely differentiated between Africans and non-Africans. The protective dysfunctional 602S allele is rare in Africa, but expands to become the dominant allele among individuals of European descent. This supports the hypothesis that this locus may be under selection from *Mycobacterium* or other pathogens that are recognized by TLR1 and its co-receptors (Wong et al. 2010). Although in many examples the predisposition alleles maintain their frequencies in Africa, other loci show contrasting distributions, where the predisposition alleles are significantly prevalent in European populations while the MRCA pre-dated human migration out of Africa. Alleles like the NOD2 associated with Crohn's disease are thought to have undergone negative selection, with the deleterious haplotype becoming almost extinct in Africa while heterozygote advantage and/or intergenic interactions maintained the present frequencies among Europeans (Nakagome et al. 2012). Even in situations in which mutations underlying completely new traits occurred subsequent to the human expansion out of Africa, understanding the patterns of gene fixation in comparison to ancestral African alleles will give insight into the mode and date of selection event(s) affecting a particular locus. The problem is to identify such event(s) with certainty.

Allele frequency patterns may form some of the evidence supporting a recent history of major human trait emergence. Selection signatures are present among diverse populations that are believed to have diverged and migrated from Africa at an early stage. There is evidence for recent positive selection in the *FOXI1* gene region in Africa, where climate might be related to this recent adaptive event in humans. Of the multiple functions of FOXI1, its role in kidney-mediated water–electrolyte homeostasis is the most obvious candidate for explaining a climate-related adaptation (Moreno-Estrada et al. 2010). Bains et al. (2013) reported significant intra-African diversity at the *CYP A5* gene, which is likely to contribute to multiple pharmacogenetic profiles across the continent. Their results revealed a significant correlation between *CYPA5* expression phenotypes and aridity data consistent with a hypothesis that the enzyme is important in salt retention adaptation.

Based on the above overview, the pre-exodus of humans from Africa is a period worthy of renewed investigation. Many of the adaptive traits have their origin in that period. The genomes of the extant African populations is the best available source to mine for origins of these traits and to make sense of their biology. The area between the Red Sea and Lake Chad is especially interesting; the stage would have expanded to include much of the current Sahara and the northwest mountains as the likely scene of much of the early-phase events of major adaptations in humans.

3.2 Population Structure Drift and Culture

Coincidental associations between genome compartments, cultural elements, and genetic variation as embodied in the tagging approach mentioned earlier are mostly the outcome of genetic drift – a potent evolutionary force that will be discussed in what follows.

Drift, or the change in allele frequency under sampling effect, works in combination with all evolutionary forces, including migration and selection. Drift creates differences in allele frequency due to chance based on sampling under different conditions and means of isolation, either through biological, cultural, or geographical barriers, conditions under which the effects of natural selection become amplified.

Drift, as one of the main forces in shaping population genetic structures, has been amply studied in the out-of-Africa context, particularly in relation to what is known as the serial founder effect (Chiaroni et al. 2009). It can be observed on population markers used to trace human migrations, where specific patterns unique to certain populations are observed. The impact of migration on population structure in relation to genetic drift has not been well documented, especially in Africa. A possible justification is that within-Africa migration was relatively limited in range, although cases of wide range expansion and isolation exist – for instance, those of the south African hunter-gatherers, the Bantu expansion, and the Afroasiatic pastoralists (González-Santos et al. 2015). Demic expansion or geographic isolation are not the prime or sole mechanisms for drift. Populations could be set apart by cultural barriers like languages. Africa is host to some of the most diverse portfolios of languages and dialects. The linguistic dimension as a cultural and genetic barrier is one of the most intuitive aspects in gaining insights into population relationships. It is used now almost routinely together with geography and ethnicity to define populations as a substitute for the archaic biblical and racial terms and definitions of human groups (Caucasian, Mongoloids, Hamites, etc.). Linguistic/genetic links, originally suggested by Cavalli-Sforza (updated and reviewed in Cavalli-Sforza 1997), have almost consistently shown relevance at a global scale or at the continental and regional scales. There are several examples in Africa, including the broad association between genetic variation and clustering of the four major linguistic families shown by Tishkoff et al. (2009), as well as the associations within linguistic families – for example, the fact that Oromo and Amhara from Ethiopia cluster together with Berber and other North Africans, corroborating the claimed relations between these populations (Canossi et al. 2010). Linguistic variation and language origin have recently been traced to Africa (Atkinson 2011; Fort and Pérez-Losada 2016), and the notion of a common ancestral language was brought to light again. Linguistic affinities depicted population genetic affinities, with the exception of a few cases of horizontal transfer (often resulting in linguistic replacement), which may occur during acculturation.

Drift may arise with migration and isolation by distance or isolation resulting from language unintelligibility, even in situations where the different groups share a geographic location, such as is the case of the Nubians in the Nuba Mountains of Kordofan, Sudan and the southern nations and nationalities in Ethiopia, where nearly 50 groups each with their own language and culture are present. In the Nuba Mountains, more than 40 different languages exist, often in close proximity, sharing a total area of no more than 50,000 km^2 and believed to foster genetic isolation of these neighboring communities. A population study employing high-resolution markers will be most interesting in testing this hypothesis.

Browning et al. (2010) provided evidence for purifying selection on the *CYP1A2* gene and also concluded that Afar, Amhara, Oromo, Maale, and Anuak Ethiopian populations are highly diverse, with a substantial amount of previously uncharacterized *CYP1A2* variation. In some studies in the same ethnic groups, we find differences between subgroups, e.g., significant genetic distance exists between Ari cultivators and Ari potters. Very close ethnic group neighbors, cultivators are also observed to have significant genetic distances, as can be exemplified between Dasenech and Ari cultivators, which have greater distances than between Europeans and Japanese.

3.3 Food Culture and Genetic Variation

As omnivores, humans have tremendous flexibility in their diet. Some human societies, such as Eskimos/Inuit, traditionally depended solely on a meat diet, while others had little access to meat. In some areas, humans have practiced agriculture and animal husbandry for thousands of years, while in others they have lived on only what they could collect or hunt. Even among agricultural societies, there is tremendous variation in the most important crops, including cereals, horticultural, and root-based cultures.

The previous examples of genetic variation in drug-metabolizing enzymes are among many where differences in parts of our genome may be influenced and correlate with variation in food culture (Stover 2006; Stover and Caudill 2008). A testimony to the profoundness of diversity in food culture is the variety of staple foods and cereals found in Africa. Compared to rice in Asia, wheat and potatoes in Europe, the Near East, and America, Africans additionally consume sorghum, wheat, millet, yam, teff, and more. Plant and animal husbandry is not a Eureka! type of discovery. It took a long time to develop and wield together the crops, technologies, and social practices necessary to build an effective agricultural system. At present, we understand that individuals, domesticates, and agricultural systems are biologically bound by intimate relations. We can cite several African domesticates like noog, enset, teff, African rice, yam, oil palm, sorghum, fonio, and pearl millet. For example, in many parts of the African savanna there is a close correlation between village sites past and present and the occurrence of the Baobab (*Adansonia* spp.) since the tree has multiple uses, including storage of water in its cisterns for the dry season. Baobab, like other similar trees, is not man-dependent, but man sometimes depends on it. As traditional African agriculture is a mosaic of crops, the populations who developed these crops also evolved to maintain mutations of various enzymes that are needed to digest and balance the nutrition required. Such observations are further corroborated by milk-dependent groups (discussed below), where multiple rare variants are a cause of a common phenotype with several different lactase persistence (LP) associated alleles present in a single ethnic group (Ingram et al. 2009). A great deal of the variation might have resulted from differences in biochemical constituents of these crops exerting evolutionary pressures on the genome. An example is the putative selective pressure by the wheat protein (gluten), known to be part of the etiology of celiac disease. The investigation of East African population groups represented by hunter-gatherers, pastoralists, rain-based agriculturists, and root-based agriculturists is likely to uncover similar scenarios. Horsfall et al. (2011) reported that variation of a short repeat sequence $(TA)^n$ (rs8175347) covering the TATA box of *UGT1A1* (UDP-glucuronosyltransferase 1A1) is associated with hyperbilirubinemia (Gilbert's syndrome) and adverse drug reactions, and is used for dosage advice for irinotecan. Several reports indicate that the low-activity (risk) alleles ($[TA]^7$ and $[TA]^8$) are very frequent in Africans, but the patterns of association with other variants in the *UGT1A* gene complex that may modulate these responses are not well known. rs8175347 and two other clinically relevant *UGT1A* variants (rs11692021 and rs10929302) were assayed in 2616 people from Europe and Africa. Low-activity $(TA)^n$ allele frequencies were highest in equatorial Africa, $(TA)^7$ being the most common in Cameroon, Ghana, southern Sudan, and in Ethiopian Anuak. Haplotypic diversity was also greatest in equatorial Africa, but in Ethiopia was very variable across ethnic groups. Resequencing of the promoter of a sample subset revealed no novel variations, but rs34547608 and rs887829 were typed and shown to be tightly associated with $(TA)^n$. The results illustrate the need for investigation of the effect of *UGT1A* variants other than $(TA)^n$ on the risk of irinotecan toxicity, as well as hyperbilirubinemia due to hemolytic anemia or human immunodeficiency virus protease inhibitors, so that appropriate pharmacogenetic advice can be given.

Differences in nutritional practices and associated nutrigenomic traits are expected to be magnified when it comes to the stark contrast between the sedentary and nomadic lifestyles, reflecting human health and disease status and possibly physical traits. The difference in susceptibility to diseases may have emanated from different patterns of exposure to microbes and vectors. Difference in susceptibility to malaria, for example, has been observed and documented in West Africa between Fulani and Mosi (Modiano et al. 2001); the former are known to have a strong historical background of pastoralist culture.

Both sedentary agriculture and pastoralist cultures have extensively shaped our genome in the relatively short period since their appearance as lifestyles, i.e., during the last 10,000–20,000 YBP. Both have shaped the social, economic, and even political history of our kind, and continue to do so. In the past, the relationship between farmers and nomadic pastoralists was always affected by rivalry because of the potential competition over resources. Some of the bloodiest conflicts that took place recently in different parts of the African continent were rooted in this troubled relationship. The socioeconomic and historical contents of these conflicts were often greatly missed in their analysis. Traditional pastoralists survive in few places in the world nowadays. In fact, some of the last enclaves of these ancient practices still exist across the continent, perhaps in the very same terrains that might have witnessed their nascence. They can be encountered in the African Sahel, where annual alternations of dry and wet seasons force them to continued mobility. Little, however, is known about the genetic structure of these populations.

Of the several well-investigated loci and genomic regions implicated in human adaptation to dietary requirements are the LP-associated genes. Lactase persistence is widely accepted to have evolved in human pastoralist societies with long histories and traditions of milk consumption. The rich spectrum of LP polymorphisms in African pastoralists poses questions on the history of the practice on the continent and the relationship with other historical pastoralists worldwide. Genetics of LP as a form of cultural–genetic adaptation in the human species extend beyond adaptation to shed light on phylogeography and genetic history. Sharing of polymorphisms between North Europeans and Africans (Fulani, for example) evokes some fundamental questions on the history of cattle domestication and the spread of pastoralist culture. This interesting haplotype sharing between Fulani and Northern Europeans was further corroborated by Y chromosome macro-haplogroup analysis (Hassan et al. 2008). The interpretation of this sharing could go both ways: Either Fulani have some Indo-European ancestry emanating from back-migration to Africa, or alternatively both the practice of pastoralism and the Indo-European groups have a recent African origin exemplified in common Y chromosome E haplogroups. Issues of ancestry of a certain cultural group are complex, given the fact that farmers and pastoralists may have different population histories, particularly when it comes to male and female lineages (see Chapter 2). One example that received attention in the African Sahel is the contrast between Fulani nomads' genetic structure and those recorded in Central Asia, between farming and pastoral groups (Černý et al. 2011). The results suggest that the emergence of pastoralism might be an earlier and/or a demographically more important event than the introduction of sedentary agriculture, at least in this part of Africa – a daring proposition given the dictum that agriculture preceded animal domestication. The striking association between the Y chromosome E haplogroup pastoralist culture and Afroasiatic family of languages has been suggested to have sprouted from a common culture that flourished in the Sahara during the last glacial epochs (Gebremeskel and Ibrahim 2014).

The notion that mutations are more diversified toward the origin of a trait is a plausible one, though it may be challenged by rare incidents in which migration followed by drift could create a similar condition. However, contour maps of the main haplotypes of LP provide interesting insights on both the history of the cultural practices and populations carrying these haplotypes/mutations (Ranciaro et al. 2014). The variants arising primarily as adaptations to cultural practices are greatly influenced by other forces like genetic drift and migration. These variants are henceforth a powerful tool in drawing population histories. Generally speaking, it seems that the practice of pastoralism emerged in the area of the Sahara and African Sahel before its spread, together with the proto-groups that belonged to the linguistic family/culture.

Another potentially interesting corollary is that between agriculture and alcohol metabolism. The example of the genetics of *ALDH* might provide insights similar to those of lactase persistence. Whereas pastoralists have biologically evolved to metabolize lactose sugar and a naturally fermented form of yogurt, the agriculturalist societies almost invariably practiced brewing and fermentation of crops as a means of extending the usability of their harvest and increasing nutritional values. A parallel example to that of LP is to identify mutations in *ALDH* potentially associated with alcohol metabolism and tolerance. Although several studies have amply documented the genetics of alcohol tolerance and toxicity, no population studies analogous to LP were carried out. The alcohol intolerance observed among pastoralists may have produced cultural and even sociopolitical ramifications that often transpire in the prosecution of alcohol consumption and susceptibility to alcohol-associated diseases (an alternative scenario would be low selective signal in the gene due to low or moderate pressure of lightly brewed food).

3.4 Inbreeding

Understanding the genetic structure of different populations is paramount to shedding light on crucial aspects of effective population size and inbreeding. These have far-reaching implications for the design of genetic studies and control of genetic diseases, as well as the understanding of population histories. The effect of inbreeding on augmenting non-communicable diseases and increasing the genetic load is well established (Martin et al. 1973; Pulanić et al. 2008). There are social and cultural dimensions that are often overlooked, despite the fact that the cultural element of inbreeding is paramount. The reason why human populations avoided inbreeding in the past may be based on observation of the remarkable cost and the genetic load that may ensue when it occurs in small populations. How some populations reverted to consanguinity, however, is not entirely clear. Preservation of wealth and other cultural elements are believed to underpin that culture (Kristensen et al. 2015), although biological contributions cannot be ruled out as a cultural trigger, e.g., drastic decrease in female population size.

Culture remains the single most influential factor in the practice of consanguinity, as shown in the practice of marriage among Africans of Afroasiatic linguistic backgrounds, mainly Arabic-speaking groups. In Sudan, *Nilo-Saharan* tribes predominantly practice and observe a strict out-breeding culture in comparison to Afroasiatic groups, some of whom might have migrated with native cultures preferring marriage between relatives. A spectacular manifestation of such contrast of culture is the accumulated deleterious effects in the frequency of HbS among the *Bagara* of western Sudan, which has increased over a relatively short period during the past 300 years to become one of the highest in Africa apparently due to consanguinity (Bereir et al. 2007; Mohammed et al. 2006). The Ethiopian equivalent to this is Harari and Siltee groups.

This interplay between cultural and genetic traits is scarcely investigated. Inbreeding technically extends beyond the cultural practice of consanguineous cousin marriage to encompass hazards even to the most conscientious communities, in case the effective sizes of these communities drop to a certain level due to isolation. There are other natural causes that include disease, especially among migrant communities. A notable example is the *out-of-Africa* populations. Their effective size is already reduced significantly compared to Africans. Indeed, the deleterious consequences of inbreeding may be less drastic if the population effective size is large enough to allow for avoidance of significant homozygosity.

Drift is influenced by both population size and inbreeding, as well as natural selection and migration. In evolutionary terms, biological selection is widely believed to have a lesser influence on human fitness, with culture predominating over time. However, there is plenty of evidence for episodes of natural selection that may be deduced in retrospect from natural populations not under drift, as evident from their genetic structure. An example of such a case is polymorphisms in the *TNFA* gene that show major population differentiation between sets of population data (Eid et al. 2010). Not many examples exist of the way natural selection affects genetic drift and vice versa. One such example is observed in the populations of Hausa and Masalit, where the first example of loss of a balancing selection under a "semi drift" condition and in association with decreased selection pressure (Salih et al. 2010) was reported.

3.5 Natural Selection and Disease

What makes the study of genetics in Africa all the more interesting is the potential to gain insights into biology through understanding features and mechanisms unique to its populations – mechanisms that might still be embedded within the genomes and genetic histories of Africans. We should bear in mind that our understanding of many biological functions has largely emanated from encountering genetic dysfunctionality often manifested in abnormal clinical phenotypes. We know now that Africans did not diversify only by their sheer number of variants in genetic loci related to neutral evolution and drift, but also through a rich history of genetic adaptation and natural selection sustained by the human species during its course of early development. The extremes of environmental conditions present in the African continent may have been responsible for the initial diversity of traits following interactions with a wide range of challenges.

Of all the forces that dictate allele frequency change, whether adaptation to physical environment or culture, including food, natural selection by diseases is probably the most potent evolutionary force that had the upper hand in shaping our current genome. These forces induced drastic and fundamental changes in the human genome following the advent of agriculture and pastoralism, as a result of the changes incurred on human population structure, demography, and surrounding environments.

It is widely believed that both agriculture and, to a lesser extent, animal domestication permitted the formation of large communities of human settlements that in subsequent epochs became the primordial form of city-states. The sheer crowding resulting from human sedentary life and increased population densities, in addition to proximity to domesticated animals and vectors, provided a conducive environment for the first disease epidemics to take place.

Another example for the influence of demography on disease emergence is schistosomiasis. *Schistosoma mansoni* is estimated through a coalescent model to have originated in East Africa 107.5–147.6 kya, a period overlapping with the earliest archaeological evidence for fishing in Africa. The parasite was found to have experienced a decline in effective population size 20–90 kya, before dispersing across the continent during the Holocene (Crellen et al. 2016).

Although a written or well-documented history of infectious disease epidemics is scarce in Africa and for the globe, one of the most notorious was in Africa. The Bible, one of the earliest human texts, portrays a vivid narrative of the plague that swept Egypt following the wrath of the God of Jews. It might not be the same microbial plague that afflicted Europe millennia later, but the fact remains that Africa was and still is the home and perhaps the cradle of a multitude of dangerous parasites and microbes: malaria, leishmaniasis, Lassa fever, ebola, smallpox, tuberculosis, and more. Technological advancements reflected in improved human health and medical care through vaccination and medications rendered the threat of wide microbial outbreaks a rather distant memory, at least in the industrialized countries. This African notoriety, however, might soon be overshadowed by the surge of chronic and lifestyle diseases that are sly and difficult to treat. Africa, together with isolated spots in Asia and South America, remains an exception in having both infectious and non-infectious diseases looming on its population. Five of the ten leading killers are communicable, prenatal, and nutritional disorders largely affecting children. These are rampant in the underdeveloped countries, especially those of Africa (Murray and Lopez 1997).

Malaria and leishmaniasis are two protozoan parasites that seem to have originated in Africa and followed the same trail of human migration out of Africa, at least for the latter (Ibrahim and Barker 2001). Qutob et al. (2012) investigated the link between genetic diversity using heterozygosity as a surrogate measure for pathogen richness, following correction for the effect of drift. They recovered a clear decline in diversity with geographic distance from Africa and confirmed the positive relationship between genetic diversity and pathogen richness for all three classical MHC class I genes – an interesting correlation that deserves further scrutiny, given its implications for disease epidemiology. The intriguing question then becomes why, given their origin and long common history with African populations, do certain parasites continue to inflict such a grave toll on populations in Africa against dictums of host–parasite co-evolution? The reasons why some populations or individuals are still vulnerable to fatal epidemics could be illuminating for aspects of population history and mechanisms of protective immunity and how the host and parasite gene and population trees intersect over time (Figure 3.2). The notion of the

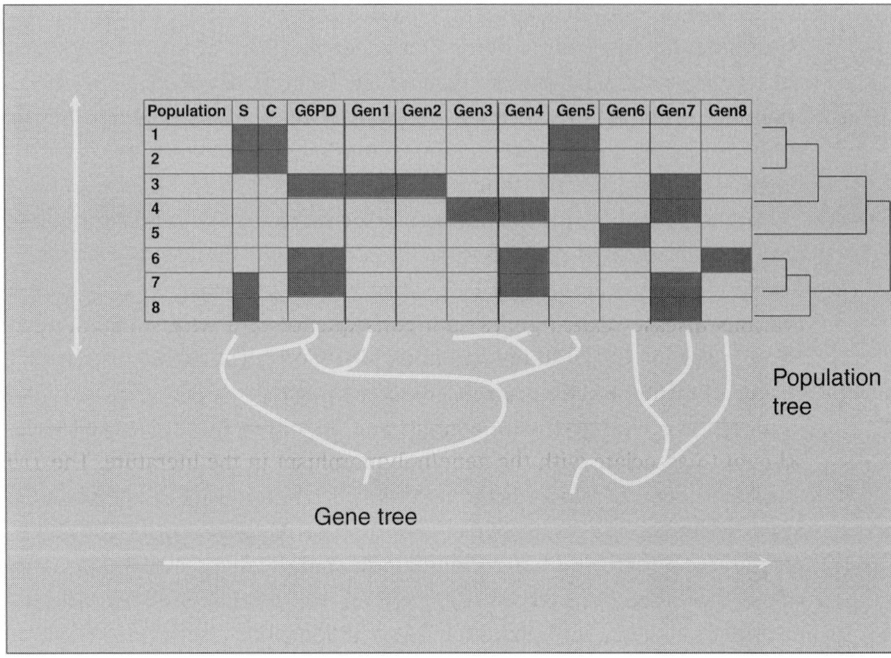

Figure 3.2 The genetic history of a disease employs information from distributions of disease susceptibility alleles (gene tree) and population trees. These are often discordant. Both trees encompass and are influenced by evolutionary forces like migration selection and drift that may be disparate, incumbent upon independent genetic trajectories in time and space.

human genome as an archive of past and recent onslaughts by infectious agents is appealing and has been utilized to shed light on the immunobiology of common human infections and their epidemiology and history. The case claimed for CCR5-Δ32 protection against or delayed onset of the clinical manifestations of HIV1 is the most renowned example (Marmor et al. 2006). Countless similar episodes might have taken place before and after human migration out of Africa. Humans would have barely succeeded in making it through their subsequent diasporas across alien and hostile environments if they were not equipped with an impressive load of physiological and immunological mechanisms, resulting from at least 60,000 years of adaptation and combating disease, and possibly much more from their hominid and primate legacies.

Dynamics and rates of natural selection and allele fixation have been a point of controversy, and remain so. The magnitude and intensity of the selective pressure, as well as the background of the genetic structure of a certain population at a specific locus, are determinants for the ensuing selective signal. What is known as extreme natural selection, mostly inflicted by infectious diseases and physical conditions, may result in either directional selection and selective sweeps, or heterozygosity and heterosis. Both are manifested in signatures engraved in the human genome. The claimed limited evidence for classic selective sweeps in Africa should be interpreted with caution, as signals of sweep may often be confused with features of population structure. Interpretation of the classic measure of Tajima's D is an example in which measures of selection could often be confused with population expansion. Negative selection has been shown globally to reduce population differentiation at amino acid-altering mutations, particularly in disease-related genes. Conversely, positive selection is seen to influence regional adaptation of human populations by increasing population differentiation in gene regions, primarily at non-synonymous and 5′-UTR variants (Barreiro et al. 2008).

Directional selection and examples of near-allele fixation could be encountered in various disease-related genes as a consequence of a wide range of epidemics. The distribution of *TNF-376* polymorphism in Nilotics with a completely different allele frequency spectrum in comparison to other populations (Eid et al. 2010) may attest to a past epidemic of malaria, meningitis, or any of the diseases that have recently been shown to associate with the gene/polymorphism in the literature. The *TNFA* gene is part of a network that influences susceptibility to several diseases of infectious and non-infectious natures. Hence, and as discussed earlier, functionality should always be considered within the context of biological systems paradigms and "genetic borrowing" (see below).

As for heterosis, malaria and hemoglobinopathies (particularly sickle cell disease) stand out as the example that received most scrutiny. Sickle cell mutation was the showcase in population genetics and will not be discussed in detail in this chapter for convenience. A parallel example on cystic fibrosis and cholera is more of a conjecture, rather than based on the physiological explanation, that is nonetheless plausible.

In tuberculosis (TB), for example, a wide range of susceptibility from fatal TB to total resistance to infection and/or progression to disease exists. Such a spectrum of host–pathogen interaction outcomes is mediated in part by the genetic constitution of individuals and populations that impact their potential or innate and adaptive immunity, including in the intermediary stages of disease (Bekele, unpublished data). These susceptibility/resistance mechanisms involve innate as well as acquired immunity conferred by previous exposure, and as the outcome determines survival, the immunity genes behind such phenotypes are subjected to natural selection (Quintana-Murci et al. 2007). In this respect, different gene polymorphisms have been found which are associated with increased susceptibility and severity of TB. Some of these polymorphisms are functional, but for many of these no functional (immunologic) changes have been demonstrated, and these associations, therefore, need further confirmation and investigation (Van Crevel et al. 2002).

Functionality of SNPs is conventionally based on coding region analysis, paying little attention to regulatory sequences. Interestingly, a common variant, at 5q33.3, was significantly associated with TB (odds ratio = 0.37, $p = 2.11 \times 10^{-8}$). This variant lies within a genomic region that includes *IL12B* and is embedded in an *H3K27Ac* histone mark (Sobota et al. 2016). Another example of natural selection on regulatory sequences is the heterosis selective signal documented with spectacular excess of heterozygosity across genes of 5q31. This was attributed to an over-dominance effect between malaria and visceral leishmaniasis, using an *in silico* approach to show that candidate SNPs in non-coding sequences may potentially influence expression, product activity, stability, and regulation of gene products (Amin et al. 2015).

The impact of such allele frequency change in genes lying in interface with the environment or responsible for immune defenses like those residing in 5q31 has to be understood in their potential biological pathways and relationship to other diseases (like atopy and others), as discussed above and below.

Of the contemporary episodes of extreme selection by infectious diseases, and the most tragic, that could have led to such allele differences was the epidemic of visceral leishmaniasis (VL) in southern Sudan (Perea et al. 1989). The VL outbreak was particularly fatal to the Nuer ethnicity, as it was decades ago for the JumJum tribe of the Blue Nile during the 1940 epidemic (Sati 1958). The Nuer were shown to have relatively smaller effective size compared to other Nilotics (Hassan et al. 2008), which could have serious consequences if susceptibility alleles/networks are higher in frequency or fixed among them.

Alternatively, we may also speculate that the severity of the VL epidemic among the Nuer and JumJum is related to their immunologic naïvety to the parasite, having moved recently into the endemic Upper Nile and Blue Nile areas while progressively encountering the peripheral transmission of the disease sylvatic cycle (Ibrahim and Amin 2015). Both oral tradition and archaeological evidence suggest a migration of the Nilotics into the area from further north in the past few millennia (Shillington 2005). The gravity and high death toll among particular ethnic groups could furthermore be interpreted in light of genetic drift. The force of drift operates more effectively in situations of isolation or inbreeding, hence leading to lower effective population size and excess susceptibility/resistance alleles that may end in the tragic situation described above.

Diseases of limited prevalence in humans historically and those known to be of inferior selective pressure (like cancer and late-onset polygenic diseases) cannot explain the allele distribution of susceptibility loci associated with these diseases. An example is the genome-wide signal of selection in the prostate stem cell antigen (*PSCA*) gene in the bladder and gastric cancers (Bhatia et al. 2011). Of the several examples in the literature for functional association of genes that are classified as underlying susceptibility to diseases, cancer has been consistently excluded as a likely source of a significant selection index. More recently, however, skin cancer has been debated as a possible selective force of skin color trait during early human evolution (Andersen et al. 2012).

In a similar fashion, the tumor suppressor gene *p53* has been shown to display an obvious cline from south to north of its *p53* Arg–Pro polymorphism with the derived allele (Arg) being more common in Asians and Europeans compared to Africans (Hirshfield et al. 2010). This polymorphism is believed to be under selection by affecting main fitness traits like fecundity (Orr 2009), leading to a relative advantage in this trait. In such a case, even if the fitness index is relatively small, it will lead to substantial allele frequency differences over time. Another example is the relationship of the skin pigmentation to vitamin D and diseases like rickets in Europe in the past and presently, and prostate and breast cancers in Americans of African ancestry (Murphy et al. 2012).

3.6 Biological Network and Genetic Borrowing

The impact of natural selection on a locus or loci extends beyond the direct phenotypic traits associated with that locus or loci because it is seldom the case that a gene decides only a single trait. Genes are mostly part of genetic networks that operate epistatically or multiplicatively.

New paradigms in biology embrace the complexity of the biological information systems embedded in these networks and feedback loops which are bound to reflect on our understanding of the trade-off between disparate phenotypes and existing "borrowing" of genetic effects that occur routinely in our bodies through shared biological networks. The human body, like other living systems, is functional on a parsimony basis, i.e., the body has a limited set of "tricks" from which it makes a living and functions. One feature of that is the way functional genes and protein networks interact and borrow effects of natural selection.

The most notorious example of direct borrowing in two infectious diseases against each other is the one mentioned earlier of HIV and plague. *Yersinia pestis*, the microbe causing plague, and the HIV virus both utilize the same chemokine receptor, CCR5. It has been speculated that a common deletion, CCR5-Δ32, was brought into the current elevated frequencies in Europe through the repeated epidemics known as the Black Death that ravaged Europe during the Middle Ages (Cohn and Weaver 2006). Although this has been questioned by some authors, who proposed an older coalescence date for the mutation (Libert et al. 1998; Stephens et al. 1998), the mutation has almost zero frequency in sub-Saharan Africa, where HIV is rampant. Interestingly, CCR5 variants are not found to be associated with HIV/AIDS risk in Africa – the origin and center of the current AIDS pandemic. The gene, therefore, may not be subjected to rapid evolutionary change as a result of the HIV epidemic in Africa (Ramaley et al. 2002), indicating the historical context of disease "focality" and resistance/susceptibility alleles, and raising questions on the existence of other relevant genetic predisposition loci in Africa.

In terms of infectious versus non-infectious diseases, there are the examples of sickle cell disease and malaria, and cystic fibrosis and cholera, and others mentioned earlier. Examples of balancing selection and trade-off in biology and medicine will remain as malaria still invokes its lethal impact across the tropics. The historical context in the origin and spread of such polymorphisms is exemplified in how sickle cell disease was brought to Sudan through the Sahel by nomadic Bagara tribes from West Africa and Lake Chad. We were able to date this to ~300 YBP by estimating the amount of introgression of certain Y chromosome haplogroups (Bereir et al. 2007). The frequency of the sickle allele in the Nilo-Saharan-speaking group and other East African groups might have been nil prior to these demographic events, raising questions about the population adaptive responses to malaria epidemics, a disease that has evidently been endemic to the Nile Valley for millennia.

These are obvious examples because of the effect size of the involved variants and the proportional gravity of the selection effects. However, there are probably many more examples of disease associations arising from putative epistatic effects that could explain eQTL (expression quantitative trait loci) signals in genome wide association studies (GWASs) of common diseases and increase the penetration of rare coding variants (Lappalainen et al. 2013).

The literature is packed with examples of alleles maintained by natural selection on standing variations which makes the basis of such trade-offs all the more interesting. One example is oculocutaneous albinism type 2 (OCA2) and leprosy (Tuli et al. 2012) – also found in high prevalence in sub-Saharan African populations. Also, Crohn's disease risk alleles on the *NOD2* locus are believed to have been maintained by natural selection on standing variation (Nakagome et al. 2012). These variations affect some common diseases that are on the rise in Africa and worldwide, like cancer and kidney diseases. *MYH9* and *APOL1*, both of which are associated with a number of rare and common syndromes, were recently identified as susceptibility genes for major forms of kidney disease (Colares et al. 2014; Oleksyk et al. 2010). *MYH9* was reported to disproportionately affect individuals of African descent (OR 3–8, $p < 10^{-8}$).

Furthermore, the commonly held notion of genetic disease categorization between Mendelian and complex diseases is now under revision, with evidence suggesting that 54 percent (524 of 968) of the Mendelian disease genes were also involved in complex diseases. Mendelian and complex disease genes were associated with more diseases and phenotypes, and were involved in more complex protein–protein interaction networks than Mendelian but not complex disease (MNC). Those genes on average encoded the longest proteins and had the highest transcript count among all gene categories. Especially, tissue specificity of MC genes was much higher than that of any other gene categories, although their expression level was similar to that of essential genes (Jin et al. 2012). In brief, without understanding the intricacies of biological network regulation and how genes/proteins interact, neither the genetic differences nor the effects of selection will be clearly understood and properly elucidated.

3.7 Epigenetics, an Added Layer of Complexity in the Selection Plateau

Until very recently, it was largely believed that susceptibility to disease, especially chronic disorders of adulthood, is exclusively mandated by genetic factors, such as those lying mainly in the protein coding part of the human genome. Recent advances, however, have identified that non-genetic factors are also key players in the predisposition and onset of adult-onset disorders. Early insults during prenatal development have been corroborated to entail long-term consequences of possible transgenerational heritability that is not associated directly with DNA sequence alterations (Heijmans et al. 2008; Tobi et al. 2009). The term "epigenetic" is used to define such changes in the phenotype or gene expression patterns underlying mechanisms other than direct DNA sequence changes. Therein, each cell in the body acquires a unique epigenetic tag that reflects its genotype, developmental history, and environmental influences acting in concert upon it. These changes then influence the function and thus the phenotype of the organism.

It has also been proven that adverse prenatal environmental conditions are associated with persistent epigenomic changes that parallel altered rates of aging and higher susceptibilities to adult-onset disorders, including schizophrenia, cancer, and type 2 diabetes. This deviant phenotype forms adaptively via developmental plasticity, which attempts to match an individual's responses to the predicted environment (Godfrey et al. 2007). Accordingly, phenotypes that match the predicted demands during adulthood will promote health, while a significant mismatch will hamper adaptability to subsequent lifecourse conditions, thus elevating predispositions to the disorders (Gluckman and Hanson 2006; Heijmans et al. 2009; Tobi et al. 2009). Prenatal circumstances have been linked to the recently emerging disorders which have entailed an epidemiological shift in their favor relative to acute infections. One such attested factor is prenatal hunger during pregnancy, related to famines in many parts of the globe, including the African continent. Such a link can only be understood through the spectacles of the vast realm of epigenetics. However, to date, the famine–epigenetic-disease paradigm has never been studied seriously, despite the several advantages it poses, and our knowledge remains meager when it comes to the dynamics of natural selection on epigenetic marks. This might be an outcome of a lack of a theoretical framework outlining the relationship between epigenetic and genetic mechanisms, in addition to the methodological limitations in the field of epigenetics.

References

Amin M, Elhassan A, Rocket K, et al. (2015). Overdominance effects between malaria and visceral leishmaniasis in the 5q31 region. *J Biol Sci* 15(4): 221–226.

Andersen KG, Shylakhter I, Tabrizi S, et al. (2012). Genome-wide scans provide evidence for positive selection of genes implicated in Lassa fever. *Philos Trans R Soc B Biol Sci* 367: 868–877.

Atkinson QD (2011). Phonemic diversity supports a serial founder effect model of language expansion from Africa. *Science* 332: 346–348.

Bains RK, Kovencuric M, Plaster CA, et al. (2013). Molecular diversity and population structure at the cytochrome P450 3A5 gene in Africa. *BMC Genet* 3: 14–34.

Barreiro LB, Laval G, Quach H, Patin E, and Quintana-Murci L (2008). Natural selection has driven population differentiation in modern humans. *Nat Genet* 40(3): 340–345.

Bereir RE, Hassan HY, Salih NA, et al. (2007). Co-introgression of Y-chromosome haplogroups and the sickle cell gene across Africa's Sahel. *Eur J Hum Genet* 15(11): 1183–1185.

Bhatia G, Patterson N, Pasaniuc B, et al. (2011). Genome-wide comparison of African-ancestry populations from CARe and other cohorts reveals signals of natural selection. *Am J Hum Genet* 89(3): 368–381.

Browning SL, Tarekegn A, Bekele E, Bradman N, and Thomas MG (2010). CYP1A2 is more variable than previously thought: a genomic biography of the gene behind the human drug-metabolizing enzyme. *Pharmacogenet Genomics* 20(11): 647–664.

Canossi A, Piancatelli D, Aureli A, et al. (2010). Correlation between genetic HLA class I and II polymorphisms and anthropological aspects in the Chaouya population from Morocco (Arabic speaking). *Tissue Antigens* **76**: 177–193.

Cavalli-Sforza LL (1997). Genes, peoples, and languages. *Proc Natl Acad Sci U S A* **94**: 7719–7724.

Černý V, Pereira L, Musilová E, et al. (2011). Genetic structure of pastoral and farmer populations in the African Sahel. *Mol Biol Evol* **28**(9): 2491–2500.

Chiaroni J, Underhill PA, and Cavalli-Sforza LL (2009). Y chromosome diversity, human expansion, drift, and cultural evolution. *PNAS* **106**(48): 20174–20179.

Cohn SK Jr. and Weaver LT (2006). The Black Death and AIDS: CCR5-Δ32 in genetics and history. *QJM* **1076**: 497–503.

Colares VS, Titan SMO, Pereira AC, et al. (2014). MYH9 and APOL1 gene polymorphisms and the risk of CKD in patients with lupus nephritis from an admixture population. *PLoS One* **9**(3): e87716.

Crellen T, Allan F, David S, et al. (2016). Whole genome resequencing of the human parasite *Schistosomamansoni* reveals population history and effects of selection. *Sci Reports* **6**: 20954.

de Magalhaes JP and Matsuda A (2012). Genome-wide patterns of genetic distances reveal candidate loci contributing to human population-specific traits. *Ann Hum Genet* **76**: 142–158.

Deshpande O, Batzoglou S, Feldman MW, and Cavalli-Sforza LL (2009). A serial founder effect model for human settlement out of Africa. *Proc Biol Sci* **276**(1655): 291–300.

Eid NA, Hussein AA, Elzein AM, et al. (2010). Candidate malaria susceptibility/protective SNPs in hospital and population-based studies: the effect of sub-structuring. *Malar J* **9**: 119.

Elhassan N, Gebremeskel EI, Elnour MA, et al. (2014). The episode of genetic drift defining the migration of humans out of Africa is derived from a large East African population size. *PloS One* **9**: e97674.

Eyre-Walker A (2002). Changing effective population size and the McDonald–Kreitman test. *Genetics* **162**(4): 2017–2024.

Fort J and Pérez-Losada J (2016). Can a linguistic serial founder effect originating in Africa explain the worldwide phonemic cline? *J R Soc Interface* **117**. pii: 20160185.

Gebremeskel EI and Ibrahim ME (2014). Y-chromosome E haplogroups: their distribution and implication to the origin of Afro-Asiatic languages and pastoralism. *Eur J Hum Genet* **12**: 1387–1392.

Gluckman PD and Hanson MA (2006). Evolution, development and timing of puberty. *Trends Endocrinol Metab* **17**(1): 7–12.

Godfrey KM, Lillycrop KA, Burdge GC, Gluckman PD, and Hanson MA (2007). Epigenetic mechanisms and the mismatch concept of the developmental origins of health and disease. *Pediatr Res* **61**(5 Pt 2): 5R–10R.

González-Santos M, Montinaro F, Oosthuizen O, et al. (2015). Genome-wide SNP analysis of Southern African populations provides new insights into the dispersal of Bantu-speaking groups. *Genome Biol Evol* **7**(9): 2560–2568.

Hassan HY, Underhill PA, Cavalli-Sforza LL, Ibrahim ME (2008). Y-chromosome variation among Sudanese: restricted gene flow, concordance with language, geography, and history. *Am J Phys Anthropol* **137**: 316–323.

Heijmans BT, Tobi EW, Stein AD, et al. (2008). Persistent epigenetic differences associated with prenatal exposure to famine in humans. *PNAS* **105**(44): 17046–17049.

Heijmans BT, Tobi EW, Lumey LH, and Slagboom PE (2009). The epigenome: archive of the prenatal environment. *Epigenetics* **4**(8): 526–531.

Hirshfield KM, Rebbeck TR, and Levine AJ (2010). Germline mutations and polymorphisms in the origins of cancers in women. *J Oncol.* DOI: 10.1155/2010/297671.

Horsfall LJ, Zeitlyn D, Tarekegn A, et al. (2011). Prevalence of clinically relevant UGT1A alleles and haplotypes in African populations. *Ann Hum Genet* **75**: 236–246.

Ibrahim M and Amin M (2015). Population genetics in Sudan. *Khartoum Med J* **8**(2): 1095–1098.

Ibrahim ME and Barker DC (2001). The origin and evolution of the *Leishmania donovani* complex as inferred from a mitochondrial cytochrome oxidase II gene sequence. *Infect Genet Evol* **1**: 61–68.

Ingram CJE, Raga TO, Tarekegn A, et al. (2009). Multiple rare variants as a cause of a common phenotype: several different lactase persistence associated alleles in a single ethnic group. *J Mol Evol* **69**: 579–588.

Jallow M, Teo YY, Small KS, et al. (2009). Genome-wide and fine-resolution association analysis of malaria in West Africa. *Nat Genet* **41**: 657–665.

Jin W, Qin P, Lou H, Jin L, and Xu S (2012). A systematic characterization of genes underlying both complex and Mendelian diseases. *Hum Mol Genet* **21**(7): 1611–1624.

Keinan A, Mullikin JC, Patterson N, and Reich D (2009). Accelerated genetic drift on chromosome X during the human dispersal out of Africa. *Nat Genet* **41**(1): 66–70.

Kristensen TN, Hoffmann AA, Pertoldi C, and Stronen AV (2015). What can livestock breeders learn from conservation genetics and vice versa? *Front Genet* **6**: 38.

Lappalainen T, Sammeth M, Friedländer MR, et al. (2013). Transcriptome and genome sequencing uncovers functional variation in humans. *Nature* **501**(7468): 506–511.

Libert F, Cochaux P, Beckman G, et al. (1998). The deltaccr5 mutation conferring protection against HIV-1 in Caucasian populations has a single and recent origin in Northeastern Europe. *Hum Mol Genet* **7**(3): 399–406.

Marmor M, Hertzmark K, Thomas SM, Halkitis PN, and Vogler M (2006). Resistance to HIV infection. *J Urban Health* **83**(1): 5–17.

Martin AO, Kurczynski TW, and Steinberg AG (1973). Familial studies of medical and anthropometric variables in a human isolate. *Am J Hum Genet* **25**(6): 581–593.

Miller M, Fadl M, Mohamed HS, et al. (2007). Y chromosome lineage- and village-specific genes on chromosomes 1p22 and 6q27 control visceral leishmaniasis in Sudan. *PloS Genet* **3**: 679–688.

Modiano D, Luoni G, Sirima BS, et al. (2001). The lower susceptibility to *Plasmodium falciparum* malaria of Fulani of Burkina Faso (West Africa) is associated with low frequencies of classic malaria-resistance genes. *Trans R Soc Trop Med Hyg* **95**(2): 149–152.

Mohammed AO, Attalla B, Bashir FM, et al. (2006). Relationship of the sickle cell gene to the ethnic and geographic groups populating the Sudan. *Community Genet* **9**(2): 113–120.

Morabia A and Abel T (2006). The WHO report "Preventing Chronic Diseases: a vital investment" and us. *Soz Praventivmed* **51**: 74.

Moreno-Estrada A-PE, Sikora EJ, Ramírez-Soriano CF, and Bosch E (2010). African signatures of recent positive selection in human FOXI1. *BMC Evol Biol* **10**: 267.

Murphy AB, Kelley B, Nyame YA, et al. (2012). Predictors of serum vitamin D levels in African American and European American men in Chicago. *Am J Mens Health* **6**(5): 420–426.

Murray CJL and Lopez AD (1997). Mortality by cause for eight regions of the world: Global Burden of Disease Study. *Lancet* **349**: 1269–1276.

Nakagome S, Mano S, Kozlowski L, et al. (2012). Crohn's disease risk alleles on the NOD2 locus have been maintained by natural selection on standing variation. *Mol Biol Evol* **29**(6): 1569–1585.

Oleksyk TK, Nelson GW, An P, Kopp JB, and Winkler CA (2010). Worldwide distribution of the MYH9 kidney disease susceptibility alleles and haplotypes: evidence of historical selection in Africa. *PloS One* **5**(7): e11474.

Orr HA (2009). Fitness and its role in evolutionary genetics. *Nat Rev Genet* **10**(8): 531.

Perea WA, Moren A, Ancelle T, and Sondorp E (1989). Epidemic visceral leishmaniasis in southern Sudan. *Lancet* **2**(8673): 1222–1223.

Poloni ES, Naciri Y, Bucho R, et al. (2009). Genetic evidence for complexity in ethnic differentiation and history in East Africa. *Ann Hum G* **73**: 582–600.

Pulanić D, Polasek O, Petrovecki M, et al. (2008). Effects of isolation and inbreeding on human quantitative traits: an example of biochemical markers of hemostasis and inflammation. *Hum Biol* **80**(5): 513–533.

Quintana-Murci L, Alcais A, Abel L, and Casanova JL (2007). Immunology in natural selection: clinical, epidemiological and evolutionary genetics of infectious disease. *Nat Immunol Disease* **18**: 1165–1171.

Qutob N, Balloux F, Raj T, et al. (2012). Signatures of historical demography and pathogen richness on MHC class I genes. *Immunogenetics* **64**(3): 165–175.

Ramaley PA, French N, Kaleebu P, et al. (2002). HIV in Africa: chemokine-receptor genes and AIDS risk. *Nature* **417**(6885): 140.

Ranciaro A, Campbell MC, Hirbo JB, et al. (2014). Genetic origins of lactase persistence and the spread of pastoralism in Africa. *Am J Hum Genet* **94**: 496–510.

Reich DE, Cargill M, Bolk S, et al. (2001). Linkage disequilibrium in the human genome. *Nature* **411** (6834): 199–204.

Salih NA, Hussain AA, Almugtaba IA, et al. (2010). Loss of balancing selection in the beta S globin locus. *BMC Med Genet* **11**: 21.

Sati MH (1958). Early phases of an outbreak of kala-azar in the Southern Fung. *Sudan Medical Journal* **1**: 98–111.

Shillington K (2005). *History of Africa.* Palgrave Macmillan.

Sobota RS, Stein CM, Kodaman N, et al. (2016). A locus at 5q33.3 confers resistance to tuberculosis in highly susceptible individuals. *Am J Hum Genet* **98**(3): 514–524.

Stephens JC, Reich DE, Goldstein DB, et al. (1998). Dating the origin of the CCR5-Δ32 AIDS-resistance allele by coalescence of haplotypes. *Am J Hum Genet* **62**: 1507–1515.

Stover PJ (2006). Influence of human genetic variation on nutritional requirements. *Am J Clin Nutr* **83**: 436S–442S.

Stover PJ and Caudill MA (2008). Genetic and epigenetic contributions to human nutrition and health: managing genome–diet interactions. *J Am Diet Assoc* **108**: 1480–1487.

Tenesa A, Navarro P, Hayes BJ, et al. (2007). Recent human effective population size estimated from linkage disequilibrium. *Genome Res* **17**: 520–526.

Tishkoff SA, Reed FA, Friedlaender FR, et al. (2009). The genetic structure and history of Africans and African Americans. *Science* **324**: 1035–1044.

Tobi EW, Lumey LH, Talens RP, et al. (2009). DNA methylation differences after exposure to prenatal famine are common and timing- and sex-specific. *Epigenetics* **4**(8): 526–531.

Tuli AM, Valenzuela RK, Kamugisha E, and Brilliant MH (2012). Albinism and disease causing pathogens in Tanzania: are alleles that are associated with OCA2 being maintained by balancing selection? *Med Hypotheses* **79**(6): 875–878.

Van Crevel R, Ottenhoff TH, and van der Meer JW (2002). Innate immunity to Mycobacterium tuberculosis. *Clin Microbiol Rev* **15**(2): 294–309.

Wong SH, Gochhait S, Malhotra D, et al. (2010). Leprosy and the adaptation of Human Toll-Like Receptor 1. *PloS Pathog* **6**: e1000979. DOI: 10.1371/journal.ppat.1000979.

4 Genetic Susceptibility to Visceral Leishmaniasis

Hiba S. Mohamed, Muntaser E. Ibrahim, and Jenefer M. Blackwell

4.1 Leishmaniasis

The leishmaniases are a group of related diseases caused by parasites of the genus *Leishmania* (Order Kinetoplastidae, family Trypansomatidae). The clinical presentations of leishmaniasis include visceral leishmaniasis (VL) or kala-azar, post-kala-azar dermal leishmaniasis (PKDL), localized cutaneous leishmaniasis (LCL), disseminated cutaneous leishmaniasis (DCL), and mucosal leishmaniasis (ML). These different clinical patterns of infection are principally related to the species of *Leishmania*. All species of *Leishmania* are transmitted by sand flies; *Phlebotomus* spp. in the Old World and *Lutzomyia longipalpis* in the New World. Humans, wild animals, and domestic animals are known to act as reservoir hosts. Visceral leishmaniasis is associated with infection by members of the *L. donovani* species complex, including *L. donovani sensu strictu*, *L. infantum*, and *L. aethiopica*, and is characterized by fever, weight loss, epistaxis, pancytopenia, anorexia, abdominal pain, cough, diarrhea, nausea and vomiting, weakness, fatigue, splenomegaly, hepatomegaly, and lymphadenopathy. Other clinical signs of VL can include edema, neurological changes, and, less frequently, psychosis and jaundice (Zijlstra and el-Hassan 2001b). As well as being a severe systemic infection, a proportion of patients treated for VL go on to present with cutaneous disease, termed PKDL (Zijlstra and el-Hassan 2001a; Zijlstra et al. 1994). Recent analysis by the WHO estimates that 200,000–400,000 new cases occur annually, 90 percent of which are in India, Bangladesh, Sudan, South Sudan, Ethiopia, and Brazil (Alvar et al. 2012). In Africa, the highest incidence rate was reported in the East African region, with estimated annual incidence of 29,400–56,700. The reported VL cases per year in the region is 8560, of which Sudan, South Sudan, and Ethiopia have the highest incidence rates, in that order. A small number of cases have been reported from Somalia, Uganda, Kenya, and Eritrea. Visceral leishmaniasis is also present in other foci in Africa, such as Morocco, Algeria, and Tunisia, with estimated annual incidence rate of 300, 130, and 110 cases, respectively (Alvar et al. 2012).

Whether an individual infected with the *Leishmania* parasite goes on to develop VL is determined by a complex interplay between parasite, host, and vector genetic factors, as well as environmental and socioeconomic risk factors (Badaro et al. 1986; Jeronimo et al. 1994; Zijlstra et al. 1994). However, immunologically we know that *Leishmania*-specific cellular immune responses are a correlate of infection control in resistant individuals, measured traditionally as a positive delayed type hypersensitivity (DTH) skin test responses *in vivo* (Jeronimo et al. 1994) or lymphocyte proliferation responses *ex vivo* (Ho et al. 1982), and more recently as an important balance between the pro-inflammatory cytokines tumor necrosis factor (TNF) and interferon-γ (IFNγ) and anti-inflammatory interleukin-10 (IL-10) (Kumar et al. 2014; Singh et al. 2012b). Clinical VL is associated with high levels of IL-10 (Nylen and Sacks 2007; Nylen et al. 2007; Zijlstra and el-Hassan 2001b), while neutralization of IL-10 in splenic aspirates of VL patients promotes parasite killing (Gautam et al. 2011). Although it has traditionally been thought that VL patients failed to make IFNγ in response to leishmanial antigens as determined in *ex vivo* peripheral blood mononuclear cell assays, recent studies in India demonstrated that CD4-positive T cells from active VL patients were able to make antigen-specific IFNγ responses in whole-blood assays (Singh et al. 2012a). Thus, it is the balance between antigen-specific pro-inflammatory and anti-inflammatory cytokine responses that are important in determining the outcome of infection with parasites of the *L. donovani* species complex.

4.2 Genetic Susceptibility to Visceral Leishmaniasis

Multiple lines of evidence support a role for host genetics in determining the clinical outcome of infection with species of *Leishmania* that cause clinical VL, in particular the spectrum from asymptomatic through subclinical to clinical VL. Studies in mice (reviewed by Blackwell [1996] and Blackwell et al. [2009]) provided early support for a strong genetic component to susceptibility to *L. donovani* infection. In this defined model system, it was possible to demonstrate that different genes control innate versus adaptive immunity, as expected for a complex disease. In humans, studies based on ethnic differences (Ibrahim et al. 1999; Khalil et al. 2002; Zijlstra et al. 1994), familial aggregation and segregation analysis (Bucheton et al. 2002; Cabello et al. 1995; Peacock et al. 2001; Zijlstra et al. 1994), and a high relative risk (λ_{2S} = 34) of disease in further siblings of affected sibling pairs (Peacock et al. 2001) support a genetic hypothesis, with longitudinal studies showing a strong interplay between environmental and host factors during outbreaks (Badaro et al. 1986; Bucheton et al. 2002; Jeronimo et al. 2004). Segregation analysis undertaken following total population surveys that measured DTH responses in Peru (Shaw et al. 1995) and Brazil (Feitosa et al. 1999) support multifactorial genetic control over a sporadic model. In a recent study of the quantitative DTH response in families from Natal, Brazil (Jeronimo et al. 2007a), familial correlations were estimated using the FCOR program of the SAGE (Statistical Analysis for Genetic Epidemiology) software package (version 4). Heritability (h^2) was estimated from the sibling correlations r using the equation $h^2 = 2r$. In these families there were 440 sibling pairs with a correlation of 0.42, 212 grandparent–grandchild pairs with a correlation of 0.265, and 90 cousin pairs with a correlation of 0.13. This is consistent with genetic control of the DTH response, where first-degree relatives have a stronger correlation than second- or third-degree relatives. Estimated heritability of the DTH immune response was

84 percent, suggesting a substantial genetic component to variation in induration size, as determined by the DTH skin test.

A number of approaches can be taken in trying to identify genetic risk factors in human population studies. These include candidate gene studies based on our knowledge of genetic risk in mice, and of immunological pathways involved in regulating pathology and disease outcome, as well as unbiased genome-wide approaches. A number of labs, including our respective labs, have undertaken both candidate gene (Bucheton et al. 2003b, 2007; Faghiri et al. 1995; Jeronimo et al. 2007b; Karplus et al. 2002; Meddeb-Garnaoui et al. 2001; Mohamed et al. 2003, 2004; Peacock et al. 2002; Salih et al. 2007) and genome-wide linkage studies (GWLS) (Bucheton et al. 2003a; Jamieson et al. 2007; Jeronimo et al. 2007a; Miller et al. 2007) for the VL phenotype (see below). In Brazil we also analyzed candidate genes/gene regions (Jeronimo et al. 2007b; Karplus et al. 2002) using the DTH response as a qualitative trait (i.e., positive or negative response) and carried out a GWLS analyzing DTH as a quantitative trait loci (QTL) (Jeronimo et al. 2007a). It should be noted, however, that most of the candidate gene allelic association studies carried out to date were underpowered to find multiple genes of small effect sizes (odds ratios [OR] < 2). Similarly, all GWLS using multi-case pedigrees were only sufficiently powered to find single major loci, or at best oligogenic control, rather than confirmed linkage, which, in the context of a GWLS, requires LOD scores >5.4 to be achieved (Lander and Kruglyak 1995). More recently, we had the opportunity to carry out a well-powered genome-wide association study (GWAS) as part of phase 2 of the Wellcome Trust Case Control Consortium (WTCCC2) (Fakiola et al. 2013). The HLA region was the only association to achieve genome-wide significance, commonly accepted as $p < 5 \times 10^{-8}$ (Dudbridge and Gusnanto 2008), across the epidemiological divides of geography and parasite species (Fakiola et al. 2013). Nevertheless, some common genes (Table 4.1) are highlighted here that have emerged by studying different populations using the traditional candidate gene and GWLS methods, including those uncovered through founder effects and population substructure in Africa (Miller et al. 2007).

4.3 Candidate Gene Studies

A number of candidate gene studies have been reported for VL (Table 4.1) (Bucheton et al. 2003b; Faghiri et al. 1995; Karplus et al. 2002; Meddeb-Garnaoui et al. 2001; Mohamed et al. 2003, 2004; Peacock et al. 2002; Salih et al. 2007). Interesting among these are those arising from murine studies, where both innate immunity under the control of the *Slc11a1* (formerly *Lsh/Ity/Bcg/Nramp1*) gene (Bradley 1974, 1977) and acquired immunity directed by the major histocompatibility complex (H-2 in mice, HLA in man) (Blackwell et al. 1980) were shown to be important. In the light of our knowledge of immune responses important in immunopathology, we also focus here on genes that drive pro-inflammatory versus anti-inflammatory responses.

Table 4.1 Genetic loci showed positive association/linkage with susceptibility to VL

Chromosomal loci	Genes	Phenotype	Population	Reference
2q35	*SLC11A1*	VL	Sudanese	Bucheton et al. 2003a
2q35	*SLC11A1*	VL	Sudanese	Mohamed et al. 2004
6p21.33	*TNFA* allele 2	VL	Brazilian	Karplus et al. 2002
6p21.32	*HLA-DRB1* *HLA-DQA1*	VL	Brazilian and Indian	Fakiola et al. 2013
17q11.1–q12	*CCL1* *CCL16*	VL	Brazilian	Jamieson et al. 2007
5q31–q33	*IL4* *IL9*	VL	Sudanese	Mohamed et al. 2003
5q31–q33	*LECT2* *TGFBI*	DTH+	Brazilian	Jeronimo et al. 2007a
5q31–q33	*TGFB1*	DTH+ VL	Brazilian	Frade et al. 2011
6q23.3	*IFNGR1*	PKDL	Sudanese	Mohamed et al. 2003 Salih et al. 2007
22q12.3	*IL2Rβ*	VL	Sudanese	Bucheton et al. 2007
2q14	*IL-1β*	VL	Iranian	Moravej et al. 2012
2q35	*CXCR2*	VL	Indian	Mehrotra et al. 2011
22q12		VL	Sudanese	Bucheton et al. 2003a
1p22 6q27		VL	Sudanese	Miller et al. 2007
6q27 7q11.22 17q11.2–q21.3		VL	Brazilian	Jamieson et al. 2007
9p21.1		VL	Brazilian	Jeronimo et al. 2007a
6q27	*PHF10*, *C6orf70*, *DLL1*, *PSMB1*, and *TBP*	VL	Sudanese and Brazilian	Fakiola et al. 2011
6q27	*DLL1*	VL	Indian	Mehrotra et al. 2012
6p25.3–p24.3 11q13.2–q23.3 8p23.1–p21.3 Xq21.1–q26.1		VL		Fakiola et al. 2010

4.3.1 *SLC11A1* (Formerly *NRAMP1*)

Strong evidence for genetic control of multiplication of *L. donovani* in mice was shown experimentally by following parasite loads in liver and spleen after intravenous inoculation of amastigotes into different inbred mouse strains (Bradley 1979). The rate of parasite multiplication during the acute phase (first two weeks) of infection showed that the parasite multiplication rate clearly divided inbred strains into two groups. Twelve strains designated as susceptible showed >100-fold parasite replication over the first two weeks, while the remaining 13 resistant strains showed less than eight-fold increase (Bradley 1974). Inter-crossing to produce F1, F2, and backcross progeny demonstrated the presence of a single gene controlling acute susceptibility or resistance to *L. donovani* infection in mice, and the gene was designated *Lsh* (Bradley 1977). Parallel studies similarly demonstrated single-gene (designated *Ity*) control of *Salmonella typhimurium* (Plant and Glynn 1974, 1976), followed by studies demonstrating single-gene (designated *Bcg*) control of *Mycobacterium bovis* infection (Gros et al. 1981). All three genes were shown to map to the same region of mouse Chromosome 1 (Bradley 1979; Plant and Glynn 1979; Plant et al. 1982; Skamene et al. 1982). The gene was positionally cloned and redesignated the natural resistance associated macrophage protein 1 (*Nramp1*), with susceptibility associated with a single nonconservative Gly to Asp substitution at position 169 within predicted transmembrane domain 4 of the predicted polytopic integral membrane protein (Vidal et al. 1993). The *Nramp1* designation is now superseded by the designation solute carrier family 11a (proton-coupled divalent metal ion transporters) member 1, or *Slc11a1* in mice and *SLC11A1* in humans, based on its function as proton/divalent cation (Fe^{2+}, Zn^{2+}, and Mn^{2+}) transporter (Goswami et al. 2001; Jabado et al. 2000). The gene has many pleiotropic effects on macrophage function, including regulation of the CXC chemokines KC, IL-1β, iNOS, MHC class II molecules, TNF, nitric oxide (NO) release, L-arginine flux, oxidative burst, and tumoricidal as well as microbial activity (reviewed by Blackwell et al. [1994, 2001] and Skamene [1994]).

In humans, a number of studies have shown association and/or linkage of *SLC11A1* with susceptibility to infectious and autoimmune diseases (reviewed by Blackwell et al. [2003]). In Sudan, our group examined polymorphisms at *SLC11A1* in 59 multicase families of VL from the high-incidence Masalit tribe (Mohamed et al. 2004). Significant linkage across *SLC11A1* was observed. The extended transmission disequilibrium test showed biased transmission of alleles at 5′ polymorphisms in the promoter, exon 3 and intron 4, and haplotypes formed by them, but not for 3′ polymorphisms at exon 15 or the 3′ UTR. Stepwise logistic regression analysis using a case/pseudo-control dataset derived from the 59 families was consistent with the main effects contributed by the intron 4 469 + 14G/C polymorphism (Mohamed et al. 2004). In another study conducted in 37 multicase families of VL from the Aringa tribe in eastern Sudan, significant linkage was found with 5′(CA) repeat polymorphism in *SLC11A1* (Bucheton et al. 2003b). However, in a primary family-based (313 cases; 176 families) followed by large replication (941 cases; 992 controls) cohort studied in India, no evidence was found for associations between VL and nine polymorphisms at *SLC11A1* (Mehrotra et al. 2011). Nor was a signal at *SLC11A1* observed in the well-powered GWAS performed across India and Brazil (Fakiola et al. 2013). It will be of interest to determine whether significance at *SLC11A1* is retained in larger well-powered replication studies in Africa.

HLA as a Candidate Gene Region

Analysis of *L. donovani* infection over the "acquired" T cell phase of the immune response (>15 days) showed that genes encoding the IA (DQ in man) and IE (DR in man) class II molecules in the major histocompatibility complex (MHC; H-2 in mice, HLA in man) exert a profound influence on the ability of even *Slc11a1*-susceptible mice to self-cure from *L. donovani* infection (Blackwell et al. 1980). Hence, there was strong evidence that polymorphism at class II molecules is itself functionally responsible for MHC-regulated control of infection. Genes encoding the class II MHC molecules lie adjacent to class III genes (*TNF*; *LTA*) encoding members of the tumor necrosis factor (TNF) family, TNF, and lymphotoxin α. In man, high levels of circulating TNF have been observed in cases with severe clinical visceral leishmaniasis (Barral-Netto et al. 1991). This suggests that regulatory polymorphisms for genes encoded in other regions of the HLA complex could also be important in determining genetic susceptibility. Hence, there were multiple reasons to target the HLA complex as a candidate disease susceptibility locus for VL.

In human VL there was no linkage or allelic association with class II/III regions of HLA when 15 polymorphic markers across 1.03 Mb (DQB1 to TNF) were genotyped in 87 multicase families of VL in northeastern Brazil (Peacock et al. 2002). Similarly, there was no evidence for significant association with HLA class I (A and B) and class II (DR) in a small study comparing 51 VL patients and 46 controls in India (Singh et al. 1997). This region was also investigated in the Tunisian population, where 156 unrelated cases of Mediterranean VL due to *L. infantum* and 154 unrelated healthy controls were genotyped for polymorphisms at *HLA-DRB1*, *-DQB1*, *TNF*, *LTA*, *HSP70-2*, and *HSP70-hom*. No significant associations were found after Bonferroni correction for multiple testing (Meddeb-Garnaoui et al. 2001). In Sudan, two SNPs at −308 bp in the promoter of *TNF* and at amino acid position 26 in exon 2 of *LTA*, together with a more distal *TNF* microsatellite, were genotyped in 59 multicase families (Mohamed 2002). No significant linkage or allelic association was observed. Examination of the −308 bp polymorphism at *TNF* in 183 Brazilian families showed that allele 1 was transmitted more frequently than expected from heterozygous parents to DTH+ offspring, while haplotypes with the *TNF* microsatellite that contained allele 2 at the −308 bp *TNF* SNP (single nucleotide polymorphism) were associated with symptomatic VL (Karplus et al. 2002). In brief, almost all of these small case-control and family-based studies from different populations – Brazil, India, Tunisia, Iran, and Sudan – failed to demonstrate linkage or allelic association between VL and HLA genes. Nevertheless, when a well-powered GWAS was undertaken across India and Brazil, definitive evidence of association at the *HLA-DRB1–HLA-DQA1* locus was observed (Fakiola et al. 2013), thus demonstrating the limitations of undertaking small underpowered studies even when they are supported by strong hypothesis-driven prior knowledge.

Chromosome 17q11.1–q12

Genetic analysis of quantitative traits in mice (lesion size, parasite loads, and organ sizes in the viscera) identified a region on murine chromosome 11 linked to susceptibility to cutaneous leishmaniasis following *L. major* infection (Mock et al. 1993; Roberts et al. 1993). This region shows conserved synteny with human chromosome 17q11.1–q12, which contains a cluster of chemokine genes as well as the gene (*NOS2A*) encoding macrophage inducible nitric oxide synthase (iNOS) and chemokine receptors. The region was shown to be linked to susceptibility to human VL caused by *L. infantum chagasi* in a GWLS carried out in Brazil, with family-based association tests indicating significant association with two chemokine genes, *CCL1* and *CCL16* (Jamieson et al. 2007). In Sudan, a study of multicase families showed no evidence for linkage between chromosome 17q11.1–q12 markers (*NOS2A*, *D17S250*, and *D17S33*) and a putative disease susceptibility locus for VL or PKDL.

Chromosome 5q31–q33

A second region of murine chromosome 11 that was linked to controlling susceptibility in later phases of infection following *L. major* infection (Roberts et al. 1993) shows conserved synteny with human chromosome 5q31–q33 that contains a cluster of T helper 2 associated cytokine genes (*IRF1–IL5–IL13–IL4–IL9–LECT2–TGFBI*). In the mouse model, extensive visceralization of *L. major* in the susceptible BALB/c strain provides a further model of VL, making this a compelling region to look for susceptibility genes for human VL. In humans, immunological studies suggest that the balance in type 1 versus type 2 cytokine responses is important in *L. donovani* infection (Kemp et al. 1993; Sundar et al. 1997). In Brazil, a detailed study of SNPs across the region was undertaken using both symptomatic VL as a phenotype, as well as a positive delayed-type hypersensitivity skin test response (DTH) as a marker for acquired resistance to disease. Twenty SNPs were analyzed across *IRF1–IL5–IL13–IL4–IL9–LECT2–TGFBI* in 102 families (323 DTH+; 190 DTH–; 123 VL individuals). Data were analyzed for association with DTH status and VL using family-based stepwise conditional logistic regression analysis. Independent associations were observed between the DTH+ phenotype and markers in separate linkage disequilibrium blocks in *LECT2* (OR 2.25; $p = 0.005$; 95 percent CI = 1.28–3.97) and *TGFBI* (OR 1.94; $p = 0.003$; 95 percent CI = 1.24–3.03). Visceral leishmaniasis child–parent trios gave no evidence of association, but the DTH– phenotype was associated with SNP rs2070874 at *IL4* (OR 3.14; $p = 0.006$; 95 percent CI = 1.38–7.14), and SNP rs30740 between *LECT2* and *TGFBI* (OR 3.00; $p = 0.042$; 95 percent CI = 1.04–8.65). The association between *TGFBI* and DTH+ has been replicated in an independent study in Brazil in which the VL phenotype also showed association (Frade et al. 2011). Together, the results indicate several genes in the immune response gene cluster at 5q23.3–q31.1 influence outcomes of *L. infantum chagasi* infection in Brazil.

In Sudan, we focused on variants in *IL4* and *IL9* in 59 multicase families of VL with/without PKDL. Three polymorphisms were typed: a variable number tandem repeat (VNTR) in intron 2 of the *IL4* gene designated IL4RP2, a microsatellite in intron 3 of *IL4* designated IL4RP1, and microsatellites in intron 4 of the *IL9* gene. Multipoint nonparametric analysis linked the *IL4*/*IL9* gene region to VL *per se* ($p = 0.002$). Transmission disequilibrium testing with robust variance estimates confirmed association in the presence of linkage between VL *per se* and variants at *IL4* ($p = 0.008$) but not *IL9*. Stepwise logistic regression analysis showed both IL4RP2 and IL4RP1 markers contributed significantly to the association, suggesting a common disease-associated haplotype (Mohamed et al. 2003). Studies utilizing larger sample sizes may help to further define the roles of genetic variants within the chromosome 5q23.3–q31.1 region in regulating different clinical outcomes of VL across the epidemiological divides of geography and parasite species.

4.3.2 *IFNG* and *IFNGR1*

IFNγ encoded by the *IFNG* gene is a key T helper 1 and natural killer cell cytokine. It activates macrophages to kill leishmania parasites (Murray et al. 1983), exerting its effects through its receptor IFNGR1. In Sudan, two studies confirmed linkage and association of *IFNGR1* gene polymorphisms with PKDL but not the VL phenotype (Mohamed et al. 2003; Salih et al. 2007). No evidence was found for linkage or association with *IFNG* gene polymorphisms.

4.3.3 *IL2RB* and *IL1B*

Evidence for association between VL susceptibility and a novel polymorphism causing the amino acid change G245R in the membrane proximal domain of the IL-2 receptor beta chain, as well as for the variant *IL2RB*+8777T, was reported in a population from eastern Sudan (Bucheton et al. 2007). However, these associations were not thought to be the sole contributors to the region of linkage at chromosome 22q12 (see below) that had previously been reported for this population (Bucheton et al. 2003a).

In an Iranian population, 95 cases and 128 unrelated healthy controls were enrolled in a study looking for associations between VL and the variants +3953T/C and −511T/C in the *IL1B* gene encoding interleukin-1 beta (Moravej et al. 2012). Higher frequencies of the *IL1B* −511CC genotype, and the C allele, were found in VL patients ($p = 0.008$ and $p = 0.00006$, respectively) compared to endemic controls. The haplotype *IL1B* −511C/+3953C was also observed to be more frequent in VL patients compared with the controls ($p = 0.0002$).

4.3.4 *CXCR1* and *CXCR2*

IL8RA and IL8RB, encoded by *CXCR1* and *CXCR2*, are receptors for interleukin (IL)-8 and other CXC chemokines involved in chemotaxis and activation of polymorphonuclear neutrophils. Variants at *CXCR1* and *CXCR2* had been associated with susceptibility to cutaneous and mucocutaneous leishmaniasis in Brazil (Castellucci et al. 2010). In India, three SNPs (rs4674259, rs2234671, rs3138060) that tag linkage disequilibrium blocks across *CXCR1/CXCR2* were genotyped in primary family-based (313 cases; 176 nuclear families; 836 individuals) and replication (941 cases; 992 controls) samples. Family-based association analysis showed association between VL and SNPs *CXCR1* rs2234671 ($p = 0.003$) and *CXCR1* rs3138060 ($p = 0.026$), but not with *CXCR2* rs4674259. Logistic regression analysis of the case–control data under an additive model showed association between VL and SNPs *CXCR2* rs4674259 ($p = 0.027$) and *CXCR1* rs3138060 ($p = 0.028$), but not with *CXCR1* rs2234671. The three-locus haplotype T_G_C across these SNPs was shown to be the risk haplotype in both family (TRANSMIT; $p = 0.014$ – see Clayton 1999) and population ($p = 0.028$) samples (combined $p = 0.002$). *CXCR2*, but not *CXCR1*, expression measured using quantitative RTPCR was downregulated in pretreatment compared to posttreatment splenic aspirates ($p = 0.021$). This well-powered primary and replication genetic study, together with functional analysis of gene expression, implicate *CXCR2* and a role for neutrophils in determining outcome of VL in India.

4.4 Genome-Wide Linkage Studies

Family-based GWLS have been used to identify regions of the genome linked to disease susceptibility. In Sudan, the first reported genome scan for VL was carried out on 36 VL families from the Aringa tribe in eastern Sudan. The results showed significant linkage (LOD score 3.5; $p = 3 \times 10^{-5}$) with markers on chromosome 22q12 (Bucheton et al. 2003a). A similar genome-wide scan study using 69 multicase families from two villages in eastern Sudan occupied by the Masalit ethnic group failed to replicate this linkage, instead identifying major Y chromosome tagged lineage-specific linkage at 1p22 (1p22, LOD = 5.65; $p = 1.72 \times 10^{-7}$) and 6q27 (6q27, LOD = 3.74; $p = 1.68 \times 10^{-5}$) controlling VL (Miller et al. 2007). The results were interpreted in relation to founder effect and a high rate of consanguinity in recent immigrant populations (Miller et al. 2007).

In Brazil, 22 multicase families of VL were enrolled in a GWLS (Jamieson et al. 2007). Three regions (6q27, 7q11.22, and 17q11.2–q21.3) remained positive (LOD scores 1.08, 1.34, 1.14; $p = 0.013, 0.007, 0.011$) for linkage following further dense mapping. In a separate study in Brazil, a region at 9p21.1 (LOD score 1.60; $p = 0.003$) was linked to VL, while regions on chromosomes 15q26.2 (LOD score 2.50; $p = 0.0003$) and 19q33.3 (LOD score 1.93; $p = 0.001$) were linked to the DTH+ phenotype (Jeronimo et al. 2007a).

Further investigations (Fakiola et al. 2011; Mehrotra et al. 2012) have been carried out to fine-map genes for the VL phenotype within the 6q27 region. Single nucleotide polymorphisms at *PHF10*, *C6orf70*, *DLL1*, *PSMB1*, and *TBP* all showed significant associations with VL in Sudan, but only *DLL1* SNPs were significant in Brazil (Fakiola et al. 2011), perhaps reflecting an ability to localize the etiological gene more precisely in the admixed Brazilian population compared to the large lineages with consanguinity and few founders used in the Sudanese study. Association between the VL phenotype and SNPs at *DLL1* was also replicated in a well-powered genetic and functional study in an Indian population (Mehrotra et al. 2012). Expression of *DLL1* measured using quantitative RTPCR was significantly lower than other 6q27 genes in splenic aspirates from VL cases (Fakiola et al. 2011), while changes in expression of *DLL1* were observed following drug cure of VL (Mehrotra et al. 2012). *DLL1* encodes Delta-like 1, the ligand for Notch3. Within the family of Notch ligands, Delta1 and Delta4 instruct antigen-driven CD4 T cell selection down the Th1 fate as opposed to the Th2 fate that is driven by Jagged1 and Jagged2 (Amsen et al. 2004; Ong et al. 2008). The effects of polarized antigen-specific Th1 and Th2 responses on the outcome of leishmaniasis are well established in murine models (Locksley and Scott 1991; Scott et al. 1989) and human disease (Pirmez et al. 1993; Sundar et al. 1997). Clinical VL, in particular, has been associated with high Th2 cytokine responses (Sundar et al. 1997), whereas Th1-generated IFNγ levels are higher in children infected with *L. infantum chagasi* who do not progress to clinical VL than in those who do (Carvalho et al. 1992). The potential for Delta1-driven Th1 differentiation to alter the course of infection (Maekawa et al. 2003) makes genetic regulation of *DLL1* expression a highly plausible explanation for the genetic associations and down-regulated splenic expression observed with this 6q27 gene.

A primary GWLS followed by refined mapping identified linkage peaks at chromosomes 2q12.2-q14.1 and 11q13.2-q23.3, but only 11q13.2-q23.3 was replicated in a second set of families (Fakiola et al. 2010). When families were analyzed according to religious group, linkage peaks at 6p25.3–p24.3 and 8p23.1-p21.3 were found to be specific to Hindu families, while a peak at Xq21.1–q26.1 was specific to Muslim families. Only the peak at Xq21.1–q26.1 was replicated across all Muslim families. No population genetic substructure related to religious group was identified, but spatial analysis confirmed geographical heterogeneity for linkages at 6p25.3–p24.3, 8p23.1–p21.3, and Xq21.1–q26.1, with specific clusters of families contributing (Fakiola et al. 2010). None of the linkage peaks identified in Sudanese and Brazilian studies were replicated in this Indian study.

4.5 Genome Wide Association Studies

Genome wide association studies are a promising approach for identifying disease susceptibility genes for common diseases. These studies involve scanning thousands of samples in family-based or case–control cohorts. In this method, microarray DNA chips are used to genotype hundreds of thousands to millions of SNP markers located across the human genome. By 2015, >2000 published GWAS had achieved $p < 5 \times 10^{-8}$ across 17 different trait classes, information about which is summarized in the NIH Catalog of Published Genome-Wide Association Studies (Hindorff et al. 2009). This now includes a number of studies of infectious disease, including the first VL GWAS conducted by the LeishGEN Consortium and the Wellcome Trust Case Control Consortium 2 (Fakiola et al. 2013). Discovery GWAS were undertaken in 989 VL cases caused by *L. donovani* and 1089 controls from India and 357 VL cases caused by *L. infantum chagasi* in 308 families (total individuals 1970) from Brazil. Strong evidence of association was found in the *HLA-DRB1–HLA-DQA1* locus in both populations, which was replicated in a further sample of 941 VL cases and

990 controls from India. The combined p value at the top SNP rs9271858 across discovery and replication cohorts was $p = 2.76 \times 10^{-17}$. The *HLA-DRB1-HLA-DQA1* HLA class II region thus contributes to VL susceptibility in both India and Brazil, suggesting shared genetic risk factors that cross the epidemiological divides of geography and parasite species.

4.6 Conclusion and Future Perspectives

Visceral leishmaniasis is an important health problem, but remains largely a neglected disease. There is no vaccine available. The first-line drugs require injections over a period of 30 days in a hospital, and drugs are expensive and unaffordable. The disease affects poor and marginalized populations in remote areas. In India, up to 60 percent of cases are now resistant to the first-line drug. Other drugs have side-effects. Different studies have showed the importance of host genetics in susceptibility to VL. However, researchers failed to replicate many genetic loci in different populations. This may be due to small sample sizes of individual studies, a wide range of ethnic groups, and absence of major gene effects in complex diseases. Nevertheless, these studies provide knowledge about genes involved in disease pathogenesis and possible biological pathways involved in disease development. Well-designed studies are required to investigate and analyze the mechanisms of genetic interactions that generate disease phenotypes and molecular pathways of disease.

References

Alvar J, Velez ID, Bern C, et al. (2012). Leishmaniasis worldwide and global estimates of its incidence. *PLoS One* **7**: e35671.

Amsen D, Blander JM, Lee GR, et al. (2004). Instruction of distinct CD4 T helper cell fates by different notch ligands on antigen-presenting cells. *Cell* **117**: 515–526.

Badaro R, Jones TC, Lorenco R, et al. (1986). A prospective study of visceral leishmaniasis in an endemic area of Brazil. *J Infect Dis* **154**: 639–649.

Barral-Netto M, Badaro R, Barral A, et al. (1991). Tumor necrosis factor (cachectin) in human visceral leishmaniasis. *J Infect Dis* **163**: 853–857.

Blackwell JM (1996). Genetic susceptibility to leishmanial infections: studies in mice and man. *Parasitology* **112**: S67-S74.

Blackwell JM, Freeman J, Bradley D (1980). Influence of H-2 complex on acquired resistance to *Leishmania donovani* infection in mice. *Nature* **283**: 72–74.

Blackwell JM, Barton CH, White JK, et al. (1994). Genetic regulation of leishmanial and mycobacterial infections: The *Lsh/Ity/Bcg* gene story continues. *Immunol Lett* **43**: 99–107.

Blackwell JM, Goswami T, Evans CAW, et al. (2001). SLC11A1 (formerly NRAMP1) and disease. *Cell Microbiol* **3**: 773–784.

Blackwell JM, Searle S, Mohamed H, and White JK (2003). Divalent cation transport and susceptibility to infectious and autoimmune disease: continuation of the *Ity/Lsh/Bcg/Nramp1/Slc11a1* gene story. *Immunol Lett* **85**: 197–203.

Blackwell JM, Fakiola M, Ibrahim ME, et al. (2009). Genetics and visceral leishmaniasis: of mice and man. *Parasite Immunol* **31**: 254–266.

Bradley DJ (1974). Genetic control of natural resistance to *Leishmania donovani*. *Nature* **250**: 353–354.

Bradley DJ (1977). Regulation of *Leishmania* populations within the host: II. Genetic control of acute susceptibility of mice to *Leishmania donovani* infection. *Clin Exp Immunol* 30: 130–140.

Bradley DJ (1979). Regulation of Leishmania populations within the host: IV. Parasite and host cell kinetics studied by radioisotope labelling. *Acta Trop* 36: 171–179.

Bucheton B, Kheir MM, El-Safi SH, et al. (2002). The interplay between environmental and host factors during an outbreak of visceral leishmaniasis in eastern Sudan. *Microbes Infect* 4: 1449–1457.

Bucheton B, Abel L, El-Safi S, et al. (2003a). A major susceptibility locus on chromosome 22q12 plays a critical role in the control of kala-azar. *Am J Hum Genet* 73: 1052–1060.

Bucheton B, Abel L, Kheir MM, et al. (2003b). Genetic control of visceral leishmaniasis in a Sudanese population: candidate gene testing indicates a linkage to the *NRAMP1* region. *Genes Immun* 4: 104–109.

Bucheton B, Argiro L, Chevillard C, et al. (2007). Identification of a novel G245R polymorphism in the IL-2 receptor beta membrane proximal domain associated with human visceral leishmaniasis. *Genes Immun* 8: 79–83.

Cabello PH, Lima AM, Azevedo ES, and Kriger H (1995). Familial aggregation of *Leishmania chagasi* infection in northeastern Brazil. *Am J Trop Med Hyg* 52: 364–365.

Carvalho EM, Barral A, Pedral-Sampaio D, et al. (1992). Immunologic markers of clinical evolution in children recently infected with *Leishmania donovani chagasi*. *J Infect Dis* 165: 535–540.

Castellucci L, Jamieson SE, Miller EN, et al. (2010). *CXCR1* and *SLC11A1* polymorphisms affect susceptibility to cutaneous leishmaniasis in Brazil: a case–control and family-based study. *BMC Med Genet* 11: 10.

Clayton D (1999). A generalization of the transmission disequilibrium test for uncertain-haplotype transmission. *Am J Hum Genet* 65: 1170–1177.

Dudbridge F and Gusnanto A (2008). Estimation of significance thresholds for genomewide association scans. *Genet Epidemiol* 32: 227–234.

Faghiri Z, Tabei SZ, Taheri F (1995). Study of the association of HLA class I antigens with kala-azar. *Hum Hered* 45: 258–261.

Fakiola M, Mishra A, Rai M, et al. (2010). Classification and regression tree and spatial analyses reveal geographic heterogeneity in genome wide linkage study of Indian visceral leishmaniasis. *PLoS One* 5: e15807.

Fakiola M, Miller EN, Fadl M, et al. (2011). Genetic and functional evidence implicating *DLL1* as the gene that influences susceptibility to visceral leishmaniasis at chromosome 6q27. *J Infect Dis* 204: 467–477.

Fakiola M, Strange A, Cordell HJ, et al. (2013). Common variants in the *HLA-DRB1-HLA-DQA1* HLA class II region are associated with susceptibility to visceral leishmaniasis. *Nat Genet* 45: 208–213.

Feitosa MF, Axevedo E, Lima AM, and Krieger H (1999). Genetic causes involved in *Leishmania chagasi* infection in northeastern Brazil. *Genet Mol Biol* 22: 1–5.

Frade AF, Oliveira LC, Costa DL, et al. (2011). *TGFB1* and *IL8* gene polymorphisms and susceptibility to visceral leishmaniasis. *Infect Genet Evol* 11: 912–916.

Gautam S, Kumar R, Maurya R, et al. (2011). IL-10 neutralization promotes parasite clearance in splenic aspirate cells from patients with visceral leishmaniasis. *J Infect Dis* 204: 1134–1137.

Goswami T, Bhattacharjee A, Babal P, et al. (2001). Natural-resistance-associated macrophage protein 1 is an H+/bivalent cation antiporter. *Biochem J* 354: 511–519.

Gros P, Skamene E, and Forget A (1981). Genetic control of natural resistance to *Mycobacterium bovis* (BCG) in mice. *J Immunol* 127: 2417–2421.

Hindorff LA, Junkins HA, Mehta JP, and Manolio TA (2009). A catalog of published genome-wide association studies. Available at: www.ebi.ac.uk/gwas (accessed February 10, 2016).

Ho M, Siongok TK, Lyerly WH, and Smith DH (1982). Prevalence and disease spectrum in a new focus of visceral leishmaniasis in Kenya. *Trans R Soc Trop Med Hyg* 76: 741–746.

Ibrahim ME, Lambson B, Yousif AO, et al. (1999). Kala-azar in a high transmission focus: an ethnic and geographic dimension. *Am J Trop Med Hyg* **61**: 941–944.

Jabado N, Jankowski A, Dougaparsad S, et al. (2000). Natural resistance to intracellular infections: natural resistance-associated macrophage protein 1 (Nramp1) functions as a pH-dependent manganese transporter at the phagosomal membrane. *J Exp Med* **192**: 1237–1248.

Jamieson SE, Miller EN, Peacock CS, et al. (2007). Genome-wide scan for visceral leishmaniasis susceptibility genes in Brazil. *Genes Immun* **8**: 84–90.

Jeronimo SM, Oliveira RM, Mackay S, et al. (1994). An urban outbreak of visceral leishmaniasis in Natal. *Brazil Trans R Soc Trop Med Hyg* **88**: 386–388.

Jeronimo SM, Duggal P, Braz RF, et al. (2004). An emerging peri-urban pattern of infection with *Leishmania chagasi*, the protozoan causing visceral leishmaniasis in northeast Brazil. *Scand J Infect Dis* **36**: 443–449.

Jeronimo SM, Duggal P, Ettinger N, et al. (2007a). Genetic predisposition to symptomatic versus self-curing infection with the protozoan *Leishmania chagasi*: a genome wide scan. *J Infect Dis* **196**: 1261–1269.

Jeronimo SM, Holst AK, Jamieson SE, et al. (2007b). Genes at human chromosome 5q31.1 regulate delayed-type hypersensitivity responses associated with *Leishmania chagasi* infection. *Genes Immun* **8**: 539–551.

Karplus TM, Jeronimo SM, Chang H, et al. (2002). Association between the tumor necrosis factor locus and the clinical outcome of *Leishmania chagasi* infection. *Infect Immun* **70**: 6919–6925.

Kemp M, Kurtzhals JA, Kharazmi A, and Theander TG (1993). Interferon-gamma and interleukin-4 in human *Leishmania donovani* infections. *Immunol Cell Biol* **71**: 583–587.

Khalil EA, Zijlstra EE, Kager PA, and El Hassan AM (2002). Epidemiology and clinical manifestations of *Leishmania donovani* infection in two villages in an endemic area in eastern Sudan. *Trop Med Int Health* **7**: 35–44.

Kumar R, Singh N, Gautam S, et al. (2014). Leishmania specific CD4 T cells release IFNgamma that limits parasite replication in patients with visceral leishmaniasis. *PLoS Negl Trop Dis* **8**: e3198.

Lander E and Kruglyak L (1995). Genetic dissection of complex traits: guidelines for interpreting and reporting linkage results. *Nat Genet* **11**: 241–247.

Locksley RM and Scott P (1991). Helper T-cell subsets in mouse leishmaniasis induction, expansion and effector function. *Immunol Today* **12**: A58–A61.

Maekawa Y, Tsukumo S, Chiba S, et al. (2003). Delta1-Notch3 interactions bias the functional differentiation of activated CD4+ T cells. *Immunity* **19**: 549–559.

Meddeb-Garnaoui A, Gritli S, Garbouj S, et al. (2001). Association analysis of HLA-class II and class III gene polymorphisms in the susceptibility to Mediterranean visceral leishmaniasis. *Hum Immunol* **62**: 509–517.

Mehrotra S, Oommen J, Mishra A, et al. (2011). No evidence for association between *SLC11A1* and visceral leishmaniasis in India. *BMC Med Genet* **12**: 71.

Mehrotra S, Fakiola M, Mishra A, et al. (2012). Genetic and functional evaluation of the role of *DLL1* in susceptibility to visceral leishmaniasis in India. *Infect Genet Evol* **12**: 1195–1201.

Miller EN, Fadl M, Mohamed HS, et al. (2007). Y chromosome lineage- and village-specific genes on chromosomes 1p22 and 6q27 that control visceral leishmaniasis in The Sudan. *PLoS Genet* **3**: 679–688.

Mock B, Blackwell J, Hilgers J, Potter M, and Nacy C (1993). Genetic control of *Leishmania major* infection in congenic, recombinant inbred and F2 populations of mice. *Eur J Immunogenet* **20**: 335–348.

Mohamed H (2002). *Genetic susceptibility to visceral leishmaniasis in Sudan.* PhD Thesis, University of Khartoum.

Mohamed HS, Ibrahim ME, Miller EN, et al. (2003). Genetic susceptibility to visceral leishmaniasis in The Sudan: linkage and association with *IL4* and *IFNGR1*. *Genes Immun* **4**: 351–355.

Mohamed HS, Ibrahim ME, Miller EN, et al. (2004). *SLC11A1* (formerly *NRAMP1*) and susceptibility to visceral leishmaniasis in The Sudan. *Eur J Hum Genet* **12**: 66–74.

Moravej A, Rasouli M, Kalani M, et al. (2012). IL-1beta (-511T/C) gene polymorphism not IL-1beta (+3953T/C) and LT-alpha (+252A/G) gene variants confers susceptibility to visceral leishmaniasis. *Mol Biol Reports* **39**: 6907–6914.

Murray HW, Rubin BY, and Rothermel CD (1983). Killing of intracellular *Leishmania donovani* by lymphokine-stimulated human mononuclear phagocytes: evidence that interferon-gamma is the activating lymphokine. *J Clin Invest* **72**: 1506–1510.

Nylen S and Sacks D (2007). Interleukin-10 and the pathogenesis of human visceral leishmaniasis. *Trends Immunol* **28**: 378–384.

Nylen S, Maurya R, Eidsmo L, et al. (2007). Splenic accumulation of IL-10 mRNA in T cells distinct from CD4+CD25+ (Foxp3) regulatory T cells in human visceral leishmaniasis. *J Exp Med* **204**: 805–817.

Ong CT, Sedy JR, Murphy KM, and Kopan R (2008). Notch and presenilin regulate cellular expansion and cytokine secretion but cannot instruct Th1/Th2 fate acquisition. *PLoS One* **3**: e2823.

Peacock CS, Collins A, Shaw MA, et al. (2001). Genetic epidemiology of visceral leishmaniasis in northeastern Brazil. *Genet Epidemiol* **20**: 383–396.

Peacock CS, Sanjeevi CB, Shaw MA, et al. (2002). Genetic analysis of multicase families of visceral leishmaniasis in northeastern Brazil: no major role for class II or class III regions of HLA. *Genes Immun* **3**: 350–358.

Pirmez C, Yamamura M, Uyemura K, Paes-Oliveira M, and Modlin RL (1993). Cytokine patterns in the pathogenesis of human leishmaniasis. *J Clin Invest* **91**: 1390–1395.

Plant J and Glynn AA (1974). Natural resistance to *Salmonella* infection, delayed hypersensitivity and Ir genes in different strains of mice. *Nature* **248**: 345–347.

Plant J and Glynn AA (1976). Genetics of resistance to infection with *Salmonella typhimurium* in mice. *J Infect Dis* **133**: 72–78.

Plant J and Glynn AA (1979). Locating salmonella resistance gene on mouse Chromosome 1. *Clin Exp Immunol* **37**: 1–6.

Plant JE, Blackwell JM, O'Brien AD, Bradley DJ, and Glynn AA (1982). Are the *Lsh* and *Ity* disease resistance genes at one locus on mouse chromosome 1? *Nature* **297**: 510–511.

Roberts M, Mock BA, and Blackwell JM (1993). Mapping of genes controlling *Leishmania major* infection in CXS recombinant inbred mice. *Eur J Immunogenet* **20**: 349–362.

Salih MA, Ibrahim ME, Blackwell JM, et al. (2007). *IFNG* and *IFNGR1* gene polymorphisms and susceptibility to post-kala-azar dermal leishmaniasis in Sudan. *Genes Immun* **8**: 75–78.

Scott P, Pearce E, Cheever AW, Coffman RL, and Sher A (1989). Role of cytokines and CD4+ T-cell subsets in the regulation of parasite immunity and disease. *Immunol Rev* **112**: 162–182.

Shaw MA, Davies CR, Llanos-Cuentas EA, and Collins A (1995). Human genetic susceptibility and infection with *Leishmania peruviana*. *Am J Hum Genet* **57**: 1159–1168.

Singh N, Sundar S, Williams F, et al. (1997). Molecular typing of HLA class I and class II antigens in Indian kala-azar patients. *Trop Med Int Health* **2**: 468–471.

Singh OP, Gidwani K, Kumar R, et al. (2012a). Reassessment of immune correlates in human visceral leishmaniasis as defined by cytokine release in whole blood. *Clin Vaccine Immunol* **19**: 961–966.

Singh OP, Stober CB, Singh AK, Blackwell JM, and Sundar S (2012b). Cytokine responses to novel antigens in an Indian population living in an area endemic for visceral leishmaniasis. *PLoS Negl Trop Dis* **6**: e1874.

Skamene E (1994). The *Bcg* gene story. *Immunobiology* **191**: 451–460.

Skamene E, Gros P, Forget A, et al. (1982). Genetic regulation of resistance to intracellular pathogens. *Nature* **297**: 506–509.

Sundar S, Reed SG, Sharma S, Mehrotra A, and Murray HW (1997). Circulating T helper 1 (Th1) cell- and Th2 cell-associated cytokines in Indian patients with visceral leishmaniasis. *Am J Trop Med Hyg* **56**: 522–525.

Vidal SM, Malo D, Vogan K, Skamene E, and Gros P (1993). Natural resistance to infection with intracellular parasites: isolation of a candidate for Bcg. *Cell* **73**, 469–485.

Zijlstra EE and el-Hassan AM (2001a). Leishmaniasis in Sudan: post kala-azar dermal leishmaniasis. *Trans R Soc Trop Med Hyg* **95**(Suppl. 1): S59–S76.

Zijlstra EE and el-Hassan AM (2001b). Leishmaniasis in Sudan: visceral leishmaniasis. *Trans R Soc Trop Med Hyg* **95**(Suppl. 1): S27–S58.

Zijlstra EE, el-Hassan AM, Ismael A, and Ghalib HW (1994). Endemic kala-azar in Eastern Sudan, a longitudinal study on the incidence of clinical and subclinical infection and post-kala-azar dermal leishmaniasis. *Am J Trop Med Hyg* **51**: 826–836.

5 Genetics of Infection in Sub-Saharan Africa

What Can the Study of Mendelian Immunodeficiency Disorders Contribute?

Melanie Newport

5.1 Introduction: The Burden of Infectious Diseases in African Populations

While non-communicable diseases have overtaken infectious diseases as the leading cause of mortality and morbidity globally, the burden of infectious diseases remains disproportionately high in sub-Saharan Africa (SSA) compared to other regions. Despite improvements in the prevention and control of certain infectious diseases, driven in particular by the sixth Millennium Development Goal – to combat HIV/AIDS, malaria, and other diseases – the burden of infectious diseases remains high. In 2017 it was estimated that 1.18 million people died from tuberculosis (TB) and 619,800 from malaria (GBD 2017 Mortality and Causes of Death Collaborators, 2018). HIV/AIDS remains a major cause of premature death in SSA (954,500 deaths in 2017), despite the availability of effective anti-retroviral treatment. Despite gains in life expectancy in SSA due to reduction in diarrheal and lower respiratory tract infections in children, 808,900 died from lower respiratory tract infections, and 553,800 from diarrheal diseases, and 153,100 from meningitis in 2017.

Meanwhile, neglected tropical infectious diseases (NTDs), a disparate group of predominantly infectious diseases that affect the world's poorest people – the so-called bottom billion – are responsible for more days lost due to disability or death (56.6 million) than malaria or TB (Hotez et al. 2006, 2009). Half of this vulnerable population live in SSA. There are currently 20 infectious diseases represented in the WHO NTD control program, including lymphatic filariasis, onchocerciasis, trachoma, dracunculiasis (guinea worm), and soil-transmitted helminths (www.who.int/neglected_diseases/diseases/en).

While public health measures such as improved nutrition, sanitation, and mass drug administration for neglected tropical infectious diseases such as onchocerciasis (river blindness) have contributed to control certain diseases, the impact of diseases for which there are no effective preventive interventions remains high. The lack of effective vaccines for HIV, malaria, and TB (the BCG vaccine is not effective in adult populations who bear the main burden of TB) is a major obstacle to the control of these conditions which is compounded by the ever-increasing difficulty of treating them due to multidrug resistance (Abubakar et al. 2013; Takala-Harrison and Laufer 2015). There is also increasing concern around the impact of emerging and re-emerging infectious diseases such as pandemic H1N1 influenza and more recently ebola virus disease (Fineberg 2014; Martinez-Romero and Garcia-Sastre 2015) such that research is still required to develop new tools to control, eliminate, or eradicate important infections. Advances in genetics and genomics that have shown some success in identifying susceptibility loci for non-communicable diseases are being applied to identify key host pathways involved in immunity to infectious diseases, with the aim of facilitating the development of much-needed new drugs and vaccines. While many of these discoveries have been undertaken in large-scale case–control studies for complex, multifactorial diseases such as hypertension and inflammatory bowel disease (Wellcome Trust Case Control Consortium 2007), this chapter will focus on the invaluable contribution of the study of rare Mendelian disorders toward unraveling the molecular pathology of complex disorders, including many infectious diseases. Given the importance of TB in SSA and the body of work in Mendelian susceptibility to mycobacterial infections, this chapter will focus on this topic to illustrate the principles. Much of the work described here relates to studies that were conducted outside Africa: This mainly reflects the more advanced clinical and technological capacity required to diagnose and investigate rare immunodeficiencies at the time the work was undertaken. There have been a number of initiatives more recently such as H3Africa (www.h3africa.org) that are now redressing this balance.

5.2 Genetic Susceptibility to Infectious Diseases: Historical Perspectives

Individual responses to the same pathogen vary considerably from asymptomatic infection through to fatal disease. Host factors, including genes, are clearly important determinants of the individual response and until the end of the nineteenth century it was widely believed that susceptibility to infectious diseases such as leprosy and TB was genetically determined (Dubos and Dubos 1952). The hereditary concept lost credibility when it was challenged by Koch's famous discovery in 1882 that TB was caused by infection with *Mycobacterium tuberculosis* (Koch 1932). However, a number of observations over subsequent years rekindled interest in the genetic component of mycobacterial susceptibility. In 1926, prior to the development of anti-mycobacterial chemotherapy, 251 children from Lubeck in Germany were accidentally immunized with *M. tuberculosis* instead of BCG (Smith 1988). Of these, 47 developed no detectable disease, 127 developed radiologically evident disease, and 77 died, suggesting host variation in susceptibility. It was noted that certain infections clustered in families (although families share more than their genes) and the prevalence of many infectious diseases varies considerably between different ethnic groups (Stead 1992; Stead et al. 1990).

The most convincing epidemiological evidence for a genetic component to susceptibility to infectious diseases comes from twin studies. In the 1930s, several studies were undertaken in twins with TB (Simmonds 1963). If one twin developed TB, the second was at greater risk of contracting the disease if the twin pair was monozygotic than if it was dizygotic, with concordances of 60 percent and 20 percent respectively (Kallman and Reisner 1943a, 1943b). Concordance was less than 10 percent in married partners. Although there has been criticism of the study designs, which may have led to inflated concordance values, more recent studies have been conducted, the older data re-evaluated, and similar conclusions drawn (Comstock 1978).

Leprosy, also caused by a mycobacterial species, *M. leprae*, was also found to be more concordant in monozygous twins than dizygous twins, again suggesting a heritable component to susceptibility to contracting disease when infected (Chakravartti and Vogel 1973).

Further epidemiological evidence suggesting a genetic component to susceptibility to mycobacterial disease is derived from family studies, mainly conducted in families with leprosy (Shields et al. 1987). However, attempts to define the nature of this component have proved inconclusive. For example, a study of 91 Filipino sibships with lepromatous leprosy (Smith 1979) and a family study of tuberculoid leprosy (Haile et al. 1985) suggested autosomal recessive inheritance of susceptibility, but other studies of leprosy probands and their first-degree relatives were more consistent with a multifactorial pattern of inheritance (Rawlinson et al. 1988; Serjeantson 1983). In Desirade, a Caribbean island with a high prevalence of leprosy, a complex segregation analysis using 27 multigenerational pedigrees gave results consistent with the presence of a recessive or codominant gene (Abel and Demenais 1988; Abel et al. 1989).

Numerous investigations have associated human leucocyte antigens (HLA, equivalent to MHC in mice) with the subtypes of leprosy or TB. It is clear that HLA type influences disease phenotype once infected, but there is no evidence that susceptibility *per se* is controlled by HLA polymorphisms (Mehra 1990; Pitchappan 1990). At the population level, the most consistent association is between leprosy and the HLA class II antigens, especially HLA-DR2, which is associated with both tuberculoid and lepromatous forms of the disease (Van Eden and De Vries 1984). However, linkage analyses could not demonstrate a susceptibility gene linked to the HLA system, suggesting a more complex genetic mechanism.

5.3 More Recent Approaches to Mapping TB Susceptibility Genes: What Have We Learned from Monogenic Disorders?

Tuberculosis may be considered to be a genetically heterogeneous disease in which several genes interact with *M. tuberculosis* genes and the environment to contribute to the overall clinical phenotype, which ranges from asymptomatic carriage (also known as latent TB) through to severe life-threatening disease that can affect any organ of the body (Newport 2013). However, the underlying genetic model is unclear, with many questions unanswered – e.g., how many genes are involved? Does every gene contribute equally to the phenotype or are there some genes that have a larger influence? How do the different genes interact with each other? Thus, identification of the genes involved is challenging. One way around this problem, which applies to many other diseases, both communicable and non-communicable, is to study families in which a particular component of a complex phenotype segregates as a monogenic disorder. Although such families are rare, they bring great insight into the important pathways that are affected in complex diseases.

For example, this approach successfully identified mutations in various genes in the mineralocorticoid system and the electrolyte transport systems in the nephron that were responsible for familial hypertension, including Liddle's syndrome (Liddle et al. 1963; Simonetti et al. 2012).

Genetic analysis of monogenic primary human immunodeficiency disorders has also shed light on the critical pathways and genes required to control mycobacterial infections in humans. Over the past 50 years there have been a number of reported cases of severe disseminated infection with weakly virulent mycobacteria in individuals without recognized predisposing immunodeficiency (Newport 2003). These included reports in the literature of disseminated non-tuberculous mycobacteria (NTM) infection occurring in more than one member of a family. In the first reported family, a Danish mother and her two daughters who lived on a small farm in close contact with poultry died from disseminated *M. avium* infection (Engbaek 1964). Uchiyama et al. reported two siblings with *M. avium intracellulare* complex infection in whom a monocyte defect was demonstrated but not clearly defined (Uchiyama et al. 1981). More recently, Holland et al. reported a heterogeneous group of patients with disseminated non-tuberculous mycobacterial infection which included three male members of one family in whom interferon-gamma (IFNγ, a major component of the host immune response to mycobacterial infection) production by mononuclear cells was defective (Holland et al. 1994).

A number of children were also described who had developed disseminated infection with *M. bovis* BCG vaccine substrains in apparently healthy infants (Casanova et al. 1995). The children were extensively investigated in France, but no known immunodeficiency was found in 50 percent of them. High rates of affected siblings and parental consanguinity suggested the existence of a novel primary immunodeficiency syndrome.

5.4 Molecular Basis of Mendelian Susceptibility to Mycobacterial Disease in Humans

The first mutation leading to disseminated mycobacterial infection was reported in a consanguineous Maltese family, where a point mutation causing a premature stop codon in the gene encoding the IFNγ receptor ligand binding chain (IFNGR1) was identified (Levin 1995; Newport et al. 1996). The affected children were homozygous for the same mutation and did not express the receptor on their immune cells, leading to a severe clinical phenotype. Three out of four children died and the fourth survived as a result of a bone marrow transplant. A different mutation was identified in the same gene to explain disseminated BCG-osis in a vaccinated infant whose parents were first-cousins (Jouanguy et al. 1996).

These families defined a new clinical syndrome: an inherited immunodeficiency that predisposed to mycobacterial infection, subsequently named autosomal recessive IFNGR1 deficiency or immunodeficiency 27A (OMIM phenotype #209950 and gene #107470). However, mutation in *IFNGR1* did not explain all cases of MSMD. Further investigation of other children with severe mycobacterial infections, including families with *M. tuberculosis* infection, led to the discovery of mutations in six other autosomal genes and two X-linked genes within the IFNγ/IL-12/-23 pathway, reviewed in more detail elsewhere (Bustamante et al. 2014). The autosomal genes encode the signal transducing chain of the IFNγ receptor (IFNGR2), the p40 subunit of IL-12 (IL12B), the beta subunit of the IL-12 receptor (IL12RB1, which is shared with the IL-23 receptor), the signal transducing and activator of transcription molecule 1 (STAT1), the ubiquitin-like modifier ISG15 (ISG15), and interferon regulatory factor 8 (IRF8). The resulting disorders from mutations in these genes can be inherited as dominantly or recessively, depending on the molecular effect of the mutation. The genes encoding nuclear factor-κB-essential modulator (NEMO) and cytochrome b(-245) beta subunit (CYBB) are located on the X chromosome. Many different mutations have been described for most of these genes, leading to a spectrum of clinical presentation and disease severity (Bustamante et al. 2014; Dupuis et al. 2000). There remain a number of patients for whom a genetic etiology has yet to be identified, suggesting that mutations in other genes await discovery.

Most of these studies described above were conducted in Europe and North America, and their direct relevance for SSA populations is yet to be ascertained. The individuals presented to highly specialized clinical settings where facilities and resources allowed the detection of opportunistic infections and extensive investigation to characterize their immune defect. A number of the families were of northern African origin and groups working in northern African countries have contributed significantly to research on Mendelian disorders, also known as orphan diseases (Barbouche et al. 2017; Boisson-Dupuis et al. 2015; Prando et al. 2013; Romdhane et al. 2011), and this has been attributed to the high rates of consanguinity in the region. The SSA region also has a high rate of consanguineous marriages, and also higher rates of homozygosity, which increases the risk of Mendelian disorders (Bittles and Black 2010). However, the clinical and diagnostic systems across most of SSA are under-resourced and it is possible that in-born errors of immunity go undiagnosed and are attributed to other diseases, and there may not be awareness of these rare diseases.

Meanwhile, genes responsible for MSMD have been tested as candidate genes in studies of TB in outbred populations from a number of different African ethnic populations (Stein et al. 2007). Interestingly, there does appear to be evidence of genetic selection on the genes encoding IFNγ and its receptor chains, IFNGR1 and IFNGR2 (Manry et al. 2011).

5.4.1 Resistance to infection conferred by single-gene defects at the population level and the role of natural selection

While the disorders described above are rare, and arise from random mutations usually in the context of related parents, there are also more common genetic variants that confer resistance, rather than susceptibility, to infection. Infectious diseases have been a major selective force on human evolution, with the best-known example being the protective effect of the sickle cell polymorphism in the beta-globin gene (HBB) against malaria in heterozygous carriers (Rees et al. 2010). The distribution of the frequency of this allele maps directly onto the distribution of *Plasmodium falciparum* malaria, with highest frequencies of both found in SSA. A point mutation in a transcription factor binding site in the promoter region of the gene encoding the Duffy antigen receptor for chemokines (DARC), which acts as a receptor for *P. vivax*, abrogates protein expression on the erythrocyte, leading to resistance to *P. vivax* malaria, again having most impact in SSA, particularly in the West African region (Tournamille et al. 1995; Welch et al. 1977). In the case of HIV infection, a 32 base pair deletion in the chemokine receptor 5 (*CCR5*) gene, exploited by the virus as a co-receptor to gain entry into target cells, prevents expression of the receptor, making cells genetically resistant to HIV infection (Samson et al. 1996). However, while this polymorphism is relatively common in people of Caucasian ethnicity, it has not been described in African populations, highlighting the need for population-specific data. Understanding the molecular basis of HIV resistance in CCR5del32 homozygotes led to the development of a new class of anti-retroviral drugs, highlighting the potential for genetics as a tool in drug discovery (Westby and van der Ryst 2005).

5.5 Future Directions

There is a lack of knowledge about Mendelian immunodeficiencies in SSA, both in terms of disease burden and their underlying molecular mechanisms. This is more likely due to the lack of resources for diagnosing and investigating families than an absence of such diseases. Given the rich depth of diversity of genetic variation within the continent and an increasing compelling evidence base that data collected in Caucasian populations cannot be extrapolated to other genetically diverse populations, it is time for this to change (Tishkoff et al. 2009). New technologies such as whole-genome and whole-exome sequencing are becoming cheaper and more refined, making studies on as few as one family or even one individual feasible. Much-neglected investment in genetics and genomics research in African populations with parallel expansion of the capacity for ground-breaking research on locally relevant diseases, led by local senior investigators (e.g., through the H3Africa consortium), is changing the landscape and, given that we still do not know the molecular basis of around 50 percent of all Mendelian disorders (Chong et al. 2015), the field of Mendelian disorders should provide rich pickings in terms of both the advancement of genomic medicine and development of research training.

References

Abel L and Demenais F (1988). Detection of major genes for susceptibility to leprosy and its subtypes in a Caribbean island: Desirade Island. *Am J Hum Genet* 42: 256–266.

Abel L, Demenais F, and Baule M-S (1989). Genetic susceptibility to leprosy on a Caribbean island: linkage analysis with five markers. *Int J Leprosy* 57: 256–266.

Abubakar I, Zignol M, Falzon D, et al. (2013). Drug-resistant tuberculosis: time for visionary political leadership. *Lancet Infect Dis.* DOI: 10.1016/S1473-3099(13)70030-6.

Barbouche MR, Mekki N, Ben-Ali M, and Ben-Mustapha I (2017). Lessons from genetic studies of primary immunodeficiencies in a highly consanguineous population. *Front Immunol* **8**: 737.

Bittles AH and Black ML (2010). Evolution in health and medicine Sackler colloquium: consanguinity, human evolution, and complex diseases. *Proc Natl Acad Sci U S A* **107**(Suppl. 1): 1779–1786.

Boisson-Dupuis S, Bustamante J, El-Baghdadi J, et al. (2015). Inherited and acquired immunodeficiencies underlying tuberculosis in childhood. *Immunol Rev* **264**: 103–120.

Bustamante J, Boisson-Dupuis S, Abel L, and Casanova JL (2014). Mendelian susceptibility to mycobacterial disease: genetic, immunological, and clinical features of inborn errors of IFN-gamma immunity. *Semin Immunol* **26**: 454–470.

Casanova JL, Jouanguy E, Lamhamedi S, Blanche S, and Fischer A (1995). Immunological conditions of children with BCG disseminated infection. *Lancet* **346**: 581.

Chakravartti MR and Vogel F (1973). A twin study on leprosy. In Becker PE, Lenz W, Vogel F, and Wendt GG, eds., *Topics in Human Genetics*, Georg Thieme, pp. 1–123.

Chong JX, Buckingham KJ, Jhangiani SN, et al. (2015). The genetic basis of Mendelian phenotypes: discoveries, challenges, and opportunities. *Am J Hum Genet.* DOI: 10.1016/j.ajhg.2015.06.009.

Comstock G (1978). Tuberculosis in twins: a re-analysis of the Prophit survey. *Am Rev Respir Dis* **117**: 621–624.

Dubos R and Dubos J (1952). *Tuberculosis, Man and Society: the White Plague.* Little, Brown and Co.

Dupuis S, Doffinger R, Picard C, et al. (2000). Human interferon-gamma-mediated immunity is a genetically controlled continuous trait that determines the outcome of mycobacterial invasion. *Immunol Rev* **178**: 129–137.

Engbaek HC (1964). Three cases in the same family of fatal infection with *M. avium. Acta Tuberc Scand* **45**: 105–117.

Fineberg HV (2014). Pandemic preparedness and response: lessons from the H1N1 influenza of 2009. *N Engl J Med* **370**: 1335–1342.

GBD 2017 Mortality and Causes of Death Collaborators (2018). Global, regional, and national age-sex-specific mortality for 282 causes of death in 195 countries and territories, 1980-2017: a systematic analysis for the Global Burden of Disease Study 2017. *Lancet* **392**: 1736–1788.

Haile RWC, Iselius L, Fine PEM, and Morton NE (1985). Segregation and linkage analysis of 72 leprosy pedigrees. *Hum Hered* **35**: 43–52.

Holland SM, Eisenstein EM, Kuhns DB, et al. (1994). Treatment of refractory disseminated nontuberculous mycobacterial infection with interferon gamma. *N Engl J Med* **330**: 1348–1355.

Hotez PJ, Molyneux DH, Fenwick A, et al. (2006). Incorporating a rapid-impact package for neglected tropical diseases with programs for HIV/AIDS, tuberculosis, and malaria. *PLoS Medicine* **3**: e102.

Hotez PJ, Fenwick A, Savioli L, and Molyneux DH (2009). Rescuing the bottom billion through control of neglected tropical diseases. *Lancet* **373**: 1570–1575.

Jouanguy E, Altare F, Lamhamedi S, et al. (1996). Interferon-gamma-receptor deficiency in an infant with fatal bacille Calmette-Guerin infection. *N Engl J Med* **335**: 1956–1959.

Kallman FJ and Reisner D (1943a). Twin studies on genetic variations in resistance to tuberculosis. *J Hered* **34**: 293–301.

Kallman FJ and Reisner D (1943b). Twin studies on the significance of genetic factors in tuberculosis. *Am Rev Tuberculosis* **47**: 549–574.

Koch R (1932). The etiology of tuberculosis. *Am Rev Tuberculosis* **25**: 285–323.

Levin M, Newport S, D'souza S, et al. (1995). Familial disseminated atypical mycobacterial infection in childhood: a mycobacterial susceptibility gene? *Lancet* **345**: 79–83.

Liddle GW, Bledsoe T, and Coppage WS (1963). A familial renal disorder simulating primary aldosteronism but with negligible aldosterone secretion. *Trans Assoc Am Phys* **76**: 199–213.
Manry J, Laval G, Patin E, et al. (2011). Evolutionary genetics evidence of an essential, nonredundant role of the IFN-gamma pathway in protective immunity. *Hum Mutat* **32**: 633–642.
Martinez-Romero C and Garcia-Sastre A (2015). Against the clock towards new Ebola virus therapies. *Virus Res.* DOI: 10.1016/j.virusres.2015.05.025.
Mehra NK (1990). Role of HLA linked factors in governing susceptibility to leprosy and tuberculosis. *Trop Med Parasitol* **41**: 352–354.
Newport MJ (2003). The genetics of nontuberculous mycobacterial infection. *Exp Rev Mol Med* **5**: 1–13.
Newport MJ (2013). The contribution of host genetics to TB disease. *Curr Respir Med Rev* **9**: 156–162.
Newport MJ, Huxley CM, Huston S, et al. (1996). A mutation in the interferon-gamma receptor gene and susceptibility to mycobacterial infections in man. *N Engl J Med* **335**: 1941–1949.
Pitchappan RM (1990). Genetics of tuberculosis susceptibility. *Trop Med Parasitol* **41**: 355–356.
Prando C, Samarina A, Bustamante J, et al. (2013). Inherited IL-12p40 deficiency: genetic, immunologic, and clinical features of 49 patients from 30 kindreds. *Medicine (Baltimore)* **92**: 109–122.
Rawlinson WD, Basten A, Britton WJ, and Serjeantson SW (1988). Leprosy and immunity: genetics and immune function in multiple case families. *Immunol Cell Biol* **66**: 9–21.
Rees DC, Williams TN, and Gladwin MT (2010). Sickle-cell disease. *Lancet* **376**: 2018–2031.
Romdhane L, Abdelhak S, Research Unit on Molecular Investigation of Genetic Orphan Diseases, et al. (2011). Genetic diseases in the Tunisian population. *Am J Med Genet A* **155A**: 238–267.
Samson M, Libert F, Doranz BJ, et al. (1996). Resistance to HIV-1 infection in Caucasian individuals bearing mutant alleles of the CCR-5 chemokine receptor gene. *Nature* **382**: 722–725.
Serjeantson SW (1983). HLA and susceptibility to leprosy. *Immunol Rev* **70**: 89–112.
Shields ED, Russell DA, and Pericak-Vance MA (1987). Genetic epidemiology of susceptibility to leprosy. *J Clin Invest* **79**: 1139–1143.
Simmonds B (1963). *Tuberculosis in Twins*. Pitman Medical Publishing.
Simonetti GD, Mohaupt MG, and Bianchetti MG (2012). Monogenic forms of hypertension. *Eur J Pediatr* **171**: 1433–1439.
Smith DG (1979). The genetic hypothesis for susceptibility to lepromatous leprosy. *Hum Genet* **50**: 163–177.
Smith FB (1988). *The Retreat of Tuberculosis 1850–1950*. Croom Helm.
Stead WW (1992). Genetics and resistance to tuberculosis. *Ann Intern Med* **116**: 937–994.
Stead WW, Senner JW, Reddick WT, and Lofgren JP (1990). Racial differences in susceptibility to infection by *Mycobacterium tuberculosis*. *N Engl J Med* **322**: 422–427.
Stein CM, Zalwango S, Chiunda AB, et al. (2007). Linkage and association analysis of candidate genes for TB and TNFalpha cytokine expression: evidence for association with IFNGR1, IL-10, and TNF receptor 1 genes. *Hum Genet* **121**: 663–673.
Takala-Harrison S and Laufer MK (2015). Antimalarial drug resistance in Africa: key lessons for the future. *Ann N Y Acad Sci* **1342**: 62–67.
Tishkoff SA, Reed FA, Friedlaender FR, et al. (2009). The genetic structure and history of Africans and African Americans. *Science* **324**: 1035–1044.
Tournamille C, Colin Y, Cartron JP, and Le Van Kim C (1995). Disruption of a GATA motif in the Duffy gene promoter abolishes erythroid gene expression in Duffy-negative individuals. *Nat Genet* **10**: 224–228.
Uchiyama N, Greene GR, Warren BJ, et al. (1981). Possible monocyte killing defect in familial atypical mycobacteriosis. *J Pediatr* **98**: 785–788.

Van Eden W and De Vries RRP (1984). Occasional review: HLA and leprosy – a re-evaluation. *Lepr Rev* **55**: 89–104.

Welch SG, McGregor IA, and Williams K (1977). The Duffy blood group and malaria prevalence in Gambian West Africans. *Trans R Soc Trop Med Hyg* **71**: 295–296.

Wellcome Trust Case Control Consortium (2007). Genome-wide association study of 14,000 cases of seven common diseases and 3,000 shared controls. *Nature* **447**: 661–678.

Westby M and van der Ryst E (2005). CCR5 antagonists: host-targeted antivirals for the treatment of HIV infection. *Antivir Chem Chemother* **16**: 339–354.

6 Pharmacogenomics and Infectious Diseases in Africa[1]

An Evolutionary Perspective

Jennifer L. Baker, Daniel Shriner, Amy R. Bentley, and Charles N. Rotimi

6.1 Introduction

Modern humans evolved in Africa approximately 200,000 years ago (Campbell and Tishkoff 2010). As groups migrated out of Africa they underwent bottlenecks leading to sharp reductions in population size and genetic diversity (Amos and Hoffman 2010; Harpending and Rogers 2000; Ramachandran et al. 2005). To this day, African populations retain the most genetic diversity globally (Auton et al. 2015). In order to survive both within and out of Africa, early human populations had to adapt to their novel environments, including new food resources, colder climates, higher altitudes, and, especially, infectious diseases (Balaresque et al. 2007; Fumagalli et al. 2011). These adaptive requirements, facilitated by natural selection, led to an increased frequency of alleles that were beneficial in that environment. Due to the fact that these adaptive requirements were driven by local environmental pressures, some of these evolutionarily advantageous alleles display geographic and ancestral specificity, as observed in the genomes of present-day humans (Fumagalli et al. 2011).

Finding genetic signals of natural selection is becoming increasingly possible as high-throughput biotechnology and novel computational methods are developed (Koboldt et al. 2013; Tyler-Smith et al. 2015). The ability to perform genome-wide association studies (GWAS) and large-scale genetic variation surveys have led to an exponential increase in the amount of publicly available genetic data (Mardis 2013) and an accompanying explosion in genomic comparisons both within and among populations (Rosenberg et al. 2011; Shriner et al. 2014; Li et al. 2011). These data support the clinical significance of identifying genetic outcomes of natural selection, such as pinpointing susceptibility loci for infectious diseases plaguing human populations (Rowell et al. 2012) or identifying responses to xenobiotic challenges (Relling and Klein 2011). These advances have opened the field of genomic medicine, which takes an individual's genetic data into account when planning clinical care (Ginsburg and Willard 2009; Muenke 2013) and is indispensable in the push for personalized, or

[1] This chapter was first published as JL Baker, D Shriner, AR Bentley, and CN Rotimi (2016) Pharmacogenomics and infectious diseases in Africa. *Springer Nature*. DOI: 10.1038/tpj.2016.78.

precision, medicine. In this capacity, genomics has the potential to inform clinicians and contribute to precision medicine on numerous fronts, such as disease prediction, treatment response, and avoidance of adverse drug reactions (ADRs) or off-target effects (Adeyemo and Rotimi 2014).

Unfortunately, most clinical studies remain overwhelmingly European-centered (Adeyemo and Rotimi 2014; Burchard et al. 2015; Chen et al. 2014; Hindorff et al. 2009; Need and Goldstein 2009; Oh et al. 2015) and the majority of current drug recommendations were established using data from clinical trials performed in Caucasian or Asian populations and may be inappropriate for African populations (Adeyemo and Rotimi 2014; Pillai et al. 2013). This sampling bias is not trivial, as estimates for both disease risk and allele frequencies vary significantly between worldwide populations, as well as between African populations (Adeyemo and Rotimi 2009; Carlson et al. 2013; Chen et al. 2012; Ntzani et al. 2012; Ramos et al. 2014). In fact, population-specific studies have established that Africa has the highest genetic diversity in the world (Tishkoff et al. 2009), as is reflected culturally in the more than 2000 languages spoken on the continent (Cashdan 2001) and genetically in regional allele frequency differences among African genomes (Gurdasani et al. 2015; Ramos et al. 2014) and in drug-metabolizing and -transporting genes in particular (Kerb et al. 2009).

Scientists have long been cognizant of the differences patients present when given the same medication. Clinical observations in the 1950s first pointed to genetic variations as the key determinants of these inconsistencies (Evans et al. 1960; Kalow and Gunn 1959; Kalow and Staron 1957). Understanding these genetic variants was the impetus for two fields. The first, pharmacogenetics, centers on variations in genes involved with (1) how a drug gets into and out of the body, termed pharmacokinetics (i.e., absorption, distribution, metabolism, and excretion [ADME]), and (2) how the drug acts on the body to produce its effects, termed pharmacodynamics (Ma and Lu 2011). There can be marked differences in a drug's pharmacokinetic profile between individuals as a result of polymorphisms in drug-metabolizing enzymes (e.g., cytochrome P450 enzymes) and transport proteins (e.g., P-glycoprotein) (Kerb et al. 2009). The term pharmacogenetics was first coined in 1959 by F. Vogel (Vogel 1959) and was the precursor to the second field, pharmacogenomics. This is a much broader investigation of genetic variations at the level of the genome, focusing on how an individual's total genetic variation affects drug response. The completion of the human genome sequence (Hattori 2004; Lander et al. 2001; Venter et al. 2001) set the stage for this field, which not only identifies variations in drug response, but also searches for molecular mechanisms, statistically assesses clinical significance, and develops screening tests with the ultimate goal of developing individualized drug therapy.

Because genetic variation in ADME genes varies significantly between populations (Li et al. 2011; Ramos et al. 2014), determining genotypes of genetic variants associated with known ADR in specific populations is critical for dosing guidelines and avoiding therapeutic failure. As a result of the large genetic heterogeneity in African populations, it is unsafe to extrapolate results from one African ancestral group to another, as a drug which is found to be in the correct plasma concentration in one group can cause reduced effectiveness or, more troublesome, complete drug toxicity in another (Tekola-Ayele et al. 2014; Zanger et al. 2008). The differential metabolism of codeine resulting from genetic variability in *CYP2D6* provides an example. Due to multiple gene duplication events, some individuals have several copies of *CYP2D6*, leading to ultrarapid metabolizing activity (Aklillu et al. 1996). These individuals quickly convert codeine into morphine, a dangerous outcome that can lead to severe ADRs, including death. The global distribution of carriers varies widely, from very low rates reported in West Africa, to 3 percent of Northern Europeans, 5–10 percent of Southern Europeans, to a high of about 30 percent of Ethiopians (Aklillu et al. 1996; Bernal et al. 1999; Bradford 2002; Sachse et al. 1997). These data, along with a strong warning from the US Food and Drug Administration against the use of codeine for the management of pain in children (FDA 2013), and the current lack of infrastructure to genotype individuals, led the Ethiopian Food, Medicine and Health Care Administration and Control Authority to ban its use entirely (Anberbir 2015).

More than 25 percent of drugs that begin clinical development ultimately fail; however, using publicly available genomic information regarding known associations between genes and diseases has been shown to significantly improve this rate (Nelson et al. 2015). This finding demonstrates the need for an accessible reference database that uses an unbiased approach to aggregate worldwide gene–disease associations. The largest such reference is the Pharmacogenomics Knowledge Base (www.pharmgkb.org), which collates genomic data from the scientific community, as well as molecular, cellular, and clinical phenotype data regarding genetic variations that affect drug response (Klein et al. 2001). In a move toward formulating procedures for the clinical implementation of pharmacogenomic data, a joint project between PharmGKB and the Pharmacogenomics Research Network resulted in the establishment of the Clinical Pharmacogenetics Implementation Consortium (CPIC) (www.pharmgkb.org/page/cpic), which publishes guidelines regarding gene–drug interactions (Relling and Klein 2011). Although African-specific observations are increasingly being made through this consortium (Johnson et al. 2011; Martin et al. 2014; Muir et al. 2014; Scott et al. 2013), specific dosing recommendations for this cohort lag far behind. In order to rectify this incongruence, it is essential to systematically investigate, monitor, and report the distribution of important pharmacogenomic variants across all of the diverse African populations (Tekola-Ayele et al. 2014).

One of the first instances of a change in a clinical procedure as a result of observed pharmacogenetic differences in drug handling was with isoniazid, an anti-tuberculosis drug introduced in 1952 (Hughes et al. 1954) that is metabolized differently depending on which N-acetyltransferase variant an individual possesses (Walker et al. 2009). Further research confirmed an "African-specific" mutation (Dandara et al. 2003) that affects drug efficacy and differs among African populations, establishing that pharmacogenetics in Africa should be discussed within the framework of the genetic diversity that exists among the many different African populations. Since that time, the literature has grown to include multiple examples of genetic polymorphisms for drug transporters and metabolizing enzymes associated with variable interethnic drug efficacy (Griese et al. 1999; Kesselring et al. 2009; Kudzi et al. 2010; Masimirembwa et al. 1996; Ngaimisi et al. 2013; Swart et al. 2012; Swart et al. 2013). A large majority of the published reports concern genes known to encode drug-metabolizing enzymes, such as the glutathione S-transferase and cytochrome P450 (CYP450) superfamilies, different variants of which can produce reduced, absent, or increased drug metabolism. In fact, cytochrome P450 enzymes metabolize more than 80 percent of therapeutic drugs (Lynch and Price 2007; Seripa et al. 2010; Swen et al. 2011). The variation that exists in these genes results in differential drug responses (Dandara et al. 2014a), the clinical manifestations of which must be identified for the safe and efficacious use of a drug.

6.2 The Burden of Infectious Diseases in Africa

Although countless factors have allowed natural selection to shape genetic variation, none are as powerful as infectious pathogens (Fumagalli et al. 2011; Rowell et al. 2012). In order for an infectious disease to exert a selective pressure on the genome, it must substantially affect morbidity and mortality prior to, or during, the reproductive age for many generations. When this happens, natural selection acts on a population's genomic variability, keeping beneficial genetic variants and removing the negatives. Host genetic variation is shaped as a result of this evolutionary process and plays a role in susceptibility to infectious disease, resulting in population-specific genetic variations, depending on the prevalence of endemic infectious diseases (Chapman and Hill 2012; Mangano and Modiano 2014). Segregation analysis detected major genes associated with schistosomiasis, leishmaniasis, leprosy, and tuberculosis (Abel and Demenais 1988; Abel et al. 1991, 1992, 1995), which were later mapped using methods that combined population- and patient-based studies (Casanova 2015). As more data become publicly available, the causative genetic factors for disease susceptibility continues to grow (Alcaïs et al. 1997, 2007; Baghdadi et al. 2006; Cobat et al. 2009; Dessein et al. 1999; Mira et al. 2004), setting the stage for a closer look at this topic as it relates to the burden of infectious diseases in Africa. If the current benefits of personalized medicine are to be extended to African populations, a more thorough characterization of the diversity of African ancestries will be necessary (de Vries et al. 2015; Gurdasani et al. 2015; Rotimi et al. 2014; Siwo et al. 2015; Wonkam and Mayosi 2014) and is the motivation for this review. In particular, we will examine Africa's burden of infectious disease, discuss known genetic adaptations to five selected diseases that are prevalent in Africa, and present examples of clinically relevant population-specific variants.

6.2.1 Malaria

Transmitted to humans by infected female mosquitos, human malaria is caused by four species of the protozoan parasite *Plasmodium*: *P. vivax*, *P. malariae*, *P. ovale*,

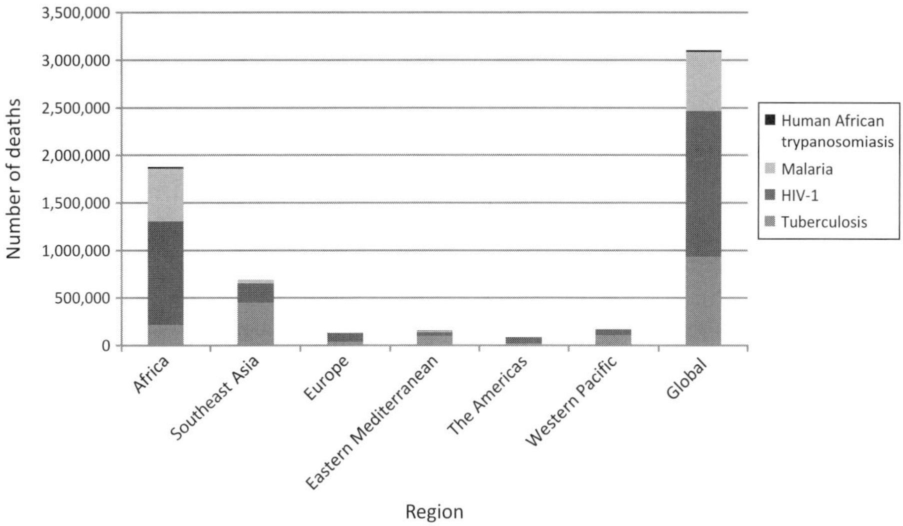

Figure 6.1 The global burden of the diseases discussed in this review as of 2012. Lassa fever is not included in the figure because the disease only occurs in Africa.

and the parasite that causes the most severe cases of malaria, *P. falciparum* (WHO 2015c). Malaria is endemic throughout most of the tropics, with *P. falciparum* producing the major burden of disease globally, followed by *P. vivax* (Guerra et al. 2008). Due to concerted global efforts, the number of deaths from malaria declined considerably between 2000 and 2015 (WHO 2015c). Despite these significant gains, malaria still poses a significant worldwide public health problem, causing approximately 438,000 fatalities each year, with 292,000 deaths occurring in African children (WHO 2015c) (Figure 6.1).

P. falciparum is the leading cause of malaria in Africa, while *P. vivax* is the dominant pathogen outside of Africa (Howes et al. 2015; Mendis et al. 2001). The Duffy antigen receptor for chemokines (ACKR1), commonly known as DARC, is the red blood cell receptor for the malaria-causing *P. vivax* (Miller et al. 1976). A Duffy-negative phenotype associated with rs2814778 abrogates DARC surface expression on red blood cells and confers resistance to *P. vivax* (Langhi and Orlando Bordin 2006). This genetic variant is found extensively in malaria-endemic regions of sub-Saharan Africa (SSA), especially in West African populations (Table 6.1) (Gething et al. 2012; Howes et al. 2011; Langhi and Orlando Bordin 2006; Mendis et al. 2001; Miller et al. 1976). However, DARC-independent host cell invasion is apparently possible, as studies in Madagascar (Ménard et al. 2010) and Brazil (Cavasini et al. 2007) have identified *P. vivax* parasites that successfully invaded Duffy-negative erythrocytes.

Upon infection, the parasite incubates in the liver and enters the bloodstream, infecting red blood cells and causing high fever, sweats, chills, and frequently death (Wassmer et al. 2015; WHO 2015c). However, the parasites do not thrive in the presence of sickle-shaped blood cells. Thus, one of the consequences of thousands of

Table 6.1 Allele frequencies across African populations for rs2814778

Population	Sample size	Frequency of C	Reference
Berber	25	0.040	(Cheung et al. 2000)
Moroccan	48	0.106	(Cheung et al. 2000)
Tunisian	44	0.148	(Cheung et al. 2000)
Tunisian	58	0.155	(Cheung et al. 2000)
Libyan	31	0.161	(Cheung et al. 2000)
Libyan	31	0.161	(Cheung et al. 2000)
Tunisian	44	0.171	(Cheung et al. 2000)
Libyan	67	0.202	(Cheung et al. 2000)
Tunisian	94	0.207	(Cheung et al. 2000)
Algerian	35	0.214	(Cheung et al. 2000)
Tunisian	42	0.214	(Cheung et al. 2000)
Berber	30	0.217	(Cheung et al. 2000)
Tunisian	44	0.227	(Cheung et al. 2000)
Mozabite	26	0.231	(Cheung et al. 2000)
Moroccan	41	0.256	(Cheung et al. 2000)
Tunisian	45	0.256	(Cheung et al. 2000)
Tunisian	12	0.500	(Cheung et al. 2000)
Amhara	24	0.521	(Gurdasani et al. 2015)
Oromo	24	0.583	(Gurdasani et al. 2015)
Wolayta	24	0.604	(Gurdasani et al. 2015)
Somali	8	0.625	(Cheung et al. 2000)
Jews, Ethiopian	34	0.779	(Cheung et al. 2000)
Zulu	100	0.800	(Gurdasani et al. 2015)
Bantu speakers (South Africa)	8	0.880	(Cheung et al. 2000)
Maasai	143	0.930	(Cheung et al. 2000)
Mbuti	37	0.973	(Cheung et al. 2000)
Maasai	19	0.974	(Cheung et al. 2000)
Sandawe	39	0.974	(Cheung et al. 2000)
Somali	24	0.979	(Gurdasani et al. 2015)
Chagga	44	0.989	(Cheung et al. 2000)
Biaka	69	0.993	(Cheung et al. 2000)
YRI	108	0.995	(Auton et al. 2015)
Baganda	100	1	(Gurdasani et al. 2015)
Bantu speakers (Kenya)	12	1	(Cheung et al. 2000)
Biaka	35	1	(Cheung et al. 2000)
ESN	99	1	(Auton et al. 2015)
Gumuz	24	1	(Gurdasani et al. 2015)
Hausa	38	1	(Cheung et al. 2000)
Ibo	47	1	(Cheung et al. 2000)
Lisongo	7	1	(Cheung et al. 2000)
LWK	99	1	(Auton et al. 2015)
MAG	113	1	(Auton et al. 2015)
Malinke	113	1	(Cheung et al. 2000)
Mandenka	24	1	(Cheung et al. 2000)

Table 6.1 (cont.)

Population	Sample size	Frequency of C	Reference
MSL	85	1	(Auton et al. 2015)
San	7	1	(Cheung et al. 2000)
Yoruba	76	1	(Cheung et al. 2000)
Zaramo	39	1	(Cheung et al. 2000)

Key: ESN, Esan in Nigeria; LWK, Luhya in Webuye, Kenya; MAG, Gambian in Western Division, Mandinka; MSL, Mende in Sierra Leone; YRI, Yoruba in Ibadan, Nigeria.

Table 6.2 Allele frequencies across African populations for rs334

Population	Sample size	Frequency of A	Reference
Amhara	24	0	(Gurdasani et al. 2015)
Gumuz	24	0	(Gurdasani et al. 2015)
Oromo	24	0	(Gurdasani et al. 2015)
Somali	24	0	(Gurdasani et al. 2015)
Wolayta	24	0	(Gurdasani et al. 2015)
Zulu	100	0.005	(Gurdasani et al. 2015)
Baganda	100	0.070	(Gurdasani et al. 2015)
LWK	99	0.101	(Auton et al. 2015)
MAG	113	0.115	(Auton et al. 2015)
ESN	99	0.121	(Auton et al. 2015)
MSL	85	0.124	(Auton et al. 2015)
YRI	108	0.139	(Auton et al. 2015)

ESN, Esan in Nigeria; LWK, Luhya in Webuye, Kenya; MAG, Gambian in Western Division, Mandinka; MSL, Mende in Sierra Leone; YRI, Yoruba in Ibadan, Nigeria.

years of human co-existence with the malaria pathogen is that inherited red blood cell disorders known as sickle cell disease (SCD) are maintained as a result of balancing selection (Pasvol et al. 1978). Sickle cell disease is a painful disease caused by abnormal hemoglobin (Hb) alleles called hemoglobin S (HbS, "sickle") at rs334 (Table 6.2). Individuals with two sickle alleles (HbSS) produce red blood cells that take the sickle shape, thus severely restricting their oxygen-carrying capacity and interrupting healthy blood flow. Despite the debilitating effects of the homozygous condition, HbS is maintained due to protection against malaria, as *P. falciparum* cannot thrive in sickled red blood cells (Allison 1954; Fumagalli et al. 2009; Hill et al. 1991; Pasvol et al. 1978). In fact, carriers of the sickle allele have 60 percent protection against overall mortality (Aidoo et al. 2002), resulting in a close geographical overlap of a high sickle allele frequency and malarial endemicity (Piel et al. 2010).

Glucose-6-phosphate dehydrogenase (G6PD) deficiency is an inherited enzyme abnormality and, similar to other red blood cell disorders, is prevalent in countries where malaria is historically endemic (Howes et al. 2013), signifying that the deficiency may confer some protection (Bienzle et al. 1972; Howes et al. 2012; Tishkoff et al. 2001). Primaquine is the only drug licensed for the prevention and, in combination with other drugs, treatment of *P. vivax*. Unfortunately, the drug can cause severe hemolytic anemia in G6PD-deficient patients (Alving et al. 1956; Domingo et al. 2013).

Malaria prophylaxis and treatment is challenging because the *Plasmodium* parasite is capable of generating drug resistance in a relatively short time (Shahinas et al. 2013). Therefore, an increasing number of African countries are heeding the recommendations of the WHO for treating malaria by administering artemisinin-based combined treatments (ACTs), which combine a form of artemisinin (either artemether or artesunate) with a longer-acting drug such as lumefantrine, amodiaquine, mefloquine, or sulfadoxine-pyrimethamine (Ogbonna and Uneke 2008; WHO 2015b). Artemisinins work by quickly removing parasites from the blood, leaving fewer parasites upon which the partner drug must act (Roederer et al. 2011). The implementation of combination therapy is intended to reduce parasite resistance, as it should be more difficult for *P. falciparum* to become resistant to two drugs with unrelated modes of action (Yeung et al. 2004). However, resistance against the limited number of effective drugs is the foremost impediment to successfully treating malaria (Hyde 2007; Mehlotra et al. 2009).

The ACTs presently recommended for treatment of uncomplicated malaria are substrates of CYP enzymes. For example, amodiaquine is regularly used in conjunction with artesunate, and, to a lesser extent, with sulfadoxine-pyrimethamine (Faucher et al. 2009). The enzyme CYP2C8 metabolizes amodiaquine, as well as another malaria medication, chloroquine (Li et al. 2002; Paganotti et al. 2012). A variant of this enzyme, CYP2C8*2, results in the poor metabolizer phenotype, where individuals with at least one copy of CYP2C8*2 experience a longer drug half-life and increased adverse side-effects (Parikh et al. 2007). Allele frequencies of CYP2C8*2 differ between East and West Africa (Paganotti et al. 2012). A comprehensive list of the known CYP450 enzymes involved in the metabolism of antimalarial drugs includes: artemether: CYP2A6, CYP2B6, CYP2C8, CYP2C9, CYP2C19, CYP3A4, and CYP3A5; lumefantrine: CYP2D6, CYP3A4, and CYP3A5; amodiaquine: CYP1A1, CYP1B1, and CYP2C8; mefloquine: CYP3A4 and CYP3A5; chloroquine: CYP2C8, CYP2D6, CYP3A4, and CYP3A5; sulfadoxine-pyrimethamine: CYP2C9 and CYP2D6; primaquine: CYP1A1, CYP1A2, CYP3A4, and CYP3A5; and quinine: CYP1A1, CYP1A2, and CYP3A5 (Bains 2013).

6.2.2 Human African Trypanosomiasis

Human African trypanosomiasis (HAT), also known as African sleeping sickness, is spread by blood-feeding tsetse flies infected with one of two protozoans, *Trypanosoma brucei rhodesiense* and *T.b. gambiense* (Capewell et al. 2011, 2015). *T.b. rhodesiense* causes East African (or Rhodesian) trypanosomiasis and accounts for 3 percent of cases annually, while two types of *T.b. gambiense*, termed groups 1 and 2, cause West African (or Gambian) trypanosomiasis and account for 97 percent of cases annually (Capewell et al. 2011; Pays et al. 2014). Primarily confined to remote rural areas of SSA (Franco et al. 2014; Welburn et al. 2016), incidences of the two illnesses generally map respectively to the east and west side of the African Rift Valley (Welburn et al. 2001), although this historically defined boundary is shifting in Uganda as infected cattle move northwards, taking *T.b. rhodesiense* with them (Wardrop et al. 2010). Fortunately, disease prevalence has been steadily dropping since systematic global data collection began in 1940, reaching an all-time low of 3796 diagnosed cases in 2015 (WHO 2015e), down from 30,000 diagnosed and an estimated 300,000 undiagnosed cases in 1995 (Simarro et al. 2011; WHO 1998).

T.b. gambiense causes a prolonged chronic disease that can take decades before it becomes fatal, while *T.b. rhodesiense* causes a more acute illness that progresses rapidly and can lead to death within weeks or months after infection (Checchi et al. 2008; Lamour et al. 2015; Odiit et al. 1997). Regardless of the infecting agent, clinical progression of the illness is divided into two stages (Checchi et al. 2008): (1) the early or hemolymphatic stage, characterized by nonspecific inflammatory symptoms similar to those associated with malaria or enteric fever, making diagnosis and treatment a challenge (Sternberg 2005); and (2) the late meningoencephalitic stage, characterized by weight loss, behavioral changes, stupor, and coma (Checchi et al. 2008, 2015). If left untreated, both can be fatal (Lamour et al. 2015).

There is an unequal incidence of HAT cases among people living in close geographical proximity to each other, suggesting a genetic component to HAT susceptibility (Welburn et al. 2016). Thus far, genes associated with the susceptibility and severity of HAT include *FAS* and *FASLG* (Lopes et al. 1999; Martins et al. 2001), *IL23R* (Ribeiro et al. 2010), *SIGLEC6* and *SIGLEC12* (Crocker et al. 2007), *RNASEL*, *CXCR6*, and *IFIH1* (Fumagalli et al. 2010b), *APOL1* (Genovese et al. 2010), and *OAS2/3.32*.

Although the burden of HAT on the African continent has plummeted in recent years, evolutionary signatures of the arms race between humans and trypanosomes are still evident in their genomes (Capewell et al. 2015; MacLean et al. 2004; Uzureau et al. 2013). For example, humans have a trypanosome lytic factor (TLF), which includes the key component APOL1. When TLF is taken up by trypanosomes, APOL1 localizes to the lysosome and forms a pore that leads to osmotic swelling and rupture of the trypanosome. However, *T.b. rhodesiense* has evolved a serum resistance-associated (SRA) virulence factor which, when expressed as a protein in TLF-resistant lines, binds and inactivates APOL1 (Capewell et al. 2015). Variants in the *APOL1* gene evade SRA inactivation, restoring protection from HAT (Thomson et al. 2014). In fact, the frequency of the HAT-protective G allele at rs73885319 is 26 percent among the 1000 Genomes AFR populations and as high as nearly 50 percent among the Esan in Nigeria (Table 6.3), while this variant is absent among non-African ancestry populations. Furthermore, using a different resistance mechanism, group 2 *T. b. gambiense* can also overcome APOL1, an adaptive strategy that reflects the ability of *T. brucei* to continually evolve and infect new hosts (Capewell et al. 2015). Curiously, the protective G allele at rs73885319 is common in western Africa and rare to absent in eastern Africa; the advantages conferred by *APOL1* variants are not

Table 6.3 Allele frequencies across African populations for rs73885319

Population	Sample size	Frequency of G	Reference
Western Africa			
Esan	99	0.495	(Auton et al. 2015)
Ibo	411	0.488	(Adeyemo et al. 2015)
Edo	14	0.464	(Adeyemo et al. 2015)
Akan	361	0.436	(Adeyemo et al. 2015)
Asante	35	0.409	(Behar et al. 2011)
Yoruba	563	0.342	(Adeyemo et al. 2015)
Ga-Adangbe	158	0.247	(Adeyemo et al. 2015)
Mandinka	113	0.243	(Auton et al. 2015)
Jola	79	0.234	(Gurdasani et al. 2015)
Ewe	45	0.189	(Adeyemo et al. 2015)
Mende	85	0.124	(Auton et al. 2015)
Wolof	78	0.122	(Gurdasani et al. 2015)
Fula	73	0.116	(Gurdasani et al. 2015)
Bulsa	22	0.114	(Behar et al. 2011)
Southern Africa			
Zulu	100	0.13	(Gurdasani et al. 2015)
Sotho	86	0.081	(Gurdasani et al. 2015)
Bantu South Africa	8	0.07	(Kopp et al. 2011)
San	7	0	(Kopp et al. 2011)
Central Africa			
Somie	65	0.164	(Behar et al. 2011)
Congo	55	0.109	(Behar et al. 2011)
Bakola	19	0.05	(Ko et al. 2013)
Biaka Pygmy	36	0.042	(Kopp et al. 2011)
Mada	19	0.03	(Ko et al. 2013)
Cameroon	64	0.008	(Behar et al. 2011)
Mbuti Pygmy	15	0	(Kopp et al. 2011)
Lemande	18	0	(Ko et al. 2013)
Fulani	19	0	(Ko et al. 2013)
Eastern Africa			
Sena	51	0.122	(Behar et al. 2011)
Malawi	50	0.12	(Behar et al. 2011)
Baganda	100	0.08	(Gurdasani et al. 2015)
Barundi	97	0.077	(Gurdasani et al. 2015)
Banyarwanda	100	0.075	(Gurdasani et al. 2015)
Kikuyu	55	0.064	(Adeyemo et al. 2015)
Luhya	99	0.056	(Auton et al. 2015)
Sandawe	19	0.05	(Ko et al. 2013)
Iraqw	19	0.05	(Ko et al. 2013)
Hadza	19	0.05	(Ko et al. 2013)
Bantu Kenya	12	0.045	(Kopp et al. 2011)
Sudanese	24	0.042	(Pagani et al. 2012)
Kalenjin	99	0.02	(Adeyemo et al. 2015)

Table 6.3 (cont.)

Population	Sample size	Frequency of G	Reference
Anuak	76	0.02	(Behar et al. 2011)
Wolayta	24	0	(Gurdasani et al. 2015)
Tygray	21	0	(Pagani et al. 2012)
Somali	38	0	(Gurdasani et al. 2015)
Sengwer	19	0	(Ko et al. 2013)
Oromo	26	0	(Gurdasani et al. 2015)
Maale	76	0	(Behar et al. 2011)
Gumuz	24	0	(Gurdasani et al. 2015)
Borana	18	0	(Ko et al. 2013)
Ari Cultivator	24	0	(Pagani et al. 2012)
Ari Blacksmith	17	0	(Pagani et al. 2012)
Amhara	42	0	(Gurdasani et al. 2015)
Afar	76	0	(Behar et al. 2011)
Northern Africa			
Mozabite	30	0.018	(Kopp et al. 2011)
Kordofan	30	0	(Behar et al. 2011)

fully explained, although it has been suggested that protection may be conferred against other pathogens besides trypanosomes (Thomson et al. 2014).

Importantly, however, the protection from infectious disease that *APOL1* variants confer appears to come at the cost of chronic disease risk. HAT-protective *APOL1* variants have been strongly associated with kidney disease and, to a lesser degree, cardiovascular disease (Ito et al. 2014; Mukamal et al. 2016). This association has been confirmed for nephropathies, including focal segmental glomerulosclerosis (Genovese et al. 2010; Kopp et al. 2011), HIV-associated nephropathy (Kasembeli et al. 2015; Kopp et al. 2011), hypertension-attributed end-stage kidney disease (Genovese et al. 2010), severe lupus nephritis (Freedman et al. 2014), and chronic kidney disease progression (Parsa et al. 2013). Analyses of *APOL1* HAT-protective/kidney disease risk variants have yielded some of the highest recorded odds ratios for a common variant: 29 and 89 for risk of HIV-associated nephropathy in African Americans (Kopp et al. 2011) and South African blacks (Kasembeli et al. 2015), respectively, and 17 for focal segmental glomerulosclerosis (Kopp et al. 2011). The magnitude of effect suggested by these associations coupled with the relatively high prevalence of the risk alleles have been suggested to underlie the huge public health burden of kidney diseases among African ancestry individuals. The *APOL1* story is a notable example of how the shaping of the genome in response to infectious disease can have an impact on chronic disease risk.

According to the CDC, four drugs are currently prescribed for the treatment of HAT: pentamidine, suramin, melarsoprol, and eflornithine (WHO 2016), and some of the pharmacogenomic properties of some of these drugs have been described. For example, pentamidine is transported within the body by human OCT1, a broad selectivity transporter encoded by the highly polymorphic *SCL22A1* gene (Arimany-Nardi et al. 2015; Shu et al. 2003), variants of which affect the efficiency of uptake, distribution, and excretion of clinically relevant drugs (Arimany-Nardi et al. 2015; Gorboulev et al. 1997; Zhang et al. 1997). Furthermore, CYP450 enzymes with roles in the metabolism of pentamidine include CYP1A1, CYP1A2, CYP2C8, CYP2C19, CYP2B6, CYP3A4, CYP3A5, and CYP4A11 (Bains 2013).

6.2.3 Human Immunodeficiency Virus

Sub-Saharan Africa accounts for 71 percent of the global burden of human immunodeficiency virus (HIV), with nearly one out of every 20 adults infected with the virus (WHO 2015a) (Figure 6.1). HIV is a retrovirus that causes the immune system to become weaker as infection progresses, eventually leading to the most advanced stage of infection, acquired immunodeficiency syndrome (AIDS). There are two types of HIV, HIV-1 and HIV-2, of which the former is the predominant virus worldwide and is highly pathogenic (McCutchan 2006). Because the virus affects the immune system, an HIV-positive individual is more susceptible to infections and complications from other illnesses, especially tuberculosis (TB), which are often the ultimate cause of death (Worodria et al. 2011).

Early research suggested that host genetic polymorphisms affect individual susceptibility to HIV-1, viral load (indicative of infectiousness), and the speed of disease progression (Gonzalez et al. 1999; Schliekelman et al. 2001; Winkler et al. 2004). While HIV-1 is not believed to have been in the human population long enough to have shaped the human genome, susceptibility to HIV-1 does depend in part on an individual's genetic architecture, shaped by the evolutionary history of exposure to other infectious diseases.

A fully functioning receptor, CCR5, is required for HIV-1 virus entry (Le et al. 2015; Liu et al. 1996). A 32 base pair deletion termed CCR5 Δ32 (rs333) leads to a frameshift in the coding sequence, resulting in a non-functional protein that is not expressed in the cell membrane (Hill and Littman 1996; Martinson et al. 1997; Samson et al. 1996). Individuals homozygous for CCR5 Δ32 do not express functional receptors and are resistant to HIV-1 infection (Huang et al. 1996; Michael et al. 1997). Additionally, heterozygosity for the deletion provides a reduction in functional CCR5, thereby conveying partial resistance to HIV-1 infection; when an infection does occur, the viral loads are lower, slowing the progression to AIDS (Eugen-Olsen et al. 1997).

Data suggest that the CCR5 Δ32 deletion arose approximately 2900 years ago from a single mutation event (Hummel et al. 2005) and was subsequently subjected to positive selection (Libert et al. 1998; Stephens et al. 1998), resulting in a geographical distribution where it is more prevalent in Europeans and largely absent from Asian and African populations (Table 6.4) (Bharti et al. 2015; Libert et al. 1998; Lu et al. 1999; Martinson et al. 1997; Salem and Batzer 2007; Samson et al. 1996).

Table 6.4 Allele frequencies across African populations for rs333

Population	Sample size	Frequency of Δ32	Reference
African Pygmies	51	0	(Cheung et al. 2000)
Baka	63	0	(Cheung et al. 2000)
Beninese	26	0	(Cheung et al. 2000)
Biaka	69	0	(Cheung et al. 2000)
Biaka	35	0	(Cheung et al. 2000)
Bidjuki	40	0	(Cheung et al. 2000)
Central African	52	0	(Cheung et al. 2000)
Chagga	45	0	(Cheung et al. 2000)
Egyptians	46	0	(Cheung et al. 2000)
ESN	99	0	(Auton et al. 2015)
Gambian	56	0	(Cheung et al. 2000)
Guinean	50	0	(Cheung et al. 2000)
Hausa	38	0	(Cheung et al. 2000)
Ibo	48	0	(Cheung et al. 2000)
Ivoirian	87	0	(Cheung et al. 2000)
Jews, Ethiopian	29	0	(Cheung et al. 2000)
Kako	21	0	(Cheung et al. 2000)
Kenyan	80	0	(Cheung et al. 2000)
Lisongo	11	0	(Cheung et al. 2000)
LWK	99	0	(Auton et al. 2015)
Maasai	11	0	(Cheung et al. 2000)
MAG	113	0	(Auton et al. 2015)
Malawian	80	0	(Cheung et al. 2000)
Mbuti	39	0	(Cheung et al. 2000)
Mbuti	20	0	(Cheung et al. 2000)
MSL	85	0	(Auton et al. 2015)
Nigerians	23	0	(Cheung et al. 2000)
San, Central Kalahari	36	0	(Cheung et al. 2000)
Sanaga	37	0	(Cheung et al. 2000)
Sao Tomeans	100	0	(Cheung et al. 2000)
Sudanese	25	0	(Cheung et al. 2000)
Xhosa	64	0	(Cheung et al. 2000)
Yoruba	78	0	(Cheung et al. 2000)
YRI	108	0	(Auton et al. 2015)
Zamane	49	0	(Cheung et al. 2000)
Zambian	96	0	(Cheung et al. 2000)
Zime	60	0	(Cheung et al. 2000)
Zulu	9	0	(Cheung et al. 2000)
Senegalese	139	0.001	(Cheung et al. 2000)
Nigerians	111	0.005	(Cheung et al. 2000)
Egyptians	154	0.006	(Cheung et al. 2000)
Tunisian	145	0.010	(Cheung et al. 2000)
Berber	42	0.012	(Cheung et al. 2000)
Moroccans	167	0.015	(Cheung et al. 2000)

Table 6.4 (cont.)

Population	Sample size	Frequency of Δ32	Reference
Egyptians	72	0.021	(Cheung et al. 2000)
Tunisian	205	0.034	(Cheung et al. 2000)
Cape Verdean	97	0.036	(Cheung et al. 2000)
Cape Verdean	127	0.039	(Cheung et al. 2000)
Berber	39	0.051	(Cheung et al. 2000)

ESN, Esan in Nigeria; LWK, Luhya in Webuye, Kenya; MAG, Gambian in Western Division, Mandinka; MSL, Mende in Sierra Leone; YRI, Yoruba in Ibadan, Nigeria

A twentieth-century disease, HIV-1 has not been around long enough for selection pressure to drive CCR5 Δ32 from 0 to 10 percent in European populations (Novembre et al. 2005). Another pathogen, *Yersinia pestis*, the cause of the bubonic plague, was proposed as an agent of positive selection for CCR5 Δ32 (Stephens et al. 1998). However, the *Variola* virus, which causes smallpox, generates a stronger selective pressure than plague on pre-reproductive members of a population and therefore appears to be a more plausible candidate (Galvani and Slatkin 2003).

In addition, both the *Variola* virus and HIV-1 use chemokine receptors to enter white blood cells (Lalani et al. 1999), while the plague bacterium has an entirely different mode of entry.

Selection pressure from two infectious agents, both flaviviruses, provides the best evidence thus far to explain why this deletion is not observed in Africa. First, individuals homozygous for CCR5 Δ32 are at higher risk for tick-borne encephalitis (Kindberg et al. 2008). Second, fully functional CCR5 reduces symptoms from infection with West Nile virus (Lim et al. 2008), named for the West Nile district of Uganda where it was first isolated in 1937 (Smithburn et al. 1940). These data suggest that infectious agents played a role in shaping the genome. Whether CCR5 Δ32 is beneficial or deleterious in the context of a given infection (e.g., HIV-1 infection versus West Nile virus) is contingent upon complex interactions between an infecting pathogen and the host immune system.

For those who have access to health care in Africa, treatment of HIV/AIDS with highly active anti-retroviral therapy (HAART) has drastically increased life expectancy for infected individuals (Bekker et al. 2014; Nsanzimana et al. 2015; Ross et al. 2015), transforming a once certainly fatal disease into a chronic condition. Because patients are now receiving continuing treatment, much of African pharmacogenetic research is focused on optimizing therapeutic responses and preventing ADRs (Aceti et al. 2015) by identifying clinically relevant genetic variants (Habtewold et al. 2011; Mukonzo et al. 2013; Ngaimisi et al. 2013; Ranade et al. 2008; Swart et al. 2012; Yimer et al. 2012).

Drug efficacy disparities between African populations treated with the same HAART regimen have been reported. For example, efavirenz is metabolized to 8-hydroxyefavirenz mainly by CYP2B6 (Ward et al. 2003). A specific variant of this enzyme, CYP2B6*6, is associated with a higher plasma efavirenz concentration. The frequency distribution of the CYP2B6*6 variant allele is significantly higher in Tanzanians (41.9 percent) than Ethiopians (31.4 percent), and has also been shown to vary between Zimbabweans (49 percent) and Ugandans (35 percent) (Mukonzo et al. 2009; Nyakutira et al. 2008).

Many CYP450 enzymes are involved in the metabolism of drugs used to fight HIV. A comprehensive list includes: efavirenz: CYP2B6, CYP3A4, and CYP3A5; saquinavir: CYP3A4 and CYP3A5; maraviroc: CYP3A4 and CYP3A5; nevirapine: CYP2B6; indinavir: CYP3A4 and CYP3A5; nelfinavir: CYP3A4 and CYP3A5; ritonavir: CYP3A4 and CYP3A5; and lopinavir: CYP3A4 and CYP3A5.

Another example is the HLA-B*5701 allele, which is strongly associated with abacavir hypersensitivity syndrome (AHS) (Phillips and Mallal 2009). Abacavir causes serious ADR in 4–8 percent of patients (Phillips and Mallal 2009). Screening for the HLA-B*5701 allele has been shown to greatly reduce the frequency of AHS (Hetherington et al. 2002; Mallal et al. 2002). There are clear frequency differences of the HLA-B*5701 allele between African populations, ranging from 0 percent in the Nigerian Yoruba to 3.3 percent in the Kenyan Luhya and 13.6 percent in the Kenyan Maasai (Rotimi and Jorde 2010). Such differences for a clinically relevant variant validates the importance of individual screening and demonstrates the inadequacy of group identity such as "African" in medical decision-making at the individual level. In fact, the US FDA currently recommends that screening for HLA-B*5701 be completed prior to administration of abacavir for all patients (Rotimi and Jorde 2010). Increasingly, genomic data can help predict immunological response to HIV/AIDS therapeutic medicine, as was observed in HIV-infected women in Kenya (Oluka et al. 2015) and Zimbabwe (Mhandire et al. 2015) taking nevirapine-based anti-retroviral therapy.

6.2.4 Lassa Fever

Lassa fever is a severe viral hemorrhagic fever caused by the Lassa virus (LASV). This RNA virus of the Arenaviridae family resides in the rodent vector *Mastomys natalensis* that lives in close contact with humans and sheds the virus in urine (Monath et al. 1974; Mylne et al. 2015). First described in 1969 in the town of Lassa, Nigeria (Buckley and Casals 1970; Frame et al. 1970), Lassa fever is endemic in the West African countries of Guinea, Liberia, Nigeria, and Sierra Leone (Bowen et al. 2000; Fichet-Calvet and Rogers 2009), although it does occur in other countries as well, such as Mali and Côte d'Ivoire (Atkin et al. 2009; Günther et al. 2000). The virus infects an estimated 300,000–500,000 people annually, resulting in thousands of deaths in the region (Andersen et al. 2012; McCormick 1986).

The virus enters the cell via its cell surface receptor, alpha-dystroglycan (DAG1), replicating in a wide variety of cell types (Cao et al. 1998). The glycosyltransferase LARGE is required for viral entry as it post-translationally modifies DAG1 (Andersen et al. 2012) by producing a protein the virus needs to infect an individual (Kunz et al. 2005). A genome-wide screen for recent selective sweeps identified a signal for positive selection at a 300 kb region exclusively within the LARGE gene in populations with West African ancestry (Sabeti et al. 2007). Further data supporting the hypothesis of this virus as a driver of selection was subsequently reported for LARGE as well as for another gene biologically connected to Lassa fever, *IL21* (Andersen et al. 2012; Fumagalli et al. 2010a). For both genes, the signal has been localized to putative regulatory regions (Andersen et al. 2012), a potentially important fact in the development of future therapies.

While there are no approved vaccines for LASV infection, the antiviral therapeutic ribavirin increases survival if given early in infection (McCormick et al. 1986). The pharmacogenomics of ribavirin are well known in relation to another infectious disease, as pegylated interferon-a and ribavirin-based regimens are the mainstay for treatment of patients with chronic hepatitis C virus genotype 1 infection. In this case, a patient's interleukin-28B (IL28B) genotype predicts drug response (Ge et al. 2009; Rauch et al. 2010; Suppiah et al. 2009; Tanaka et al. 2009); the same genotype is predicted to affect drug response when given to patients with Lassa fever, although to date the literature is devoid of any Lassa fever pharmacogenomic studies.

6.2.5 Tuberculosis

Africa carries the highest overall burden of TB, with 281 cases per 100,000 population in 2014, while the global average is 133 per 100,000 population. Approximately one-third of the global population has latent, asymptomatic TB (WHO 2015d), and although those individuals only have a 10 percent chance of developing active TB disease (Wamala et al. 2016), risk is higher in individuals with compromised immune systems due to conditions such as diabetes or HIV infection. Caused by the bacteria *Mycobacterium tuberculosis*, TB begins with fever, weight loss, and coughing. Coughing progressively becomes worse with sputum and bloody coughs, chest pain, and ultimately death (Smith 2003; WHO 2015d).

TB is preventable through avoidance of infected individuals, who pass the infection on by coughing, and in most cases it is curable with a standard six-month antimicrobial drug course. Unfortunately, multidrug-resistant tuberculosis (MDR-TB) is on the rise globally. MDR-TB does not respond to the first-line anti-TB drugs, isoniazid and rifampicin, and patients are often cured with second-line drugs. However, in some instances, extensively drug-resistant TB (XDR-TB) can develop, which responds to even fewer available medicines, demonstrating the importance of developing new TB therapies (Segal and Hill 2003).

Human genetic variation affects susceptibility to mycobacterial infections, and candidate gene association studies have suggested roles for several genes and pathways (Möller and Hoal 2010). A recent genome-wide linkage study in Gambia and South Africa found suggestive linkage on 15q and Xq (Bellamy et al. 2000). Many studies have reported associations between susceptibility and resistance to TB and several HLA loci and/or alleles (Dubaniewicz et al. 2000; Kettaneh et al. 2006; Mishra et al. 2014; Yuliwulandari et al. 2010), and allele frequencies of these markers are known to vary between ethnic groups (Lombard et al. 2006). One recent study from Uganda suggests the HLA-DQB1*03:03 allele may be associated with resistance to TB (Wamala et al. 2016), but much work needs to be done between African populations to assess population-specific allele frequency differences and/or corresponding pharmacogenomic outcomes. Moreover, the contributions of rare variants with potentially large effects or multiple genes of small effect warrant systematic investigation (Möller and Hoal 2010).

Isoniazid, rifampicin, and pyrazinamide are the most commonly used drugs to fight TB and, as we observed in previous examples, the highly polymorphic CYP450 enzymes are involved in metabolism of these drugs. For example, CYP1A2, CYP2C19, and CYP2E1 affect the metabolism of isoniazid. Metabolism of rifampin involves the actions of CYP1A2, CYP2B6, CYP2C8, CYP2C9, CYP3A4, and CYP3A5. CYP2C9, CYP2C19, and CYP2E1 affect the metabolism of ethambutol and the metabolism of pyrazinamide is affected by CYP1A2 and CYP3A4 (Bains 2013).

Isoniazid, rifampicin, and pyrazinamide can cause serious ADRs in some individuals (Yew and Leung 2006). The most serious is anti-tuberculosis drug-induced hepatotoxicity (ADIH), which can lead to treatment failure and interruption, drug resistance, morbidity, and mortality (Tostmann et al. 2008; Wang et al. 2012). An increased risk of ADR is associated with SNPs rs1799929 and rs1495741 in *NAT2* (Wang et al. 2012). Ethiopian populations have significantly higher frequencies of these variants when compared with African ancestry 1000 Genomes populations (Tables 6.5 and 6.6) (Tekola-Ayele et al. 2014). In a 2011 study (Chigutsa et al. 2011), researchers examined the distribution of the *SLCO1B1* rs4149032 polymorphism that is associated with low blood concentrations of rifampicin (Table 6.7). They found that the variant allele occurred at a lower frequency in Caucasians or Asians than in African populations. The presence of rs4149032, and subsequent lower rifampicin concentration, means a higher dosage of the drug is required for African populations

Table 6.5 Allele frequencies across African populations for rs1799929

Population	Sample size	Frequency of T	Reference
San	40	0.140	(Cheung et al. 2000)
Yoruba	20	0.140	(Cheung et al. 2000)
YRI	108	0.171	(Auton et al. 2015)
Igbo	99	0.182	(Gurdasani et al. 2015)
ESN	99	0.187	(Auton et al. 2015)
MSL	85	0.188	(Auton et al. 2015)
Ga-Adangbe	100	0.200	(Gurdasani et al. 2015)
Jola	79	0.215	(Gurdasani et al. 2015)
Zulu	100	0.220	(Gurdasani et al. 2015)
Hausa	20	0.250	(Cheung et al. 2000)
Mandinka	87	0.259	(Gurdasani et al. 2015)
Sotho	86	0.262	(Gurdasani et al. 2015)
Wolof	78	0.269	(Gurdasani et al. 2015)
Fula	74	0.270	(Gurdasani et al. 2015)
Luo	16	0.270	(Cheung et al. 2000)

Table 6.5 (cont.)

Population	Sample size	Frequency of T	Reference
Barundi	97	0.278	(Gurdasani et al. 2015)
MAG	113	0.283	(Auton et al. 2015)
Baganda	100	0.305	(Gurdasani et al. 2015)
Gumuz	24	0.313	(Gurdasani et al. 2015)
LWK	99	0.328	(Auton et al. 2015)
Ibo	19	0.340	(Cheung et al. 2000)
Kalenjin	100	0.345	(Gurdasani et al. 2015)
Banyarwanda	100	0.350	(Gurdasani et al. 2015)
Kikuyu	99	0.424	(Gurdasani et al. 2015)
Somali	39	0.436	(Gurdasani et al. 2015)
Maasai	12	0.460	(Cheung et al. 2000)
Amhara	42	0.500	(Gurdasani et al. 2015)
Oromo	26	0.558	(Gurdasani et al. 2015)
Wolayta	24	0.604	(Gurdasani et al. 2015)

ESN, Esan in Nigeria; LWK, Luhya in Webuye, Kenya; MAG, Gambian in Western Division, Mandinka; MSL, Mende in Sierra Leone; YRI, Yoruba in Ibadan, Nigeria.

Table 6.6 Allele frequencies across African populations for rs1495741

Population	Sample size	Frequency of A	Reference
Biaka	31	0.190	(Cheung et al. 2000)
San	6	0.250	(Cheung et al. 2000)
Zulu	100	0.495	(Gurdasani et al. 2015)
Mbuti	15	0.500	(Cheung et al. 2000)
Yoruba	24	0.500	(Cheung et al. 2000)
Igbo	99	0.505	(Gurdasani et al. 2015)
YRI	108	0.514	(Auton et al. 2015)
Jola	79	0.525	(Gurdasani et al. 2015)
Sotho	86	0.535	(Gurdasani et al. 2015)
Bantu-speakers (Kenya)	12	0.540	(Cheung et al. 2000)
MSL	85	0.541	(Auton et al. 2015)
Bantu-speakers (South Africa)	8	0.560	(Cheung et al. 2000)
Mandenka	24	0.560	(Cheung et al. 2000)
ESN	99	0.561	(Auton et al. 2015)
MAG	113	0.566	(Auton et al. 2015)
Ga-Adangbe	100	0.590	(Gurdasani et al. 2015)
Wolof	78	0.596	(Gurdasani et al. 2015)
Baganda	100	0.600	(Gurdasani et al. 2015)
Mandinka	88	0.619	(Gurdasani et al. 2015)
Fula	74	0.622	(Gurdasani et al. 2015)

Table 6.6 (cont.)

Population	Sample size	Frequency of A	Reference
Gumuz	24	0.646	(Gurdasani et al. 2015)
Banyarwanda	100	0.655	(Gurdasani et al. 2015)
LWK	99	0.662	(Auton et al. 2015)
Barundi	97	0.670	(Gurdasani et al. 2015)
Kalenjin	100	0.730	(Gurdasani et al. 2015)
Kikuyu	99	0.737	(Gurdasani et al. 2015)
Mozabite	30	0.780	(Cheung et al. 2000)
Amhara	42	0.821	(Gurdasani et al. 2015)
Oromo	26	0.827	(Gurdasani et al. 2015)
Somali	39	0.859	(Gurdasani et al. 2015)
Wolayta	24	0.938	(Gurdasani et al. 2015)

ESN, Esan in Nigeria; LWK, Luhya in Webuye, Kenya; MAG, Gambian in Western Division, Mandinka; MSL, Mende in Sierra Leone; YRI, Yoruba in Ibadan, Nigeria.

Table 6.7 Allele frequencies across African populations for rs4149032

Population	Sample size	Frequency of C	Reference
ESN	99	0.187	(Auton et al. 2015)
Sotho	86	0.198	(Gurdasani et al. 2015)
Mandinka	88	0.205	(Gurdasani et al. 2015)
Gumuz	24	0.208	(Gurdasani et al. 2015)
MAG	113	0.226	(Auton et al. 2015)
Igbo	98	0.235	(Gurdasani et al. 2015)
Baganda	100	0.255	(Gurdasani et al. 2015)
MSL	85	0.265	(Auton et al. 2015)
LWK	99	0.278	(Auton et al. 2015)
YRI	108	0.282	(Auton et al. 2015)
Jola	79	0.285	(Gurdasani et al. 2015)
Kalenjin	100	0.330	(Gurdasani et al. 2015)
Barundi	97	0.351	(Gurdasani et al. 2015)
Kikuyu	99	0.364	(Gurdasani et al. 2015)
Banyarwanda	99	0.389	(Gurdasani et al. 2015)
Somali	39	0.423	(Gurdasani et al. 2015)
Wolayta	24	0.438	(Gurdasani et al. 2015)
Amhara	42	0.488	(Gurdasani et al. 2015)
Oromo	26	0.596	(Gurdasani et al. 2015)

ESN, Esan in Nigeria; LWK, Luhya in Webuye, Kenya; MAG, Gambian in Western Division, Mandinka; MSL, Mende in Sierra Leone; YRI, Yoruba in Ibadan, Nigeria.

to obtain the target concentration (Chigutsa et al. 2011). These two examples demonstrate how African genetic diversity extends to clinically important genetic variants and supports the urgency of working to identify ancestry-specific pharmacogenomic variants.

6.3 Traditional Herbal Medicine

Any discussion of treating infectious illnesses in Africa would not be complete without a mention of the long historical tradition of herbal medicine. These medicines should not be discounted outright as they have provided treatment and relief for many generations. Although not systematically studied in the same way as conventional therapies, increasingly their beneficial qualities are recognized, as was the case in 2015 when one half of the Nobel Prize in Medicine was awarded for research on artemisinin, a herb originally employed in Chinese traditional medicine (Tu 2011), and which has been credited with significantly reducing the mortality rates for patients infected with malaria (Nobelprize.org 2015). In many parts of the African continent, herbal medicine is accepted as part of the general health care system and has been widely reported to ease side-effects from conventional therapies such as chemotherapy (Thomford et al. 2015). There is a misconception, however, that because a treatment is a natural product, consuming it is harmless. The same enzyme systems metabolize both conventional and traditional medicines, and therefore, it is entirely possible for pharmacokinetic and pharmacodynamic interactions to occur when both therapies are combined (Butterweck and Derendorf 2008; Tarirai et al. 2010). For these reasons, it is imperative that the pharmacokinetic and pharmacodynamic profiles of interactions between conventional and traditional herbal medicines are subjected to the same scrutiny as conventional drug combinations to determine safe and reliable dosage standards (Thomford et al. 2015).

6.4 Progressing Toward Genomic Medicine in Africa

The past few decades have seen an explosion of interest in pharmacogenomic research by physicians, scientists, and the pharmaceutical industry, but as is the case in large-scale genomic studies, African populations are underrepresented (Mhandire et al. 2015), leading to slower application of pharmacological data in Africa. Factors outside and within Africa contributing to this disparity include cultural challenges, fewer genomic scientists, modest research infrastructure, and the absence of consistent national policies and funding (Tekola-Ayele and Rotimi 2015).

A significant step toward changing this paradigm began in 2010 with the creation of the Human Heredity and Health in Africa (H3Africa) Initiative, with funding from the NIH and the Wellcome Trust (National Institutes of Health 2010; Rotimi et al. 2014). The initiative, which currently involves 26 research groups and more than 500 investigators, is developing capacity to conduct health-related genomics research in Africa (Adoga et al. 2014; Dandara et al. 2014b; de Vries et al. 2015; Rotimi et al.

2014). The African Network for Drugs and Diagnostics Innovation is another initiative whose principal objective is to support African-led research and development into highly prevalent diseases on the continent (www.andi-africa.org). Sponsored by the United Nations Economic Commission for Africa and the WHO, the network also seeks to establish centers of research excellence on the continent. The African genomic data generated from these projects will help shift pharmacogenomic studies in Africa from a concentration on replicating research on candidate SNPs or genes discovered in Caucasian or Asian populations (Masimirembwa et al. 2014) to a more African-centric hypothesis-generating paradigm.

Included in these initiatives are plans to deposit African samples in large-scale national and regional biorepositories, allowing for the collation of continent-specific genotypic and phenotypic data (Matimba et al. 2008; Sirugo et al. 2004). A natural outcome of these efforts has been an abundance of ongoing discussions that transcend geographic and cultural barriers concerning the social, ethical, and legal issues that surround the collection, use, and storage of biospecimens in research (Abayomi et al. 2013; Barchi et al. 2015; Chanda-Kapata et al. 2015; Chen and Pang 2015; de Vries et al. 2015; Jao et al. 2015; Moodley et al. 2014; Rotimi et al. 2014; Staunton and Moodley 2013; Tindana et al. 2014). These dialogs are necessary to ensure the concerns of all parties are heard as these countries develop policy consensus and, ultimately, to guarantee lasting success.

6.5 Conclusion

This review details the complex history of infectious diseases in Africa and demonstrates how they have shaped African genomes. Documenting the vast genetic variation observed among African populations demonstrates the inadequacy of such group labels as "blacks" or "Africans" in biomedical research, especially in the context of pharmacogenomics and medicine. It also calls for the need to engage more diverse populations across the continent to better document the scope and extent of genetic diversity in Africans to ensure they reap the benefits of the global efforts to use genomics to improve the precision of medicine for individuals. Given the extent of disease in Africa, combined with the monetary and human costs incurred from ADRs or ineffectiveness, the necessity for economical approaches to limit disease burden on the continent is evident. Although individualized screening prior to the selection of therapy is currently cost-prohibitive, comprehensive pharmacogenetic profiling of many African populations will produce data that will improve patient care by identifying populations at risk for developing drug toxicity or non-responsiveness until such time as truly personalized medicine can become a reality.

Acknowledgments

The contents of this publication are solely the responsibility of the authors and do not necessarily represent the official view of the National Institutes of Health. This research was supported by the Intramural Research Program of the Center for Research on Genomics and Global Health (CRGGH). The CRGGH is supported by the National Human Genome Research Institute, the National Institute of Diabetes and Digestive and Kidney Diseases, the Center for Information Technology, and the Office of the Director at the National Institutes of Health (1ZIAHG200362).

References

Abayomi A, Christoffels A, Grewal R, et al. (2013). Challenges of biobanking in South Africa to facilitate indigenous research in an environment burdened with human immunodeficiency virus, tuberculosis, and emerging noncommunicable diseases. *Biopreserv Biobanking* **11**(6): 347–354.

Abel L and Demenais F (1988). Detection of major genes for susceptibility to leprosy and its subtypes in a Caribbean island: Desirade island. *Am J Hum Genet* **42**(2): 256–266.

Abel L, Demenais F, Prata A, et al. (1991). Evidence for the segregation of a major gene in human susceptibility/resistance to infection by *Schistosoma mansoni*. *Am J Hum Genet* **48**(5): 959–970.

Abel L, Cot M, Mulder L, et al. (1992). Segregation analysis detects a major gene controlling blood infection levels in human malaria. *Am J Hum Genet* **50**(6): 1308–1317.

Abel L, Vu DL, Oberti J, et al. (1995). Complex segregation analysis of leprosy in Southern Vietnam. *Genet Epidemiol* **12**(1): 63–82.

Aceti A, Gianserra L, Lambiase L, et al. (2015). Pharmacogenetics as a tool to tailor antiretroviral therapy: a review. *World J Virol* **4**(43): 198–208.

Adeyemo A and Rotimi C (2009). Genetic variants associated with complex human diseases show wide variation across multiple populations. *Public Health Genomics* **13**(2): 72–79.

Adeyemo A and Rotimi C (2014). What does genomic medicine mean for diverse populations? *Mol Genet Genomic Med* **2**(1): 3–6.

Adeyemo AA, Tekola-Ayele F, Doumatey AP, et al. (2015). Evaluation of genome wide association study associated type 2 diabetes susceptibility loci in Sub Saharan Africans. *Front Genet* **6**(355). DOI: 10.3389/fgene.2015.00335.

Adoga MP, Fatumo S, and Agwale SM (2014). H3Africa: a tipping point for a revolution in bioinformatics, genomics and health research in Africa. *Source Code Biol Med* **9**(1): 10.

Aidoo M, Terlouw DJ, Kolczak MS, et al. (2002). Protective effects of the sickle cell gene against malaria morbidity and mortality. *The Lancet* **359**: 1311–1312.

Aklillu E, Persson I, Bertilsson L, et al. (1996). Frequent distribution of ultrarapid metabolizers of debrisoquine in an Ethiopian population carrying duplicated and multiduplicated functional CYP2D6 alleles. *J Pharmacol Exp Ther* **278**(1): 441–446.

Alcaïs A, Abel L, David C, et al. (1997). Evidence for a major gene controlling susceptibility to tegumentary leishmaniasis in a recently exposed Bolivian population. *Am J Hum Genet* **61**(4): 968–979.

Alcais A, Alter A, Antoni G, et al. (2007). Stepwise replication identifies a low-producing lymphotoxin-alpha allele as a major risk factor for early-onset leprosy. *Nat Genet* **39**(4): 517–522.

Allison AC (1954). Protection afforded by sickle-cell trait against subtertian malarial infection. *BMJ* **1**: 290–294.

Alving AS, Carson PE, Flanagan CL, and Ickes CE (1956). Enzymatic deficiency in primaquine-sensitive erythrocytes. *Science* **124**(3220): 484–485.

Amos W and Hoffman JI (2010). Evidence that two main bottleneck events shaped modern human genetic diversity. *Proc Roy Soc B Biol Sci* **277**(1678): 131–137.

Anberbir Y (2015). Ethiopia: authority issues red alert on codeine drug. Available at: http://allafrica.com/stories/201511241318.html.

Andersen KG, Shylakhter I, Tabrizi S, et al. (2012). Genome-wide scans provide evidence for positive selection of genes implicated in Lassa fever. *Philos Trans R Soc London, Ser B* **367**(1590): 868–877.

Arimany-Nardi C, Koepsell H, and Pastor-Anglada M (2015). Role of *SLC22A1* polymorphic variants in drug disposition, therapeutic responses, and drug–drug interactions. *Pharmacogenomics J* **15** (6): 473–487.

Atkin S, Anaraki S, Gothard P, et al. (2009). The first case of Lassa fever imported from Mali to the United Kingdom, February 2009. *Euro Surveillance: Bulletin Europeen sur les maladies transmissibles* **14**(10): 12–14.

Auton A, Brooks LD, Durbin RM, et al. (2015). A global reference for human genetic variation. *Nature* **526**(7571): 68–74.

Baghdadi JE, Orlova M, Alter A, et al. (2006). An autosomal dominant major gene confers predisposition to pulmonary tuberculosis in adults. *J Exp Med* **203**(7): 1679–1684.

Bains RK (2013). African variation at cytochrome P450 genes: evolutionary aspects and the implications for the treatment of infectious diseases. *Evol Med Pub Health* **2013**(1): 118–134.

Balaresque PL, Ballereau SJ, and Jobling MA (2007). Challenges in human genetic diversity: demographic history and adaptation. *Hum Mol Genet* **16**(R2): 134–139.

Barchi F, Matlhagela K, Jones N, et al. (2015). "The keeping is the problem": a qualitative study of IRB-member perspectives in Botswana on the collection, use, and storage of human biological samples for research. *BMC Medical Ethics* **16**(1): 54.

Behar D, Kedem E, Rosset S, et al. (2011). Absence of APOL1 risk variants protects against HIV-associated nephropathy in the Ethiopian population. *Am J Nephrol* **34**(5): 452–459.

Bekker LG, Venter F, Cohen K-G, et al. (2014). Provision of antiretroviral therapy in South Africa: the nuts and bolts. *Antiviral Therapy* **19**(Suppl. 3): 105–116.

Bellamy R, Beyers N, McAdam KP, et al. (2000). Genetic susceptibility to tuberculosis in Africans: a genome-wide scan. *PNAS* **97**(14): 8005–8009.

Bernal ML, Sinues B, Johansson I, et al. (1999). Ten percent of North Spanish individuals carry duplicated or triplicated CYP2D6 genes associated with ultrarapid metabolism of debrisoquine. *Pharmacogenet Genomics* **9**(5): 657–660.

Bharti D, Kumar A, Mahla RS, et al. (2015). Low prevalence of CCR5-Delta32, CCR2-64I and SDF1-3′A alleles in the Baiga and Gond tribes of Central India. *Springerplus* **4**: 451.

Bienzle U, Ayeni O, Lucas AO, et al. (1972). Glucose-6-phosphate dehydrogenase and malaria. *The Lancet* **299**(7742): 107–110.

Bowen MD, Rollin PE, Ksiazek TG, et al. (2000). Genetic diversity among Lassa virus strains. *J Virol* **74**(15): 6992–7004.

Bradford LD (2002). CYP2D6 allele frequency in European Caucasians, Asians, Africans and their descendants. *Pharmacogenomics* **3**(2): 229–243.

Buckley SM and Casals J (1970). Lassa fever, a new virus disease of man from West Africa. *Am J Trop Med Hyg* **19**(4): 680–691.

Burchard EG, Oh SS, Foreman MG, Celedón JC, et al. (2015). Moving toward true inclusion of racial/ethnic minorities in federally funded studies: a key step for achieving respiratory health equality in the United States. *Am J Resp Crit Care Med* **191**(5): 514–521.

Butterweck V and Derendorf H (2008). Potential of pharmacokinetic profiling for detecting herbal interactions with drugs. *Clin Pharmacokinet* **47**(6): 383–397.

Campbell MC and Tishkoff SA (2010). The evolution of human genetic and phenotypic variation in Africa. *Curr Biol* **20**(4): R166–R173.

Cao W, Henry MD, Borrow P, et al. (1998). Identification of alpha-dystroglycan as a receptor for lymphocytic choriomeningitis virus and Lassa fever virus. *Science* **282**(5396): 2079–2081.

Capewell P, Veitch NJ, Turner CM, et al. (2011). Differences between *Trypanosoma brucei gambiense* groups 1 and 2 in their resistance to killing by trypanolytic factor 1. *PLoS Negl Trop Dis* **5**(9). DOI: 10.1371/journal.pntd.0001287.

Capewell P, Cooper A, Clucas C, et al. (2015). A co-evolutionary arms race: trypanosomes shaping the human genome, humans shaping the trypanosome genome. *Parasitology* **142**: 108–119.

Carlson CS, Matise TC, North KE, et al. (2013). Generalization and dilution of association results from European GWAS in populations of non-European ancestry: the PAGE Study. *PLoS Biology* **11**(9). DOI: 10.1371/journal.pbio.1001661.

Casanova J-L (2015). Human genetic basis of interindividual variability in the course of infection. *PNAS* **112**(51): E7118–E7127.

Cashdan E (2001). Ethnic diversity and its environmental determinants: effects of climate, pathogens, and habitat diversity. *Amer Anthrop* **103**(4): 968–991.

Cavasini CE, Mattos LC, Couto AA, et al. (2007). *Plasmodium vivax* infection among Duffy antigen-negative individuals from the Brazilian Amazon region: an exception? *Trans Roy Soc Trop Med Hyg* **101**(10): 1042–1044.

Chanda-Kapata P, Kapata N, Moraes AN, et al. (2015). Genomic research in Zambia: confronting the ethics, policy and regulatory frontiers in the 21st century. *Health Res Policy Syst* **13**(1): 60.

Chapman SJ and Hill AV (2012). Human genetic susceptibility to infectious disease. *Nat Rev Genet* **13**(3): 175–188.

Checchi F, Filipe JA, Haydon DT, et al. (2008). Estimates of the duration of the early and late stage of gambiense sleeping sickness. *BMC Infectious Diseases* **8**(1): 16.

Checchi F, Funk S, Chandramohan D, et al. (2015). Updated estimate of the duration of the meningo-encephalitic stage in gambiense human African trypanosomiasis. *BMC Res Notes* **8**(1): 292.

Chen H and Pang T (2015). A call for global governance of biobanks. *Bull WHO* **93**(2): 113–117.

Chen MS, Jr, Lara PN, Dang JH, et al. (2014). Twenty years post-NIH Revitalization Act: renewing the case for enhancing minority participation in cancer clinical trials. *Cancer* **120**(7): 1091–1096.

Chen R, Corona E, Sikora M, et al. (2012). Type 2 diabetes risk alleles demonstrate extreme directional differentiation among human populations, compared to other diseases. *PLoS Genet* **8** (4). DOI: 10.1371/journal.pgen.1002621.

Cheung KH, Osier MV, Kidd JR, et al. (2000). ALFRED: an allele frequency database for diverse populations and DNA polymorphisms. *Nucleic Acids Res* **28**(1): 361–363.

Chigutsa E, Visser ME, Swart EC, et al. (2011). The SLCO1B1 rs4149032 polymorphism is highly prevalent in South Africans and is associated with reduced rifampin concentrations: dosing implications. *Antimicrob Agents Chemother* **55**(9): 4122–4127.

Cobat A, Gallant CJ, Simkin L, et al. (2009). Two loci control tuberculin skin test reactivity in an area hyperendemic for tuberculosis. *J Exp Med* **206**(12): 2583–2591.

Crocker PR, Paulson JC, and Varki A (2007). Siglecs and their roles in the immune system. *Nat Rev Immunol* **7**(4): 255–266.

Dandara C, Masimirembwa CM, Magimba A, et al. (2003). Arylamine N-acetyltransferase (NAT2) genotypes in Africans: the identification of a new allele with nucleotide changes 481C>T and 590G>A. *Pharmacogenetics* **13**(1): 55–58.

Dandara C, Swart M, Mpeta B, et al. (2014a). Cytochrome P450 pharmacogenetics in African populations: implications for public health. *Expert Opin Drug Metab Toxicol* **10**(6): 769–785.

Dandara C, Huzair F, Borda-Rodriguez A, et al. (2014b). H3Africa and the African life sciences ecosystem: building sustainable innovation. *OMICS* **18**(12): 733–739.

Dessein AJ, Hillaire D, Elwali NE, et al. (1999). Severe hepatic fibrosis in *Schistosoma mansoni* infection is controlled by a major locus that is closely linked to the interferon-gamma receptor gene. *Am J Hum Genet* **65**(3): 709–721.

Domingo GJ, Satyagraha AW, Anvikar A, et al. (2013). G6PD testing in support of treatment and elimination of malaria: recommendations for evaluation of G6PD tests. *Malar J* **12**(391): 1–12.

Dubaniewicz A, Lewko B, Moszkowska G, et al. (2000). Molecular subtypes of the HLA-DR antigens in pulmonary tuberculosis. *Int J Infect Dis* **4**(3): 129–133.

Eugen-Olsen J, Iversen AK, Garred P, et al. (1997). Heterozygosity for a deletion in the CKR-5 gene leads to prolonged AIDS-free survival and slower CD4 T-cell decline in a cohort of HIV-seropositive individuals. *AIDS* **11**: 305–310.

Evans DA, Manley KA, and McKusick VA (1960). Genetic control of isoniazid metabolism in man. *BMJ* 2(5197): 485–491.

Faucher JF, Aubouy A, Adeothy A, et al. (2009). Comparison of sulfadoxine-pyrimethamine, unsupervised artemether-lumefantrine, and unsupervised artesunate-amodiaquine fixed-dose formulation for uncomplicated *Plasmodium falciparum* malaria in Benin: a randomized effectiveness noninferiority trial. *J Infect Dis* **200**(1): 57–65.

FDA (2013). FDA drug safety communication. Safety announcement. Available at: www.fda.gov/Drugs/DrugSafety/ucm376389.htm.

Fichet-Calvet E and Rogers DJ (2009). Risk maps of Lassa fever in West Africa. *PLoS NeglTrop Dis* 3 (3). DOI: 10.1371/journal.pntd.0000388.

Frame JD, Baldwin JM, Jr, Gocke DJ, et al. (1970). Lassa fever, a new virus disease of man from West Africa. *Am J Trop Med Hyg* 19(4): 670–676.

Franco JR, Simarro PP, Diarra A, et al. (2014). Epidemiology of human African trypanosomiasis. *Clin Epidemiol* 6(1): 257–275.

Freedman BI, Langefeld CD, Andringa KK, et al. (2014). End-stage renal disease in African Americans with lupus nephritis is associated with APOL1. *Arthritis Rheumatol* **66**(2): 390–396.

Fumagalli M, Cagliani R, Pozzoli U, et al. (2009). Widespread balancing selection and pathogen-driven selection at blood group antigen genes. *Genome Res* 19(2): 199–212.

Fumagalli M, Cagliani R, Riva S, et al. (2010a). Population genetics of IFIH1: ancient population structure, local selection, and implications for susceptibility to type 1 diabetes. *Mol Biol Evol* **27** (11): 2555–2566.

Fumagalli M, Pozzoli U, Cagliani R, et al. (2010b). Genome-wide identification of susceptibility alleles for viral infections through a population genetics approach. *PLoS Genet* **6**(2). DOI: 10.1371/journal.pgen.1000849.

Fumagalli M, Sironi M, Pozzoli U, et al. (2011). Signatures of environmental genetic adaptation pinpoint pathogens as the main selective pressure through human evolution. *PLoS Genet* **7**(11): e1002355.

Galvani AP and Slatkin M (2003). Evaluating plague and smallpox as historical selective pressures for the CCR5-Delta 32 HIV-resistance allele. *PNAS* **100**(25): 15276–15279.

Ge D, Fellay J, Thompson AJ, et al. (2009). Genetic variation in IL28B predicts hepatitis C treatment-induced viral clearance. *Nature* **461**(7262): 399–401.

Genovese G, Friedman DJ, Ross MD, et al. (2010). Association of trypanolytic ApoL1 variants with kidney disease in African Americans. *Science* **329**(5993): 841–845.

Gething PW, Elyazar IR, Moyes CL, et al. (2012). A long neglected world malaria map: *Plasmodium vivax* endemicity in 2010. *PLoS Negl Trop Dis* **6**(9). DOI: 10.1371/journal.pntd.0001814.

Ginsburg GS and Willard HF (2009). Genomic and personalized medicine: foundations and applications. *Translational Res* **154**(6): 277–287.

Gonzalez E, Bamshad M, Sato N, et al. (1999). Race-specific HIV-1 disease-modifying effects associated with CCR5 haplotypes. *PNAS* **96**(21): 12004–12009.

Gorboulev V, Ulzheimer JC, Akhoundova A, et al. (1997). Cloning and characterization of two human polyspecific organic cation transporters. *DNA Cell Biol* **16**(7): 871–881.

Griese EU, Asante-Poku S, Ofori-Adjei D, et al. (1999). Analysis of the CYP2D6 gene mutations and their consequences for enzyme function in a West African population. *Pharmacogenetics* **9**(6): 715–723.

Guerra CA, Gikandi PW, Tatem AJ, et al. (2008). The limits and intensity of *Plasmodium falciparum* transmission: implications for malaria control and elimination worldwide. *PLoS Medicine* **5**(2): 300–311.

Günther S, Emmerich P, Laue T, et al. (2000). Imported Lassa fever in Germany: molecular characterization of a new Lassa virus strain. *Emerg Infect Dis* 6(5): 466–476.

Gurdasani D, Carstensen T, Tekola-Ayele F, et al. (2015). The African Genome Variation Project shapes medical genetics in Africa. *Nature* 517(7534): 327–332.

Habtewold A, Amogne W, Makonnen E, et al. (2011). Long-term effect of efavirenz autoinduction on plasma/peripheral blood mononuclear cell drug exposure and CD4 count is influenced by UGT2B7 and CYP2B6 genotypes among HIV patients. *J Antimicrob Chemother* 66(10): 2350–2361.

Harpending H and Rogers A (2000). Genetic perspectives on human origins and differentiation. *Annu Rev Genomics Hum Genet* 1: 361–385.

Hattori M (2004). International Human Genome Sequencing Consortium: finishing the euchromatic sequence of the human genome. *Nature* 431(7011): 931–945.

Hetherington S, Hughes AR, Mosteller M, et al. (2002). Genetic variations in HLA-B region and hypersensitivity reactions to abacavir. *Lancet* 359(9312): 1121–1122.

Hill AV, Allsopp CE, Kwiatkowski D, et al. (1991). Common west African HLA antigens are associated with protection from severe malaria. *Nature* 352(6336): 595–600.

Hill CM and Littman DR (1996). Natural resistance to HIV? *Nature* 382(6593): 668–669.

Hindorff LA, Sethupathy P, Junkins HA, et al. (2009). Potential etiologic and functional implications of genome-wide association loci for human diseases and traits. *PNAS* 106(23): 9362–9367.

Howes RE, Patil AP, Piel FB, et al. (2011). The global distribution of the Duffy blood group. *Nature Comm* 2: 266.

Howes RE, Piel FB, Patil AP, et al. (2012). G6PD deficiency prevalence and estimates of affected populations in malaria endemic countries: a geostatistical model-based map. *PLoS Medicine* 9 (11). DOI: 10.1371/journal.pmed.1001339.

Howes RE, Battle KE, Satyagraha AW, et al. (2013). G6PD deficiency: global distribution, genetic variants and primaquine therapy. *Adv Parasitol.* DOI: 10.1016/B978-0-12-407826-0.00004-7.

Howes RE, Reiner RC, Jr, Battle KE, et al. (2015). *Plasmodium vivax* transmission in Africa. *PLoS Negl Trop Dis* 9(11): 1–27.

Huang Y, Paxton WA, Wolinsky SM, et al. (1996). The role of a mutant CCR5 allele in HIV-1 transmission and disease progression. *Nat Med* 2(11): 1240–1243.

Hughes HB, Biehl JP, Jones AP, et al. (1954). Metabolism of isoniazid in man as related to the occurrence of peripheral neuritis. *Am Rev Tuberculosis* 70(2): 266–273.

Hummel S, Schmidt D, Kremeyer B, et al. (2005). Detection of the CCR5-Delta32 HIV resistance gene in Bronze Age skeletons. *Genes Immun* 6(4): 371–374.

Hyde JE (2007). Drug-resistant malaria: an insight. *The FEBS Journal* 274(18): 4688–4698.

Ito K, Bick AG, Flannick J, et al. (2014). Increased burden of cardiovascular disease in carriers of APOL1 genetic variants. *Circulation Res* 114(5): 845–850.

Jao I, Kombe F, Mwalukore S, et al. (2015). Research stakeholders' views on benefits and challenges for public health research data sharing in Kenya: the importance of trust and social relations. *PLoS One* 10(9): 1–18.

Johnson JA, Gong L, Whirl-Carrillo M, et al. (2011). Clinical Pharmacogenetics Implementation Consortium guidelines for CYP2C9 and VKORC1 genotypes and warfarin dosing. *Clin Pharmacol Therapeut* 90(4): 625–629.

Kalow W and Gunn DR (1959). Some statistical data on atypical cholinesterase of human serum. *Ann Hum Genet* 23(3): 239–250.

Kalow W and Staron N (1957). On distribution and inheritance of atypical forms of human serum cholinesterase, as indicated by dibucaine numbers. *Can J Biochem Physiol* 35(12): 1305–1320.

Kasembeli AN, Duarte R, Ramsay M, et al. (2015). APOL1 risk variants are strongly associated with HIV-associated nephropathy in Black South Africans. *J Am Soc Nephrol* 26(11): 2882–2990.

Kerb R, Fux R, Mörike K, et al. (2009). Pharmacogenetics of antimalarial drugs: effect on metabolism and transport. *Lancet Infect Dis* 9(12): 760–774.

Kesselring AM, Wit FW, Sabin CA, et al. (2009). Risk factors for treatment-limiting toxicities in patients starting nevirapine-containing antiretroviral therapy. *AIDS* 23(13): 1689–1699.

Kettaneh A, Seng L, Tiev KP, et al. (2006). Human leukocyte antigens and susceptibility to tuberculosis: a meta-analysis of case-control studies. *Int J Tuberc Lung Dis* 10(7): 717–725.

Kindberg E, Mickiene A, Ax C, et al. (2008). A deletion in the chemokine receptor 5 (CCR5) gene is associated with tickborne encephalitis. *Int J Infect Dis* 197(2): 266–269.

Klein TE, Chang JT, Cho MK, et al. (2001). Integrating genotype and phenotype information: an overview of the PharmGKB project. Pharmacogenetics Research Network and Knowledge Base. *Pharmacogenomics J* 1(3): 167–170.

Ko W, Rajan P, Gomez F, et al. (2013). Identifying Darwinian selection acting on different human APOL1 variants among diverse African populations. *Am J Hum Genet* 93(1): 54–66.

Koboldt DC, Steinberg KM, Larson DE, et al. (2013). The next-generation sequencing revolution and its impact on genomics. *Cell* 155(1): 27–38.

Kopp JB, Nelson GW, Sampath K, et al. (2011). APOL1 genetic variants in focal segmental glomerulosclerosis and HIV-associated nephropathy. *J Am Soc Nephrol* 22(11): 2129–2137.

Kudzi W, Dodoo ANO, and Mills JJ (2010). Genetic polymorphisms in MDR1, CYP3A4 and CYP3A5 genes in a Ghanaian population: a plausible explanation for altered metabolism of ivermectin in humans? *BMC Med Genet* 11(1): 111.

Kunz S, Rojek JM, Kanagawa M, et al. (2005). Posttranslational modification of alpha-dystroglycan, the cellular receptor for arenaviruses, by the glycosyltransferase LARGE is critical for virus binding. *J Virol* 79(22): 14282–14296.

Lalani AS, Masters J, Zeng W, et al. (1999). Use of chemokine receptors by poxviruses. *Science* 286 (5446): 1968–1971.

Lamour SD, Gomez-Romero M, Vorkas PA, et al. (2015). Discovery of infection associated metabolic markers in human African trypanosomiasis. *PLoS Negl Trop Dis* 9(10): 1–17.

Lander ES, Linton LM, Birren B, et al. (2001). Initial sequencing and analysis of the human genome. *Nature* 409(6822): 860–921.

Langhi DM and Orlando Bordin J (2006). Duffy blood group and malaria. *Hematology* 11(5–6): 389–398.

Le AQ, Taylor J, Dong W, et al. (2015). Differential evolution of a CXCR4-using HIV-1 strain in CCR5wt/wt and CCR5Δ32/Δ32 hosts revealed by longitudinal deep sequencing and phylogenetic reconstruction. *Scientific Reports* 5: 17607.

Li J, Zhang L, Zhou H, et al. (2011). Global patterns of genetic diversity and signals of natural selection for human ADME genes. *Hum Mol Genet* 20(3): 528–540.

Li XQ, Bjorkman A, Andersson TB, et al. (2002). Amodiaquine clearance and its metabolism to N-desethylamodiaquine is mediated by CYP2C8: a new high affinity and turnover enzyme-specific probe substrate. *J Pharmacol Exp Ther* 300(2): 399–407.

Libert F, Cochaux P, Beckman G, et al. (1998). The deltaCCR5 mutation conferring protection against HIV-1 in Caucasian populations has a single and recent origin in Northeastern Europe. *Hum Mol Genet* 7(3): 399–406.

Lim JK, Louie CY, Glaser C, et al. (2008). Genetic deficiency of chemokine receptor CCR5 is a strong risk factor for symptomatic West Nile virus infection: a meta-analysis of 4 cohorts in the US epidemic. *Int J Infect Dis* 197(2): 262–265.

Liu R, Paxton WA, Choe S, et al. (1996). Homozygous defect in HIV-1 coreceptor accounts for resistance of some multiply-exposed individuals to HIV-1 infection. *Cell* 86(3): 367–377.

Lombard Z, Brune AE, Hoal EG, et al. (2006). HLA class II disease associations in southern Africa. *Tissue Antigens* 67(2): 97–110.

Lopes MF, Nunes M, Henriques-Pons P, et al. (1999). Increased susceptibility of Fas ligand-deficient gld mice to *Trypanosoma cruzi* infection due to a Th2-biased host immune response. *Eur J Immunol* 29(1): 81–89.

Lu Y, Nerurkar VR, Dashwood WM, et al. (1999). Genotype and allele frequency of a 32-base pair deletion mutation in the CCR5 gene in various ethnic groups: absence of mutation among Asians and Pacific Islanders. *Int J Infect Dis* 3(4): 186–191.

Lynch T and Price A (2007). The effect of cytochrome P450 metabolism on drug response, interactions, and adverse effects. *Am Fam Physician* **76**(3): 391–396.

Ma Q and Lu AY (2011). Pharmacogenetics, pharmacogenomics, and individualized medicine. *Pharmacol Rev* **63**(2): 437–459.

MacLean L, Chisi JE, Odiit M, et al. (2004). Severity of human African trypanosomiasis in East Africa is associated with geographic location, parasite genotype, and host inflammatory cytokine response profile. *Infect Immun* **72**(12): 7040–7044.

Mallal S, Nolan D, Witt C, et al. (2002). Association between presence of HLA-B*5701, HLA-DR7, and HLA-DQ3 and hypersensitivity to HIV-1 reverse-transcriptase inhibitor abacavir. *Lancet* **359**: 727–732.

Mangano VD and Modiano D (2014). Host genetics and parasitic infections. *Clin Microbiol Infect* **20**(12): 1265–1275.

Mardis ER (2013). Next-generation sequencing platforms. *Annu Rev Anal Chem* **6**: 287–303.

Martin MA, Klein TE, Dong BJ, et al. (2014). Clinical Pharmacogenetics Implementation Consortium guidelines for HLA-B genotype and abacavir dosing: 2014 update. *Pharmacol Ther, Part C* **95**(5): 499–500.

Martins GA, Petkova SB, MacHado FS, et al. (2001). Fas–FasL interaction modulates nitric oxide production in *Trypanosoma cruzi*-infected mice. *Immunology* **103**(1): 122–129.

Martinson JJ, Chapman NH, Rees DC, et al. (1997). Global distribution of the CCR5 gene 32-basepair deletion. *Nat Genet* **16**(1): 100–103.

Masimirembwa C, Hasler J, Bertilssons L, et al. (1996). Phenotype and genotype analysis of debrisoquine hydroxylase (CYP2D6) in a black Zimbabwean population: reduced enzyme activity and evaluation of metabolic correlation of CYP2D6 probe drugs. *Eur J Clin Pharmacol* **51**(2): 117–122.

Masimirembwa C, Dandara C, and Hasler J (2014). Population Diversity and Pharmacogenomics in Africa. In Padmanabhan S, ed., *Handbook of Pharmacogenomics and Stratified Medicine*. Elsevier, pp. 971–998.

Matimba A, Oluka MN, Ebeshi BU, et al. (2008). Establishment of a biobank and pharmacogenetics database of African populations. *Eur J Hum Genet* **16**(7): 780–783.

McCormick JB (1986). Clinical, epidemiologic, and therapeutic aspects of Lassa fever. *Med Microbiol Immunol* **175**(2–3): 153–155.

McCormick JB, King IJ, Webb PA, et al. (1986). Lassa fever: effective therapy with ribavirin. *N Engl J Med* **314**(1): 20–26.

McCutchan FE (2006). Global epidemiology of HIV. *J Med Virol* **78**(Suppl. 1): S7–S12.

Mehlotra RK, Henry-Halldin CN, and Zimmerman PA (2009). Application of pharmacogenomics to malaria: a holistic approach for successful chemotherapy. *Pharmacogenomics* **10**(3): 435–449.

Ménard D, Barnadas C, Bouchier C, et al. (2010). *Plasmodium vivax* clinical malaria is commonly observed in Duffy-negative Malagasy people. *PNAS* **107**(13): 5967–5971.

Mendis K, Sina BJ, Marchesini P, et al. (2001). The neglected burden of *Plasmodium vivax* malaria. *Am J Trop Med Hyg* **64**(1–2 Suppl.): 97–106.

Mhandire D, Lacerda M, Castel S, et al. (2015). Effects of CYP2B6 and CYP1A2 genetic variation on nevirapine plasma concentration and pharmacodynamics as measured by CD4 cell count in Zimbabwean HIV-infected patients. *OMICS* **19**(9): 553–562.

Michael NL, Chang G, Louie LG, et al. (1997). The role of viral phenotype and CCR-5 gene defects in HIV-1 transmission and disease progression. *Nat Med* **3**(3): 338–340.

Miller LH, Mason SJ, Clyde DF, et al. (1976). The resistance factor to *Plasmodium vivax* in blacks: the Duffy-Blood-Group genotype *FyFy*. *N Engl J Med* **295**(6): 302–304.

Mira MT, Alcais A, Nguyen VT, et al. (2004). Susceptibility to leprosy is associated with PARK2 and PACRG. *Nature* **427**(6975): 636–640.

Mishra G, Kumar N, Kaur G, et al. (2014). Distribution of HLA-A, B and DRB1 alleles in Sahariya tribe of North Central India: an association with pulmonary tuberculosis. *Infect Genet Evol* **22**: 175–182.

Möller M and Hoal EG (2010). Current findings, challenges and novel approaches in human genetic susceptibility to tuberculosis. *Tuberculosis* **90**(2): 71–83.

Monath TP, Newhouse VF, Kemp GE, et al. (1974). Lassa virus isolation from *Mastomys natalensis* rodents during an epidemic in Sierra Leone. *Science* **185**(4147): 263–265.

Moodley K, Sibanda N, February K, et al. (2014). "It's my blood": ethical complexities in the use, storage and export of biological samples – perspectives from South African research participants. *BMC Med Ethics* **15**(1): 4.

Muenke M (2013). Individualized genomics and the future of translational medicine. *Mol Genet Genomic Med* **1**(1): 1–3.

Muir AJ, Gong L, Johnson SG, et al. (2014). Clinical Pharmacogenetics Implementation Consortium (CPIC) guidelines for IFNL3 (IL28B) genotype and PEG interferon-α-based regimens. *Pharmacol Ther, Part C* **95**(2): 141–146.

Mukamal KJ, Tremaglio J, Friedman DJ, et al. (2016). APOL1 genotype, kidney and cardiovascular disease, and death in older adults. *Arterioscler Thromb Vasc Biol* **36**(2): 398–403.

Mukonzo JK, Roshammar D, Waako P, et al. (2009). A novel polymorphism in ABCB1 gene, CYP2B6*6 and sex predict single-dose efavirenz population pharmacokinetics in Ugandans. *Br J Clin Pharmacol* **68**(5): 690–699.

Mukonzo JK, Okwera A, Nakasujja N, et al. (2013). Influence of efavirenz pharmacokinetics and pharmacogenetics on neuropsychological disorders in Ugandan HIV-positive patients with or without tuberculosis: a prospective cohort study. *BMC Infect Dis* **13**(1): 261.

Mylne AQ, Pigott DM, Longbottom J, et al. (2015). Mapping the zoonotic niche of Lassa fever in Africa. *Trans R Soc Trop Med Hyg* **109**(8): 483–492.

National Institute of Health (2010). NIH and Wellcome Trust announce partnership to support population-based genome studies in Africa. Available at: www.nih.gov/news-events/news-releases/nih-wellcome-trust-announce-partnership-support-population-based-genome-studies-africa.

Need AC and Goldstein DB (2009). Next generation disparities in human genomics: concerns and remedies. *Trends Genet* **25**(11): 489–494.

Nelson MR, Tipney H, Painter JL, et al. (2015). The support of human genetic evidence for approved drug indications. *Nat Genet* **47**(8): 856–860.

Ngaimisi E, Habtewold A, Minzi O, et al. (2013). Importance of ethnicity, CYP2B6 and ABCB1 genotype for efavirenz pharmacokinetics and treatment outcomes: a parallel-group prospective cohort study in two Sub-Saharan Africa populations. *PLoS One* **8**(7). DOI: 10.1371/journal.pone.0067946.

Nobelprize.org (2015). The Nobel Prize in Physiology or Medicine 2015. Press Release. Nobel Media AB. Available at: www.nobelprize.org/nobel_prizes/medicine/laureates/2015/press.html.

Novembre J, Galvani AP, and Slatkin M (2005). The geographic spread of the CCR5 Δ32 HIV-resistance allele. *PLoS Biol* **3**(11): 1954–1962.

Nsanzimana S, Remera E, Kanters S, et al. (2015). Life expectancy among HIV-positive patients in Rwanda: a retrospective observational cohort study. *Lancet Global Health* **3**(3): e169–e177.

Ntzani EE, Liberopoulos G, Manolio TA, et al. (2012). Consistency of genome-wide associations across major ancestral groups. *Hum Genet* **131**(7): 1057–1071.

Nyakutira C, Roshammar D, Chigutsa E, et al. (2008). High prevalence of the CYP2B6 516G→T(*6) variant and effect on the population pharmacokinetics of efavirenz in HIV/AIDS outpatients in Zimbabwe. *Eur J Clin Pharmacol* **64**(4): 357–365.

Odiit M, Kansiime F, and Enyaru JC (1997). Duration of symptoms and case fatality of sleeping sickness caused by *Trypanosoma brucei rhodesiense* in Tororo, Uganda. *East Afr Med J* **74**(12): 792–795.

Ogbonna A and Uneke CJ (2008). Artemisinin-based combination therapy for uncomplicated malaria in sub-Saharan Africa: the efficacy, safety, resistance and policy implementation since Abuja 2000. *Trans R Soc Trop Med Hyg* **102**(7): 621–627.

Oh SS, Galanter J, Thakur N, et al. (2015). Diversity in clinical and biomedical research: a promise yet to be fulfilled. *PLoS Med* **12**(12): 1–9.

Oluka MN, Okalebo FA, Guantai AN, et al. (2015). Cytochrome P450 2B6 genetic variants are associated with plasma nevirapine levels and clinical response in HIV-1 infected Kenyan women: a prospective cohort study. *AIDS Research and Therapy* **12**: 10.

Pagani L, Kivisild T, Tarekegn A, et al. (2012). Ethiopian genetic diversity reveals linguistic stratification and complex influences on the Ethiopian gene pool. *Am J Hum Genet* **91**(1): 83–96.

Paganotti GM, Gramolelli S, Tabacchi F, et al. (2012). Distribution of human CYP2C8*2 allele in three different African populations. *Malar J* **11**(1): 125.

Parikh S, Ouedraogo JB, Goldstein JA, et al. (2007). Amodiaquine metabolism is impaired by common polymorphisms in CYP2C8: implications for malaria treatment in Africa. *Pharmacol Ther, Part C* **82**(2): 197–203.

Parsa A, Kao WH, Xie D, et al. (2013). APOL1 risk variants, race, and progression of chronic kidney disease. *N Engl J Med* **369**(23): 2183–2196.

Pasvol G, Weatherall DJ, and Wilson RJ (1978). Cellular mechanism for the protective effect of haemoglobin S against *P. falciparum* malaria. *Nature* **274**(5672): 701–703.

Pays E, Vanhollebeke B, Uzureau P, Lecordier L, and Perez-Morga D (2014). The molecular arms race between African trypanosomes and humans. *Nat Rev Microbiol* **12**(8): 575–584.

Phillips E and Mallal S (2009). Successful translation of pharmacogenetics into the clinic: the abacavir example. *Mol Diagn Ther* **13**(1): 1–9.

Piel FB, Patil AP, Howes RE, et al. (2010). Global distribution of the sickle cell gene and geographical confirmation of the malaria hypothesis. *Nat Commun* **1**: 104.

Pillai G, Davies G, Denti P, et al. (2013). Pharmacometrics: opportunity for reducing disease burden in the developing world: the case of Africa. *CPT Pharmacometrics Syst Pharmacol* **2**: e69.

Ramachandran S, Deshpande O, Roseman CC, et al. (2005). Support from the relationship of genetic and geographic distance in human populations for a serial founder effect originating in Africa. *PNAS* **102**(44): 15942–15947.

Ramos E, Doumatey A, Elkahloun AG, et al. (2014). Pharmacogenomics, ancestry and clinical decision making for global populations. *Pharmacogenomics J* **14**(3): 217–222.

Ranade K, Geese WJ, Noor M, et al. (2008). Genetic analysis implicates resistin in HIV lipodystrophy. *AIDS* **22**(13): 1561–1568.

Rauch A, Kutalik Z, Descombes P, et al. (2010). Genetic variation in IL28B is associated with chronic hepatitis C and treatment failure: a genome-wide association study. *Gastroenterol* **138**(4): 1338–1345.e7.

Relling MV and Klein TE (2011). CPIC: Clinical Pharmacogenetics Implementation Consortium of the Pharmacogenomics Research Network. *Clin Pharmacol Therapeut* **89**(3): 464–467.

Ribeiro CM, Pontes MJ, Bird S, et al. (2010). Trypanosomiasis-induced Th17-like immune responses in carp. *PLoS One* **5**(9). DOI: 10.1371/journal.pone.0013012.

Roederer MW, McLeod H, and Juliano JJ (2011). Can pharmacogenomics improve malaria drug policy? *Bull WHO* **89**(11): 838-845.

Rosenberg NA, Pritchard JK, Weber JL, et al. (2011). Genetic structure of human populations. *Science* **2381**(2002): 2381–2385.

Ross EL, Weinstein MC, Schackman BR, et al. (2015). The clinical role and cost-effectiveness of long-acting antiretroviral therapy. *Clin Infect Dis* **60**(7): 1102–1110.

Rotimi CN and Jorde LB (2010). Ancestry and disease in the age of genomic medicine. *N Engl J Med* **363**(16): 1551–1558.

Rotimi C, Abayomi A, Abimiku A, et al. (2014). Research capacity: enabling the genomic revolution in Africa. *Science* **344**(6190): 1346–1348.

Rowell JL, Dowling NF, Yu W, Yesupriya A, Zhang L, and Gwinn M (2012). Trends in population-based studies of human genetics in infectious diseases. *PLoS One* **7**(2): e25431.

Sabeti PC, Varilly P, Fry B, et al. (2007). Genome-wide detection and characterization of positive selection in human populations. *Nature* **449**(7164): 913–918.

Sachse C, Brockmoller J, Bauer S, et al. (1997). Cytochrome P450 2D6 variants in a Caucasian population: allele frequencies and phenotypic consequences. *Am J Hum Genet* **60**(2): 284–295.

Salem AH and Batzer MA (2007). Distribution of the HIV resistance CCR5-Delta32 allele among Egyptians and Syrians. *Mutat Res* **616**(1–2): 175–180.

Samson M, Libert F, Doranz BJ, et al. (1996). Resistance to HIV-1 infection in Caucasian individuals bearing mutant alleles of the CCR-5 chemokine receptor gene. *Nature* **382**(6593): 722–725.

Schliekelman P, Garner C, and Slatkin M (2001). Natural selection and resistance to HIV. *Nature* **411** (6837): 545–546.

Scott SA, Sangkuhl K, Stein CM, et al. (2013). Clinical Pharmacogenetics Implementation Consortium guidelines for CYP2C19 genotype and clopidogrel therapy: 2013 update. *Pharmacol Ther, Part C* **94**(3): 317–323.

Segal S and Hill AVS (2003). Genetic susceptibility to infectious disease. *Trends Microbiol* **11**(9): 445–448.

Seripa D, Pilotto A, Panza F, et al. (2010). Pharmacogenetics of cytochrome P450 (CYP) in the elderly. *Ageing Res Rev* **9**(4): 457–474.

Shahinas D, Folefoc A, and Pillai DR (2013). Targeting *Plasmodium falciparum* Hsp90: towards reversing antimalarial resistance. *Pathogens* **2**(1): 33–54.

Shriner D, Tekola-Ayele F, Adeyemo A, et al. (2014). Genome-wide genotype and sequence-based reconstruction of the 140,000 year history of modern human ancestry. *Scientific Reports* **4**: 6055.

Shu Y, Leabman MK, Feng B, et al. (2003). Evolutionary conservation predicts function of variants of the human organic cation transporter, OCT1. *PNAS* **100**(10): 5902–5907.

Simarro PP, Diarra A, Ruiz Postigo JA, Franco JR, and Jannin JG (2011). The human African trypanosomiasis control and surveillance programme of the World Health Organization 2000–2009: the way forward. *PLoS Negl Trop Dis* **5**(2). DOI: 10.1371/journal.pntd.0001007.

Sirugo G, Schim M, van der Loeff SO, et al. (2004). A national DNA bank in The Gambia, West Africa, and genomic research in developing countries. *Nat Genet* **36**(8): 785–786.

Siwo GH, Williams SM, and Moore JH (2015). The future of genomic medicine education in Africa. *Genome Med* **7**(1): 47.

Smith I (2003). *Mycobacterium tuberculosis* pathogenesis and molecular determinants of virulence. *Clin Microbiol Rev* **16**(3): 463–496.

Smithburn K, Hughes T, Burke A, et al. (1940). A neurotropic virus isolated from the blood of a native of Uganda. *Am J Trop Med Hyg* **s1-20**(4): 471–492.

Staunton C and Moodley K (2013). Challenges in biobank governance in Sub-Saharan Africa. *BMC Med Ethics* **14**(1): 35.

Stephens JC, Reich DE, Goldstein DB, et al. (1998). Dating the origin of the CCR5-Delta32 AIDS-resistance allele by the coalescence of haplotypes. *Am J Hum Genet* **62**(6): 1507–1515.

Sternberg JM (2005). Human African trypanosomiasis: clinical presentation and immune response. *Parasite Immunol* **26**(11–12): 469–476.

Suppiah V, Moldovan M, Ahlenstiel G, et al. (2009). IL28B is associated with response to chronic hepatitis C interferon-alpha and ribavirin therapy. *Nat Genet* **41**(10): 1100–1104.

Swart M, Whitehorn H, Ren Y, et al. (2012). PXR and CAR single nucleotide polymorphisms influence plasma efavirenz levels in South African HIV/AIDS patients. *BMC Med Genet* 13(1): 112.

Swart M, Skelton M, Ren Y, et al. (2013). High predictive value of CYP2B6 SNPs for steady-state plasma efavirenz levels in South African HIV/AIDS patients. *Pharmacogenet Genomics* 23(8): 415–427.

Swen JJ, Nijenhuis M, de Boer A, et al. (2011). Pharmacogenetics: from bench to byte – an update of guidelines. *Pharmacol Ther, Part C* 89(5): 662–673.

Tanaka Y, Nishida N, and Sugiyama M (2009). Genome-wide association of IL28B with response to pegylated interferon-α and ribavirin therapy for chronic hepatitis C. *Nat Genet* 41(10): 1105–1109.

Tarirai C, Viljoen AM, and Hamman JH (2010). Herb–drug pharmacokinetic interactions reviewed. *Expert Opin Drug Metab Toxicol* 6(12): 1515–1538.

Tekola-Ayele F and Rotimi CN (2015). Translational genomics in low- and middle-income countries: opportunities and challenges. *Public Health Genomics* 18(4): 242–247.

Tekola-Ayele F, Adeyemo A, Aseffa A, et al. (2014). Clinical and pharmacogenomic implications of genetic variation in a Southern Ethiopian population. *Pharmacogenomics J* 15(1): 101–108.

Thomford NE, Dzobo K, Chopera D, et al. (2015). Pharmacogenomics implications of using herbal medicinal plants on African populations in health transition. *Pharmaceuticals* 8(3): 637–663.

Thomson R, Genovese G, Canon C, et al. (2014). Evolution of the primate trypanolytic factor APOL1. *PNAS* 111(20): E2130–E2139.

Tindana P, Molyneux CS, Bull S, et al. (2014). Ethical issues in the export, storage and reuse of human biological samples in biomedical research: perspectives of key stakeholders in Ghana and Kenya. *BMC Med Ethics* 15(1): 1–11.

Tishkoff SA, Reed FA, and Friedlaender FR (2001). Haplotype diversity and linkage disequilibrium at human G6PD: recent origin of alleles that confer malarial resistance. *Science* 293(5529): 455–462.

Tishkoff SA, Reed FA, Friedlaender FR, et al. (2009). The genetic structure and history of Africans and African Americans. *Science* 324(5930): 1035–1044.

Tostmann A, Boeree MJ, Aarnoutse RE, de Lange WC, van der Ven AJ, and Dekhuijzen R (2008). Antituberculosis drug-induced hepatotoxicity: concise up-to-date review. *J Gastroenterol Hepatol* 23(2): 192–202.

Tu Y (2011). The discovery of artemisinin (qinghaosu) and gifts from Chinese medicine. *Nat Med* 17 (10): 1217–1220.

Tyler-Smith C, Yang H, Landweber LF, et al. (2015). Where next for genetics and genomics? *PLoS Biol* 13(7): 1–7.

Uzureau P, Uzureau S, Lecordier L, et al. (2013). Mechanism of *Trypanosoma brucei gambiense* resistance to human serum. *Nature* 501(7467): 430–434.

Venter JC, Adams MD, Myers EW, et al. (2001). The sequence of the human genome. *Science* 291 (5507): 1304.

Vogel F (1959). Moderne Probleme der Humangenetik. *Ergeb Inn Med Kinderheilkd* 12: 52–125.

de Vries J, Tindana P, Littler KJ, et al. (2015). The H3Africa policy framework: negotiating fairness in genomics. *Trends Genet* 31(3): 117–119.

Walker K, Ginsberg G, Hattis D, et al. (2009). Genetic polymorphism in N-Acetyltransferase (NAT): population distribution of NAT1 and NAT2 activity. *J Toxicol Environ Health B Crit Rev* 12(5–6): 440–472.

Wamala D, Buteme HK, Kirimunda S, et al. (2016). Association between human leukocyte antigen class II and pulmonary tuberculosis due to *Mycobacterium tuberculosis* in Uganda. *BMC Infect Dis* 16: 23.

Wang PY, Xie SY, Hao Q, et al. (2012). NAT2 polymorphisms and susceptibility to anti-tuberculosis drug-induced liver injury: a meta-analysis. *Int J Tuberc Lung Dis* **16**(5): 589–595.

Ward BA, Gorski JC, Jones DR, Hall SD, Flockhart DA, and Desta Z (2003). The cytochrome P450 2B6 (CYP2B6) is the main catalyst of efavirenz primary and secondary metabolism: implication for HIV/AIDS therapy and utility of efavirenz as a substrate marker of CYP2B6 catalytic activity. *J Pharmacol Exp Ther* **306**(1): 287–300.

Wardrop NA, Atkinson PM, Gething PW, et al. (2010). Bayesian geostatistical analysis and prediction of Rhodesian human African trypanosomiasis. *PLoS Negl Trop Dis* **4**(12): 1–10.

Wassmer SC, Taylor TE, Rathod PK, et al. (2015). Investigating the pathogenesis of severe malaria: a multidisciplinary and cross-geographical approach. *Am J Trop Med Hyg* **93**(Suppl. 3): 42–56.

Welburn SC, Fevre EM, Coleman PG, et al. (2001). Sleeping sickness: a tale of two diseases. *Parasitol Today* **17**(1): 19–24.

Welburn SC, Molyneux DH, and Maudlin I (2016). Beyond tsetse: implications for research and control of human African trypanosomiasis epidemics. *Trends Parasitol* **32**(3): 230–241.

Winkler C, An P, and O'Brien SJ (2004). Patterns of ethnic diversity among the genes that influence AIDS. *Hum Mol Genet* **13**(Suppl. 1): R9–R19.

Wonkam A and Mayosi BM (2014). Genomic medicine in Africa: promise, problems and prospects. *Genome Med* **6**(2): 11.

World Health Organization (1998). *Control and Surveillance of African Trypanosomiasis: Report of a WHO Expert Committee.* World Health Organization Technical Report. Available at: www.ncbi.nlm.nih.gov/pubmed/10070249.

World Health Organization (2015a). Global Health Observatory (GHO) data, HIV/AIDS. Available at: www.who.int/gho/hiv/en.

World Health Organization (2015b). Treatment of severe malaria. In *Guidelines for the Treatment of Malaria*, World Health Organization.

World Health Organization (2015c). *World Malaria Report 2015.* World Health Organization. Available at: www.who.int/malaria/publications/world-malaria-report-2015/report/en/.

World Health Organization (2015d). *Global Tuberculosis Report 2015.* New York: World Health Organization. Available at: www.who.int/tb/publications/global_report/en/.

World Health Organization (2015e). Cases of sleeping sickness drop to lowest level in 75 years. Available at: www.who.int/trypanosomiasis_african/cases_drop_to_lowest_since_75_years/en.

World Health Organization (2016). Trypanosomiasis: symptoms, diagnosis and treatment. Available at: www.who.int/trypanosomiasis_african/diagnosis/en/#.WONHZhhKIF4.mendeley&title= Symptoms, diagnosis and treatment (accessed April 4, 2017).

Worodria W, Massinga-Loembe M, Mazakpwe D, et al. (2011). Incidence and predictors of mortality and the effect of tuberculosis immune reconstitution inflammatory syndrome in a cohort of TB/HIV patients commencing antiretroviral therapy. *J Acquir Immune Defic Syndr* **58**(1): 32–37.

Yeung S, Pongtavornpinyo W, Hastings IM, et al. (2004). Antimalarial drug resistance, artemisinin-based combination therapy, and the contribution of modeling to elucidating policy choices. *Am J Trop Med Hyg* **71**(2 Suppl.): 179–186.

Yew WW and Leung CC (2006). Antituberculosis drugs and hepatotoxicity. *Respirology* **11**(6): 699–707.

Yimer G, Amogne W, Habtewold A, et al. (2012). High plasma efavirenz level and CYP2B6*6 are associated with efavirenz-based HAART-induced liver injury in the treatment of naïve HIV patients from Ethiopia: a prospective cohort study. *Pharmacogenomics J* **12**(6): 499–506.

Yuliwulandari R, Sachrowardi Q, Nakajima H, et al. (2010). Association of HLA-A, -B, and -DRB1 with pulmonary tuberculosis in western Javanese Indonesia. *Hum Immunol* **71**(7): 697–701.

Zanger UM, Turpeinen M, Klein K, and Schwab M (2008). Functional pharmacogenetics/genomics of human cytochromes P450 involved in drug biotransformation. *Anal Bioanal Chem* **392**(6): 1093–1108.

Zhang L, Dresser MJ, Gray AT, et al. (1997). Cloning and functional expression of a human liver organic cation transporter. *Mol Pharmacol* **51**(6): 913–921.

7 A Glimpse into Pharmacogenomics in Africa

Collet Dandara and Alice Matimba

7.1 Introduction

7.1.1 Pharmacogenetics and Pharmacogenomics

Interindividual and interpopulation differences may be used to guide and optimize disease treatment based on genetics and this is referred to as pharmacogenetics or pharmacogenomics. *Pharmacogenetics* is a branch of pharmacology that focuses on how single to few genetic markers affect drug efficacy and safety. Pharmacogenetics evaluates the effects of genetic variations on drug absorption, distribution, metabolism, and excretion (ADME), thereby broadly affecting its pharmacokinetic (PK) and pharmacodynamic (PD) properties (Hansen et al. 2009; Kitzmiller et al. 2011; Malik et al. 2010). **Pharmacogenomics**, on the other hand, is the study of how the variation across the whole genome influences response to therapeutic drugs and encompasses genes affecting host susceptibility to disease as well as disease pathways and mechanisms that affect gene expression and function. Both terms may be used interchangeably and will be referred to as pharmacogenomics in this chapter.

Taking advantage of the post-human genome advances in technologies for genome characterization, there has been an exponential increase in genomic data due to high-throughput genome sequencing; thus, more novel variants that affect drug response are being reported at a faster rate despite the low number of pharmacogenomics studies based in Africa (Choudhury et al. 2017; Collins et al. 2016; Gurdasani et al. 2015). The technological enablers include genome-wide association studies (GWAS), candidate specific mass arrays, targeted whole-exome or whole-genome analysis using next-generation sequencing, biobanking, and big data, which are not yet being fully utilized or exploited by African researchers.

7.1.2 Genomic Diversity in African Populations

Africa is the cradle of humankind, with an estimated population of 1.2 billion people. It is the second largest and second most populous continent, comprising 54 sovereign states, the most populous being Nigeria, home to nearly 200 million people, and the least populous being the Seychelles, with fewer than 100,000 people. African people speak a wide range of languages (>1500), which underlie an incredibly rich and

Note: A list of abbreviations appears in the appendix at the end of this chapter.

diverse genetic background (Gurdasani et al. 2015). The genetic diversity observed in African populations translates similarly to the diversity in response to medicines which exists among African populations and when compared with populations of the rest of the world (Masimirembwa and Matimba 2012). In this chapter, the African populations discussed primarily refer to black Africans inhabiting the eastern, southern, and western parts of the continent, commonly referred to as sub-Saharan Africa (SSA).

7.1.3 Goal of Chapter

Despite the known importance of pharmacogenomics in medicine, uptake and implementation have been slow (Moaddeb and Haga 2013), particularly in Africa. The reasons include lack of resources/infrastructure, challenging health care systems, limited capacity for research, and poor knowledge of pharmacogenomics among health professionals (Dandara et al. 2014; Kudzi et al. 2015; Masimirembwa et al. 2014). The goal of this chapter is to describe the role of genetics in pharmacokinetic and pharmacodynamic variation, highlight the relevance of pharmacogenomics in medicine, and its importance in the context of the burden of disease in Africa. The potential role of pharmacogenomics in improving therapy in Africa will be explored, drawing on established examples in its application in selected infectious and non-communicable diseases. Advances in application of pharmacogenomics for understanding gene function, disease mechanisms, and therefore clinical applicability will also be described, highlighting the role of African genomic diversity in pharmacogenomics.

7.2 Role and Applications of Pharmacogenomics in Medicine

7.2.1 Genetic Variation and Drug Response in Disease Management and Drug Development

The most characterized form of genetic variation is the single nucleotide polymorphism (SNP). This consists of substitution of single base pairs in the DNA sequence. Other common forms of genetic or genomic variations are structural variants such as insertions, deletions, inversions, and copy number variants (CNVs) affecting whole genes or genomic regions. These variations impact on gene and protein expression, and protein structure and function due to amino acid changes. As outlined by Chandra (2017), these genetic differences among individuals can affect disease management with regard to (1) the rate of disease occurrence, progression, or recurrence; (2) the drug or drug class most likely to provide benefit; (3) the therapeutic dose; and (4) the nature and extent of beneficial responses to treatment and the likelihood of drug toxicity. With relevance to pharmaceutical drug development, variation in genes encoding proteins involved in drug absorption, distribution, metabolism, excretion (ADME) play a role in the drug's pharmacokinetic variability. Variation in genes encoding for intended or unintended drug targets and other pathways related to the drug's pharmacologic effect may affect pharmacodynamic variability. Other genetic effects include predisposition to toxicities such as immune reactions, and disease progression.

Early studies characterizing genetic variation in drug response reported prolonged paralysis after administration of muscle relaxant succinylcholine due to pseudocholinesterase enzyme deficiency and deficient N-acetylation of isoniazid by N-acetyl transferase 2 (NAT2) (Kalow 1956; Price-Evans et al. 1960; Roden et al. 2011). Subsequently, pharmacogenetics was further demonstrated through ethnic variation observed by the high incidence of hemolytic anemia among African Americans receiving the antimalarial drug primaquine, who were deficient in glucose-6-phosphate dehydrogenase (G6PD) (Beutler et al. 1955). Over the years, pharmacogenetics studies have focused on genes affecting individual variation in the ADME of pharmaceutical drugs. The cytochrome P450s (CYP) are the main enzymes involved in ADME and are highly polymorphic and most studied (see Section 7.3). In the last few decades, small- to large-scale studies using massive parallel sequencing and GWAS have contributed to documentation of genetic profiles that affect specific drug efficacy and toxicity across populations for potential application in drug development (Gurdasani et al. 2015; Johansson and Ingelman-Sundberg 2008; Weinshilboum and Wang 2017). However, fewer such studies have been conducted in Africa, providing a strong case for further African-focused research to unravel the high level of micro diversity and wide variation in the prevalence of functional and clinically relevant alleles.

7.2.2 Pharmacogenomics in Genomic Medicine

Researchers have established associations between patients' genetic make-up, disease progression, disease management, and treatment outcomes. Development of pipelines to understand the underlying mechanisms and translation into clinical practice is underway. More than 60 years after the term pharmacogenetics was coined, there are now over 150 drugs with pharmacogenomic information recommended for drug labeling (www.fda.gov/Drugs/ScienceResearch/ucm572698.htm). There is evidence of clinical utility of pharmacogenomics for many drugs, encompassing all areas of medical disciplines which can be incorporated into electronic health records to tailor or individualize drug therapy (Weinshilboum and Wang 2017). There are over 250 drug–gene pairs with FDA recommendations for optimizing treatment, largely in oncology, psychiatry, infectious diseases, and neurology (Figure 7.1). The Clinical Pharmacogenetics Implementation Consortium (CPIC) has developed pharmacogenetic dosing guidelines for over 30 drugs, which can be adopted into clinical practice and used to guide clinicians in translation of patient-specific genotype information into clinical phenotypes or drug dosing groups (www.pharmgkb.org/guidelines; https://cpicpgx.org; Caudle et al. 2014; Relling and Klein 2011). Therefore, pharmacogenetics appears to be the most successful of human genomics applications in medicine. Examples are presented in the following paragraphs.

Pharmacogenetic screening has been shown to be clinically and economically effective by preventing hypersensitivity reactions to abacavir. Abacavir hypersensitivity is attributed to a genetic variant in the human leucocyte antigen (HLA) gene,

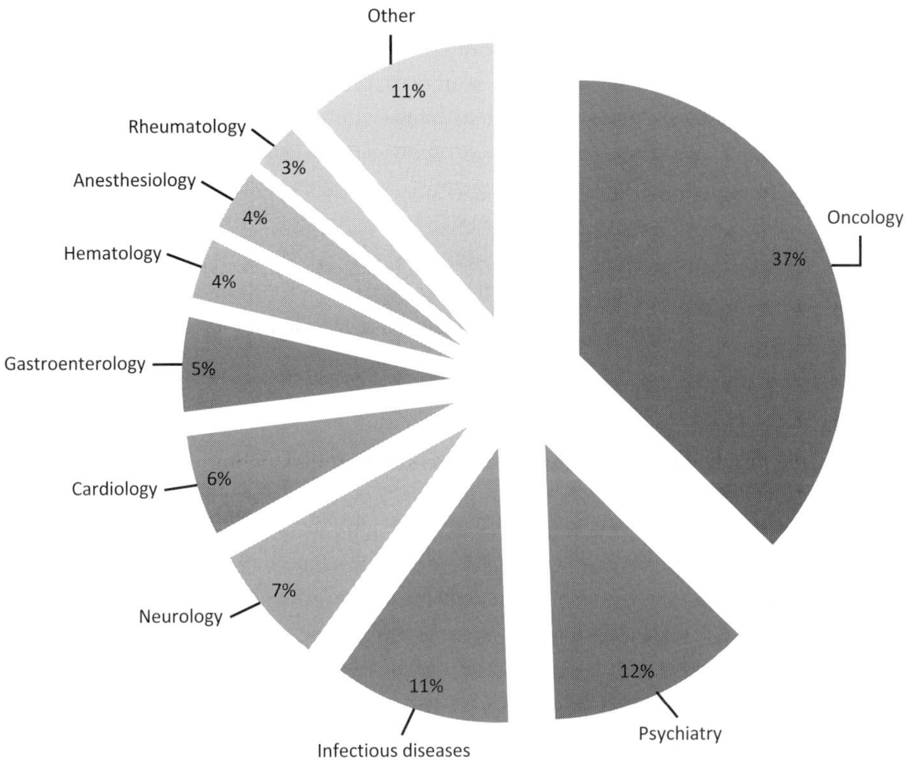

Figure 7.1 Proportion of drug–gene pairs according to disease category with US FDA recommendations for genotype-guided dosing. Based on the FDA list of Pharmacogenomic Biomarkers in Drug Labeling, last updated July 2017.

which is prevalent in Europeans but rare in populations of African ancestry (Ganesh et al. 2013; Guo et al. 2013; Ruiz-Iruela et al. 2016; Small et al. 2017). Similarly, genotype-guided dosing algorithms may improve safety and clinical outcomes in the use of anticoagulants, such as warfarin (Pirmohamed et al. 2013; Shi et al. 2015; Tang et al. 2015). However, more studies are needed in African populations due to both quantitative and qualitative differences in variants reported in genes encoding the main drug-metabolizing enzyme (DME) CYP2C9 and the drug target enzyme vitamin K oxidoreductase (VKORC1) (Deng et al. 2017; Kaye et al. 2017; Perera et al. 2013). Pharmacogenetic testing is also recommended for a range of antipsychotic drugs largely metabolized by CYP2D6 and CYP2C19, whose genes are highly polymorphic, and exhibit population-specific variants.

In oncology, screening for somatic mutations is beneficial in selecting patients who will benefit most from drug therapy targeting specific tumors (Schmidt et al. 2016). For example, mutations in the oncogene *KRAS* were associated with response to cetuximab, a monoclonal antibody used in treatment of colorectal cancer. Companion diagnostic testing is now recommended prior to initiation of cetuximab therapy (Evaluation of Genomic Applications in Practice and Prevention Working Group 2013). In another example, the human epidermal growth factor receptor 2

(HER2) overexpression due to gene duplication is indicated for use of the drug trastuzumab, and targeted treatment, also employing companion diagnostics, has improved survival of patients with breast cancer (Bringolf et al. 2017).

In childhood acute lymphoblastic leukemia (ALL), germline polymorphisms in the thiopurine methyl S-transferase gene (*TPMT*) and other genes in the thiopurine metabolic pathway were associated with dose-limiting hematopoietic toxicities leading to genotype-based dosing being recommended (Maxwell and Cole 2017; Relling et al. 2013; Yang and Bhojwani 2013). The TPMT enzyme is key to metabolism of the thiopurine drugs mercaptopurine and thioguanine. Recently, using discovery GWAS and replication cohorts from clinical trials, various novel loss-of-function variants in *NUDT15*, a gene encoding nucleotide triphosphate diphosphatase, were associated with toxicity in Asian, European, and African ancestry populations, making a strong case for *NUDT15* incorporation into thiopurine dosing recommendations (Maxwell and Cole 2017; Moriyama et al. 2017; Yang et al. 2015). These discoveries support more comprehensive pharmacogenetics-based thiopurine dose adjustments across diverse populations.

For HIV management in Africa, dose adjustments have been recommended for the non-nucleoside reverse transcriptase inhibitor (NNRTI) efavirenz, which works by blocking the function of reverse transcriptase enzyme in HIV treatment. Due to highly prevalent genetic variations in DMEs, primarily CYP2B6, African HIV patients require a lower dose than Caucasians in order to reduce the risk of the drug's adverse effects (Dhoro et al. 2015; Mukonzo et al. 2014).

The above examples of pharmacogenomics applications contribute to the development of precision medicine, which is premised on the integration of several aspects, such as a patient's demographic, clinical, genomic, environmental, and disease factors, to arrive at the best possible intervention to cure or manage. Therefore, pharmacogenetics and pharmacogenomics are at the center of precision medicine, especially aiding in drug discovery and development by pharmaceutical companies, and in the selection of drug dose combinations for patients based on their genetic make up to optimize drug efficacy and avoid adverse drug reactions (ADRs).

7.3 Genetic Variation in ADME Genes: The Basis for Differences in Drug Response

7.3.1 Pharmacokinetics and Pharmacodynamics

The use of therapeutic drugs, including herbal medicines, is the bedrock of modern medical therapy. However, as much as drugs or herbal medicines are able to cure diseases, these molecules need to be removed from the body because their accumulation can lead to toxicity. Every drug entering the body goes through the ADME process, which can be summed up as what the body does to the drug and determines the effective therapeutic outcome (Figure 7.2). Therapeutic drugs are either taken in their active form or in inactive form as pro-drugs. In both cases, the drugs are biotransformed (or metabolized) to enable their excretion or conversion to active forms.

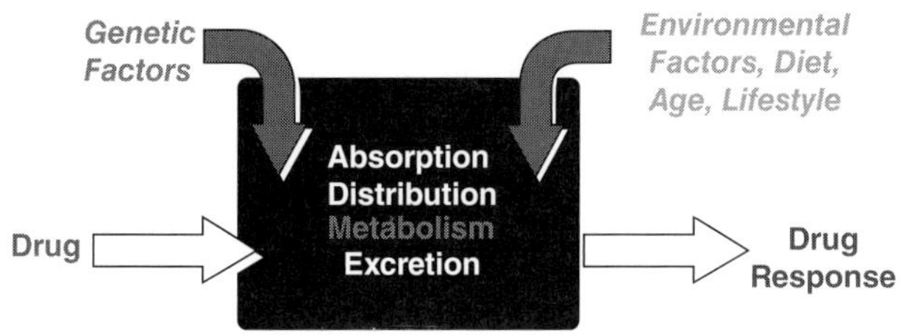

Figure 7.2 Factors affecting variation in drug response. Published with permission without modification from the September issue of *Medical Laboratory Observer* by Chandra (2017).

This process of biotransformation is catalyzed by DMEs, while the distribution of most drugs in the body is catalyzed by transporters. However, in this process of biotransformation and transport, a drug can cause physiological and biochemical changes in the body, which is referred to as pharmacodynamics (PD). Pharmacodynamics determines how well the target cells, such as heart tissue or neurons, respond to a drug. Thus, an intricate balance between the PK and PD is required so that the drug achieves the maximum intended effect, with minimal unintended adverse effects on a patient (Chandra 2017).

Pharmacokinetics and pharmacodynamics are driven by proteins in the form of enzymes, transporters, and drug receptors. Proteins are coded for by genes that can exhibit genetic variation/polymorphisms. This genetic variation affects functions or processes through which drugs act on the body, or are removed from the body. The innate genetic polymorphisms in an individual can thus cause a shift in the balance of PK and PD, resulting in an alteration in the way the body and the drug (or its metabolites) interact with each other. Genes that encode proteins important in drug ADME pathways and contain variants that are significantly correlated with individual variability in therapeutic response may be referred to as "pharmacogenes." Where there is sufficient evidence of clinical importance, these are categorized as *very important pharmacogenes* (VIPs); details can be found in the Pharmacogenomics Knowledge Base (PharmGKB: www.pharmgkb.org).

7.3.2 Genetic Variation in Genes Encoding ADME Proteins

Early studies in pharmacogenetics focused on candidate genes that were known to play a role in well-delineated drug metabolism pathways. Functional effects of genetic variants in DMEs result in phenotype classifications – poor metabolizers (PM), intermediate metabolizers (IM), extensive metabolizers (EM), or ultrarapid metabolizers (UM). The most extensively characterized types of genetic variation for DMEs are SNPs and CNVs. While many genetic variants can result in no effects on the protein, the following effects may result:

1. Abolishing the production of the required protein/enzyme due to introduction of stop codons, frame shifts in the sequence, or gene deletions. These result in null or loss of function alleles.

2. Production of a protein with reduced activity compared to the normal protein due to amino acid changes which may interfere with structure or functional sites. These are referred to as reduced-function alleles.
3. Production of a protein that has higher activity.
4. Production of higher amounts of protein due to duplicated genes (CNVs).

Generally, for a drug that is administered as an active form (ready to act), PMs accumulate higher drug plasma concentrations of the drug when given standard doses, sometimes reaching toxic levels, whereas EMs could reach favorable levels that are therapeutically effective with few or no side-effects, and UMs may fail to attain the required drug plasma concentration levels to be effective (Figure 7.3).

The most studied ADME genes include those coding for DMEs, particularly cytochrome P450s (CYP), which are responsible for the metabolism of at least 70 percent

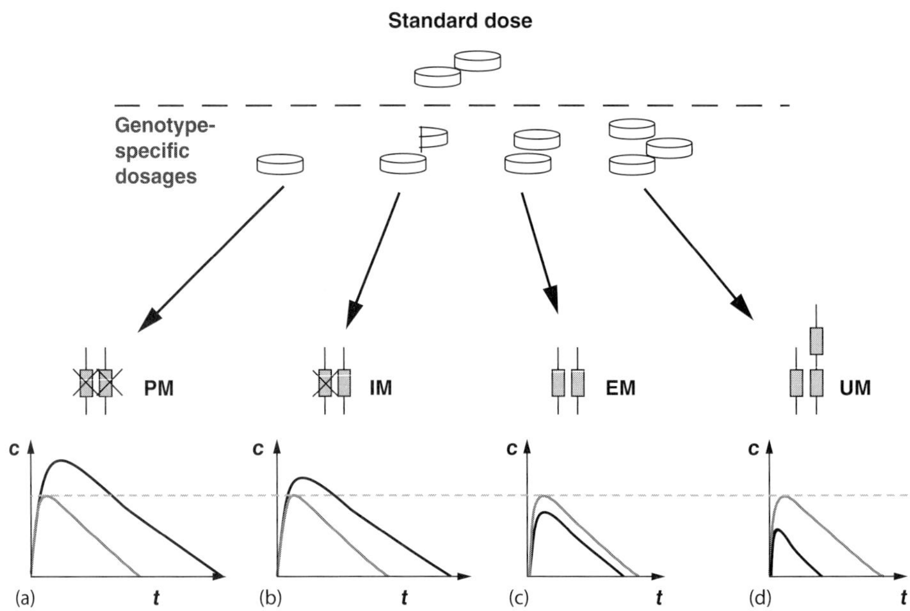

Figure 7.3 Pharmacogenetic effects of genotypes in a DME. The area between the darker and lighter lines for each graph indicates the difference in area under curve for each respective genotype. The gray dashed line indicates upper effective concentration required. (a) Individuals with no capacity to metabolize a drug (PM), thus its accumulation; this could be due to homozygosity or heterozygosity for defective alleles that severely affects or abolishes enzyme activity (null alleles) and may require lower doses. (b) Heterozygotes made up of either active and inactive alleles or different combinations of low-activity alleles that reduce activity with different effect sizes but do not completely abolish activity (IM). (c) Individuals with normal activity (EM), who would be homozygous for active alleles and would require normal doses. (d) Individuals carrying duplicated or increased copy numbers of normal active alleles, which leads to faster metabolism of the drug and could lead to failure to reach therapeutic drug plasma concentration levels (c in the graphs) and may require higher doses. Reprinted from Seeringer A, Kirchheiner J. Pharmacogenetics-guided dose modifications of antidepressants. *Clin Lab Med.* 2008 Dec; 28(4): 619–626 with permission from Elsevier.

of all therapeutic drugs and herbal medicines (Bains et al. 2013; Lynch and Price 2007; Rendic and Guengerich 2010). Over 2000 genetic variants have been reported in the *CYP* genes, including *CYP1A2*, *CYP2B6*, *CYP2C8*, *CYP2C9*, *CYP2C19*, *CYP2D6*, *CYP3A4*, and *CYP3A5*, with significant differences in allele frequencies across world populations (Masimirembwa and Matimba 2012; Preissner et al. 2013).

The CYP enzymes are highly polymorphic and yet are crucial in both the biotransformation of administered drugs and the bioactivation of some drugs that are taken in inactive forms. Thus, null alleles that result in absence of enzyme activity could mean a drug not being activated and/or adverse drug effects due to increased drug plasma concentration levels. These effects become critical for drugs that have a narrow therapeutic index (NTI), whereby small differences are observed between the minimum effective concentrations and the minimum toxic concentrations in the blood. Therapeutic drug monitoring combined with pharmacogenomics could be useful for optimizing treatment with NTI drugs such as the anticoagulant warfarin, asthma drug theophylline, and anti-cancer taxane drugs, all of which are metabolized by highly polymorphic CYPs.

The distribution of genetic variants that results in the different genotypes differs across populations. For example, a genetic variant that results in the genotype shown in Figure 7.3 could be prominent in one population but absent in another, thus, this would present as different effects of the same drug in two populations. Swen and co-workers reported on therapeutic dose recommendations for gene–drug combinations, in which it was recommended that CYP2D6 PMs receive less than 50 percent of standard doses, while CYP2D6 UMs were recommended a doubling of the standard dose for both groups to achieve the same therapeutic effects or outcome (Swen et al. 2011).

Traditionally, pharmacogenetic variation has been limited to DMEs, with special emphasis on CYPs. Successive studies have established the importance of variation in transporters in drug response. The ABC and solute carrier (SLC) transporters are two superfamilies of transport proteins; these are ubiquitous membrane-bound transport proteins with 49 and 360 genes identified, respectively (Ahmed et al. 2016; He et al. 2009; Vasiliou et al. 2009). These transporters play a key role in modulating drug transport. Although inconsistent, polymorphisms in *ABCB1* (MDR) which encodes for P-glycoprotein, an ATP-dependent drug efflux pump, were associated with higher plasma drug levels in use of anti-retroviral (ARV) and anti-cancer drugs and in dose adjustment for cyclosporine in transplant patients (Ankathil 2017; Lee et al. 2015; Swart et al. 2012a). Solute carriers catalyze transport of a range of drugs, and polymorphisms in *SLC22A1* were associated with higher plasma levels of the ARV lamivudine and anti-diabetic drug metformin (Moss et al. 2015; Sundelin et al. 2017). Polymorphisms in *SLCO1B1* were associated with higher plasma drug levels due to decreased transport activities resulting in reduced drug efficacy and adverse effects (Daka et al. 2015; Meyer et al. 2015). However, further studies are needed to determine markers relevant for clinical application in African populations.

7.3.3 Other Genetic Mechanisms

Most studies focus on polymorphisms in the coding region of the gene, the exons. Other polymorphisms may be located in non-coding sites such as introns, as well as in regulatory elements upstream or downstream of the gene. Although SNPs and CNVs are the most common types of polymorphisms, other mechanisms may affect gene and protein expression and impact drug metabolic pathways. For example, epigenetics involves structural modifications such as histone methylation and acetylation, enabling or disabling gene expression due to environmental, chemical, or physiological factors. Other mechanisms include gene-silencing molecules such as small molecular RNA, which bind mRNA molecules thereby interfering with protein expression. Although focus lies on the chromosomal genome sequence, variation in mitochondrial DNA has also been shown to affect drug response. Here we focus on SNP and CNV variants; the other mechanisms will be explored in future work.

7.3.4 Allele Nomenclature

Although a single variant may be referred to as an allele, other variants may be linked on the same sequence. This is referred to as the haplotype. Therefore, an allele may refer to a single variant or a series of linked variants. An allele and haplotype nomenclature system is required for consistent and unambiguous description of sequence variants and is essential to report and exchange information in reference to variants even when described using different reference sequences. Recommendations for sequence variant nomenclature can be found at http://varnomen.hgvs.org, and are described by den Dunnen et al. (2016). Alleles and haplotypes for ADME genes have historically used the *star* (*) nomenclature system. The default or wild type (fully functional allele) is often denoted as **1*, while other variants are denoted **2*, **3*, often in numerical order as more novel variants are discovered (Kalman et al. 2016; Robarge et al. 2007). Various databases are available for allele nomenclature for pharmacogenes, including CYPs (Gaedigk et al. 2017a; www.pharmvar.org/genes); UGT glucuronyltransferases (UGT) (www.pharmacogenomics.pha.ulaval.ca); TPMT (www.imh.liu.se/tpmtalleles?l=en); NAT (http://nat.mbg.duth.gr); and HLA (http://hla.alleles.org/nomenclature/index.html). An allele can be described by the presence of a single genetic variant or by a combination of several linked variants referred to as a haplotype. For example, the CYPC19*3 allele constitutes three SNPs – 636G>A; 991A>G; 1251A>C, while 636A>G defines the haplotype or allele. When an individual carries one copy of an allele, this is referred to as heterozygous; when carrying alleles on both chromosomes this is referred to as homozygous.

In a population, the allele frequency is calculated as the number of incidences of the allele appearing at a locus. An individual carries two alleles per locus, based on the two sets of chromosomes in the human genome. Therefore, a heterozygote individual will carry two versions of an allele, and a population of 100 individuals will therefore mean a total of 200 alleles in that population. A pharmacogenetics polymorphism refers to an allele that occurs at a frequency of at least 1 percent in a population; it is considered a rare variant if the frequency is less than 1 percent. An allele that is polymorphic in one population may be rare in a different population.

7.3.5 Genotype–Phenotype Association

The importance of a genotype is determined by its correlation with biologically or clinically important phenotypes. Phenotype characterization in pharmacogenomics encompasses drug response phenotypes, primarily therapeutic effects and toxicity. However, quantifiable biological phenotypes used include gene expression, protein levels, and enzyme activities. Pharmacological phenotypes include drug plasma levels, bioavailability, and metabolite ratios. Clinical phenotypes are based on treatment outcomes and side-effects. For example, in cancer chemotherapy, phenotypes of interest include hepatotoxicity, remission, relapse, death, and survival. Genetic association studies often involve single or a few markers at a time, while pharmacogenomics genome-wide association studies (GWAS) analyze markers across the genome and their correlation with individual or combinations of phenotypes.

7.4 Significance of Variations in Cytochrome P450 Genes

Cytochrome P450s are the principal enzymes involved in ADME. It is reported that they metabolize nearly 70 percent of therapeutically used drugs. CYPs have broad and overlapping substrate preferences and specificity, undergo inhibition by some drugs, and are inducible (Table 7.1). CYPs are highly polymorphic, resulting in alleles with variable activities whose distributions across geographical populations show both qualitative and quantitative differences. Here we discuss *CYP2D6*, *CYP3A4*, and *CYP3A5*, which are the most polymorphic CYP genes.

7.4.1 CYP2D6 Genetic Variation

CYP2D6 metabolizes some of the most commonly used drugs, many of which have an NTI, including antidepressants and antipsychotics. Earlier studies among European populations characterizing the function of CYP2D6 relied on phenotyping using probe drugs such as debrisoquine and codeine, which are principally metabolized by CYP2D6. Later, with advances in molecular techniques, particularly polymerase chain reaction (PCR), it was possible to isolate the *CYP2D6* gene, sequence it, and identify specific sequence variants (or alleles) that correlate with drug response phenotypes. The *CYP2D6* gene is located on chromosome 22 and is characterized by SNPs and CNVs (www.pharmvar.org/gene/CYP2D6). Currently, there are more than 105 alleles (haplotypes) with respect to the *CYP2D6* gene, with nearly 30 SNPs still waiting for their haplotypes to be assigned a haplotype or (*) allele. The majority of the alleles or SNPs in the *CYP2D6* gene result in either absent or decreased function of the CYP2D6 enzyme. For example, the reduced-function *CYP2D6*2 allele* is based on two variants, g.2850C>T and g.4180G>C, and overall codes for the EM

Table 7.1 Substrates, drugs, and inhibitors for cytochrome P450 enzymes CYP3A4/5, CYP2D6, and CYP2C19

Substrates	Inhibitors and inducers
CYP3A4/5	
Alfentanil, aprepitant, budesonide, buspirone, conivaptan, darifenacin, darunavir, dasatinib, dronedarone, eletriptan, eplerenone, everolimus, felodipine, indinavir, fluticasone, lopinavir, lovastatin, lurasidone, maraviroc, midazolam, nisoldipine, quetiapine, saquinavir, sildenafil, simvastatin, sirolimus, tolvaptan, tipranavir, triazolam, vardenafil, artemether, lumefantrine, mefloquine, primaquine, chroquine Drugs with narrow therapeutic index Alfentanil, astemizole*, cisapride*, cyclosporine, dihydroergotamine, ergotamine, fentanyl, pimozide, quinidine, sirolimus, tacrolimus, terfenadine*	**Inhibitors:** atazanavir, clarithromycin, indinavir, itraconazole, ketoconazole, nefazodone, nelfinavir, ritonavir, saquinavir, telithromycin **Inducers:** rifampin, carbamazepine, avasimibe, phenytoin, St. John's wort *Moderate inducers*: bosentan, efavirenz, etravirine, modafinil, nafcillin *Weak inducers*: amprenavir, aprepitant, armodafinil, echinacea**, pioglitazone, prednisone, rufinamide
CYP2D6	
Atomoxetine, desipramine, dextromethorphan, eliglustat, nebivolol, nortriptyline, perphenazine, tolterodine, venlafaxine, amitriptyline, encainide, imipramine, metoprolol, propafenone, propranolol, tramadol, trimipramine	**Inhibitors:** bupropion, fluoxetine, paroxetine, quinidine, terbinafine, cimetidine, cinacalcet, duloxetine, fluvoxamine, mirabegron **Inducers:** rifampicin
CYP2C19	
Diazepam, lansoprazole, rabeprazole, voriconazole, amitriptyline, clomipramine, imipramine, mephenytoin, omeprazole, pantoprazole, clopidogrel, proguanil, propranolol, limonene, gliclazide (sulfonylurea), citalopram	**Inhibitors:** chloramphenicol, cimetidine, ketoconazole, lansoprazole, omeprazole, oxcarbazepine, pantoprazole **Inducers:** carbamazepine, rifampicin

*Withdrawn from the United States market for safety reasons; **herbal product.
Extracted from Zanger and Schwab, 2013; and the FDA Drug Development and Drug Interactions: Table of Substrates, Inhibitors and Inducers (www.fda.gov/Drugs/DevelopmentApprovalProcess/DevelopmentResources/DrugInteractionsLabeling/ucm093664.htm#table3-3).

phenotype. Null alleles include *CYP2D6*3* due to *g.2549delA* and *CYP2D6*4* due to *1846G>A* functional variants or tag SNPs in their respective haplotypes. *CYP2D6*4* is prevalent in Caucasian populations and rare among Africans. Individuals homozygous for these alleles are likely to be classified as PMs. The Asian-specific variant includes *CYP2D6*10*, while *CYP2D6*17* and **29* are prevalent among black Africans (Bapiro et al. 2002; Masimirembwa et al. 1996; Wennerholm et al. 2001, 2002; Wright et al. 2010). Reduced CYP2D6 resulting in higher prevalence of IM phenotypes in Africans is attributed to these variants with minor allele frequencies of up to 45 percent (Gaedigk et al. 2017b).

Copy number variants in *CYP2D6* include deletions, duplications, or multiplications. Individuals carrying the deletion allele *CYP2D6*5* have reduced capacity to metabolize drugs that are substrates for CYP2D6 (Gaedigk et al. 1991). Individuals carrying a *CYP2D6* duplication *CYP2D6*2XN* are classified as UM phenotypes. This allele is more prevalent among Ethiopians and rare in other populations in the southern and western parts of Africa (Aklillu et al. 2002). Although 1–2 percent are classified as UM phenotypes in the general African population, over 20 percent carry the *CYP2D6*2N* allele in the Ethiopian and northern African populations, suggesting a high number of UM phenotypes and demonstrating the population genetic diversity in Africa. This variability in the distribution of the CYP2D6 UM phenotype can be observed in the disposition of drugs that are substrates of CYP2D6.

Due to the complex polymorphic nature of *CYP2D6*, activity scores have been used to guide phenotype characterization. For example, the **2* variant was assigned an activity score of 2, while **3*, **4*, and **5* activity scores are 0, and **10*, **17*, and **29* were assigned activity scores of 0.5 (Gaedigk et al. 2008). Genotype-based algorithms were developed which are able to accurately predict extreme phenotypes such as PMs and UMs (Twist et al. 2016). At the population level, a recent study used allele frequencies, diplotypes, and phenotype analysis to show that there is considerable variation across world populations, with genotype predicted phenotype frequencies in the range 0.4–5.4 percent, 0.4–11 percent, 1–21 percent, and 67–90 percent for PM, IM, UM, and EM status, respectively (Gaedigk et al. 2017b).

The relevance of CYP2D6 activity differs on a drug-by-drug basis, depending on whether CYP2D6 is involved in the activation or inactivation of a particular drug. The opioid analgesic codeine, used to relieve mild to moderately severe pain, is activated to morphine (the more active metabolite) by CYP2D6. Utility of CYP2D6 pharmacogenomics testing for codeine is described in the CPIC guidelines by Crews et al. (2014). CYP2D6 UMs metabolize codeine to morphine more rapidly even with normal doses, and these individuals may experience the symptoms of morphine overdose, which include extreme sleepiness, confusion, and shallow breathing. Alternative opioids such as hydrocodone or oxycodone, which are less affected by *CYP2D6* genotypes, are recommended in such cases, while considering their limitations (Crews et al. 2014, 2015).

The major classes of drug that are impacted by *CYP2D6* polymorphisms include antidepressants and antipsychotics. Antidepressants that are substrates of CYP2D6 include amitriptyline, clomipramine, desipramine, duloxetine, fluoxetine, fluvoxamine, imipramine, nortriptyline, paroxetine, and venlafaxine. Antipsychotics that are substrates of CYP2D6 include aripiprazole, flupentixol, haloperidol, olanzapine, perazine, perphenazine, pimozide, risperidone, thioridazine, and zuclopenthixol (Kitzmiller et al. 2011). Standard doses of these drugs among patients with CYP2D6 PM status result in toxic drug levels.

Tamoxifen, a selective estrogen receptor modulator (SERM) used in the treatment of both early and advanced estrogen receptor-positive (ER+) breast cancer in pre- and post-menopausal women as well as prevention of breast cancer in women at high risk, is also a substrate of CYP2D6. Tamoxifen is considered a pro-drug, whose efficacy is dependent on active metabolites, especially endoxifen. Patients with reduced CYP2D6 enzymatic activity (i.e., PMs) produce lower endoxifen levels. A strong gene–dose effect for an association between CYP2D6 activity and endoxifen concentrations was reported in various ethnic groups (Helland et al. 2017; Saladores et al. 2015). In turn, endoxifen levels were associated with improved outcomes in women in a retrospective study with pre- and post-menopausal women who had been treated adjuvantly with tamoxifen therapy (Helland et al. 2017). Although some studies have shown that genetic variation in *CYP2D6*, resulting in PM status, has been associated with poor clinical outcome of breast cancer patients treated with tamoxifen (Zembutsu 2015), inclusion of other genetic variants in predicting therapeutic outcomes is required.

Since CYP2D6 PM status varies across populations, such phenotypic diversity could be incorporated in the determination of the effectiveness of tamoxifen and other CYP2D6 substrate drugs. Similarly, some African populations have high prevalence of the UM genotypes. Therefore, steps to determine *CYP2D6* genetic profiles that put individuals at risk of therapeutic failure or adverse effects should be taken and this may include consideration of using alternative drugs that are not CYP2D6 substrates. The CPIC publishes guidelines for *CYP2D6* genotypes and dosing of different drugs as guides for clinicians in the use of these drugs in Caucasians. As there are limited studies conducted in African cohorts, these guidelines, however, do not clearly address variants that are prominent or useful among African populations, such as *CYP2D6*17* and *CYP2D6*29*, possibly due to limited clinically validated evidence in African populations. It is therefore imperative that an African consortium along similar lines to CPIC should also be established to constantly review genetic variants that affect drug responses in African populations. In 2016, an African Pharmacogenomics consortium was proposed, and it is hoped that it will catalyze the curation of the relevant data and come up with similar recommendations (Dandara et al. 2016).

7.4.2 Genetic Variation in *CYP3A4* and *CYP3A5*

The CYP3A family is the most abundant subfamily of CYP isoforms and is essential in the metabolism of many drugs, including endogenous compounds. Of the four isoforms (CYP3A4, 3A5, 3A7, and 3A43), CYP3A4 and 3A5 are the most important. Population variability in CYP3A4 and CYP3A5 activity is high, sometimes >100-fold. The variation in activity or expression levels is attributed to both environmental factors and genetic polymorphisms, primarily affecting gene expression. The most commonly reported SNP CYP3A4 g.-392G>A (*CYP3A4*1B*) affects expression levels of CYP3A4, and the frequency of this allele or variant varies among different populations (2–80 percent). Studies have reported *CYP3A4*1B* allele frequencies of 4 percent in Europeans and up to 82 percent among Africans. Recently, another SNP resulting in an allele

*CYP3A4*22* has been associated with decreased expression and activity (up to five-fold less). However, this variant is common among populations of European origin (2 percent) and rare among African populations (Fohner et al. 2013).

Of the two major CYP3A enzymes, CYP3A5 presents with the greatest variability in protein expression. This variability is attributed to three main alleles coding for CYP3A5*3, CYP3A5*6, and CYP3A5*7 due to 6986A>G, 14690G>A27131_27132insT polymorphisms, respectively. CYP3A5*3/*3 homozygosity is associated with absence of CYP3A5 enzyme activity. Among Europeans, the allele frequencies for CYP3A5*3 range from a minimum of 80 percent and reach 90 percent. Among African populations, the genotype frequency differs from as low as 0 to as high as 53 percent for populations in different regions (Bains et al. 2013). Frequencies in Africa were reported in Ghana (0.04), Senegal (0.09), Malawi (0.05), Nigeria (0.02), and Ethiopia (0.53). Another interesting example is the difference between ethnic groups in the same country – for example, among Cameroonian ethnic groups where the homozygous *CYP3A5*3/*3* genotype varies from 0 to 17 percent (Bains et al. 2013; Dandara et al. 2014).

The distribution of CYP3A4 and CYP3A5 variants is of significant pharmacogenomics relevance for African populations because the enzymes are involved in the metabolism of a wide range of drugs (Table 7.1), some of which are used for the treatment of the most commonly occurring diseases in Africa, HIV infection (efavirenz, saquinavir, maraviroc, indinavir, nelfinavir, ritonavir, and lopinavir), malaria (artemether, lumefantrine, mefloquine, primaquine, and chloroquine), and tuberculosis (rifampicin) (Dooley et al. 2008). Thus, variation in expression or activities of these enzymes is likely to result in variable effectiveness of those drugs.

In patients with comorbidities, co-use of drugs that are substrates of the same enzyme leads to either inhibition of one pathway, which is associated with increased drug plasma concentration and toxicity, or induction, which increases the amount of enzyme available and therefore reduces the drug plasma to levels that are below the therapeutic range and therefore renders the drugs ineffective. Some of the drugs metabolized by CYP3A4/5, such as alfentanil, cyclosporine, quinidine, and tacrolimus, have NTIs (Table 7.1). Indeed, in a study among Ugandans (Byakika-Kibwika et al. 2012), co-administration of artemether/lumefantrine with efavirenz was associated with reduced artemether levels (Figure 7.4). This is a challenge particularly in African populations where, in addition to known therapeutic drugs, the populations concomitantly also use herbal medicines, some of which have inhibitory or inductive effects on these DMEs. The differences in the distribution patterns of genetic variants that affect enzyme activity also indicate that drug–drug interactions do not have the same magnitudes of effect in different population regions (Dooley et al. 2008). Genetic variation in *CYP3A4* and *CYP3A5* is poorly characterized in African populations, thus there is a need for concerted efforts to sequence and characterize African populations and conduct association studies.

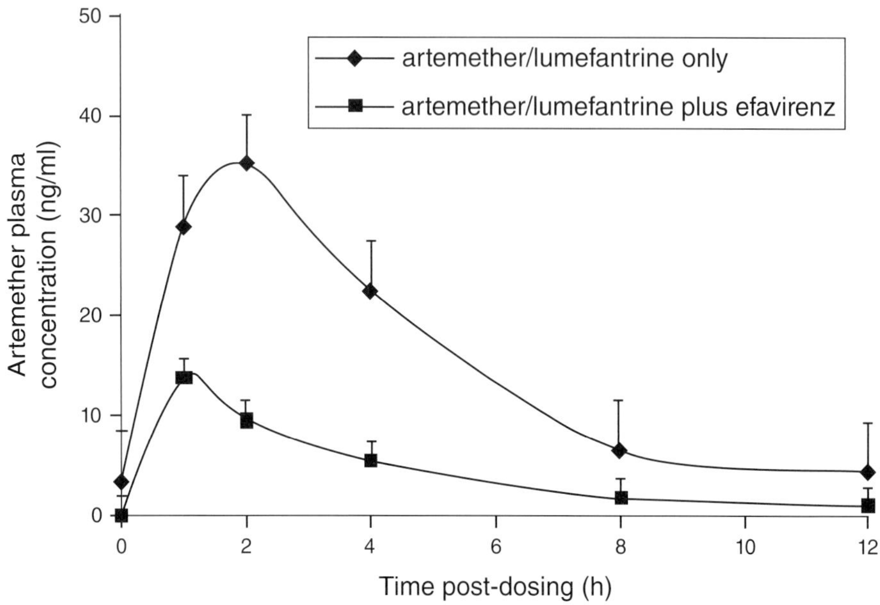

Figure 7.4 Mean plasma concentration versus time of artemether with and without efavirenz. Vertical bars represent standard errors. Adapted from Byakika-Kibwika et al. (2012).

7.5 Pharmacogenomics in Africa

Research to determine variation in drug response across African populations has largely consisted of candidate gene approaches using targeted sequencing and genotyping methods. Here we highlight population pharmacogenomics variation and describe examples of limitations in application of pharmacogenomics based on replicating non-African studies.

7.5.1 Pharmacogenomic Variation in Africa

Initially pharmacogenetics studies in African populations focused on replicating studies done in Caucasian or Asian populations, primarily genotyping of DME gene markers. The most commonly studied genes included the cytochrome P450s (CYP), N-acetyltransferases (NAT), thiopurine S-methyltransferase, urudine glucuronotransferases (UGT) and glutathione S-transferases (GST). Table 7.2 provides an analysis of a selected number of SNPs in genes coding for some of these DMEs in world populations, comparing Africans to non-Africans. A separate analysis using principal component analysis of world populations showed the clustering of the major world populations: Asian, Caucasian, and African (Rajman et al. 2017). Important take-home messages emanating from this include: (1) there is noticeable geographical variation in the distribution of genetic variants in DME genes both quantitatively and qualitatively; (2) compared to other world populations, there are many gaps in African populations in terms of knowing the frequencies of certain

Table 7.2 Allele frequencies (percent) of genetic polymorphisms of DMEs and transporters

	CYP2C9		CYP2C19		CYP2D6					NAT2		GSTM1	CYP2B6		CYP3A5	VKORC1	CYP 3A4	CYP 1A2	ABCB1
	*2	*3	*2	*3	*4	*5	*10	*17	*29	*7	*14	del/del	*c.516T*	*c.983C*	*3	*g.-1639A*	*1B	*g-163A*	*3435T*
Asian	2–9	3.9	30	10	1	6	51	0	0	13	0	55	15–21	0	65–89	66	0.7	66	35
Chinese	0	4.7	37	8	1	6	51	0	0	16	0	58	15	0	66	94.2	0.3	34	35
Japanese	0	2.3	35	11	1	3	43	0	0	10	0	44	18	0	61	90.1	0	–	–
Korean	0	3.6	21	12	2	6	51	0	0	11	0	53	14	0				–	–
Caucasian	15	5.8	15	0	25	5	2	0	0	2	0	50	27	0	96	39.8	3	72	43
Swede	11	6.6	17	0	23	5	1	0	0	2	0	51	24	0	93	30–40	4	–	–
German	14	5.1	18	0	20	2	2	0	0	4	0	51	29	0	93	37	2	–	–
American			14	0	–	–	–	0	–	2	0	54	22	0	96		3	–	–
African	2.2	1.8	16	1	2	4	6	30	15	5	13	30		7.5	31.8	10	>60	>50	10–90
African American	2.8	2	25	0	7	6	4	15	5	2	9	28	38	0	37	11	68	54	84
Tanzanian	0	0	18	1	2	4	4	18	20	3	13	33	10	2	18–26		–	49	–
Zimbabwe	0	0	13	0	2	4	6	34	17	6	14	24	45	9	77.6	44	–	57	–
Black South African	0	0	21	0	3	5	12	24	6	5	11	23	43	7	13.8	0	77	61	89
Ghanaian	0	0	6	–	7	6	3	28	–	–	–	39	49	7.6	11–27	–	78	–	11
Ethiopian	0.4	0.2	14	2	4	3	9	9	–	–	–	–	–		21–67	–	–	–	–
Kikuyu	0	0	16	0	1	–	–	33	14	–	–	28	34	7		–	–	–	–
Luo	0	0	18	0	4	–	6	23	16	3	14	29	37			–	–	–	–
Maasai	0	0	11	0	8	–	5	18	8	4	9	16	35	1		–	–	52	84
Igbo	0	0	29	0	8	–	10	14	20	4	11	23	38		13	–	79	–	–
Yoruba	0	0	10	0	3	–	7	22	10	3	8	31	42	12	16	–	79	57	88
San	0	0	12	–	9	1	5	22	2				40			–	–	–	–

genetic variants; and (3) African populations show greater diversity in the genetic variation for the pharmacogenes. Thus, more efforts should be directed toward genome characterization of African populations. Although not purely focused on pharmacogenomics, research in characterization of African genomes will be useful in further understanding pharmacogenomics variation across ethnic groups in Africa.

Most studies in Africa have focused on determination of common genetic variants and resequencing of some pharmacogenes. However, there are now dedicated efforts toward understanding the role of pharmacogenetic variation in the management of diseases contributing to the high disease burden faced in sub-Saharan Africa, which includes HIV/AIDS, tuberculosis (TB), and malaria (Carr et al. 2017; Dandara et al. 2011; Swart et al. 2013). In addition, drugs used in the management of non-communicable diseases also exhibit interindividual and population variability. There exist a handful of studies on genes affecting the metabolism of drugs used in cancer and cardiovascular and metabolic disease treatments. Advances and opportunities for pharmacogenomics application are discussed in the next section.

7.5.2 Potential Application of Pharmacogenomics in Africa

Infectious Diseases

The most relevant infectious diseases in Africa – HIV, TB, and malaria – contribute to a significant burden in most sub-Saharan African countries. Their management is complicated by various factors, including high prevalence rates and lack of adequate health care and increased susceptibility to non-communicable disease conditions. Pharmacogenomics applications can provide benefits in predicting drug resistance and complications of therapy.

HIV/AIDS

Acquired immune deficiency syndrome (AIDS) is a culmination of diseases caused by the retrovirus human immunodeficiency virus (HIV). HIV attacks the immune system, making an individual more susceptible to other infections and infection-related cancers. HIV/AIDS is one of the major public health problems in sub-Saharan Africa, requiring intensive preventive and treatment programs, and close monitoring to prevent its spread (Kharsany and Karim 2016). Due to high levels of HIV infectivity and severe symptoms ultimately causing death, the race for a cure or preventive vaccine is of high priority. To date, at least 35 drugs have been approved for HIV/AIDS management and are effective in slowing down replication of HIV, easing symptoms and improving quality of life of patients, turning HIV/AIDS into a chronic condition instead of a rapidly terminal illness (Akinboro et al. 2014; Moss 2013; Mutabazi-Mwesigire et al. 2014). These drugs fall into several classes depending on their mechanism of action, including nucleoside reverse transcriptase inhibitors (NRTIs), non-nucleoside reverse transcriptase inhibitors (NNRTIs), protease inhibitors (integrase), and fusion inhibitors (Table 7.3).

Of the 36.7 million people worldwide living with HIV/AIDS at the end of 2016, approximately 60 percent were living in eastern and southern Africa, with South Africa, Mozambique, Kenya, Zambia, Tanzania, Uganda, Zimbabwe, Malawi, and

Table 7.3 List of FDA-approved anti-retroviral drugs and their classification (as of December 2017)

Brand	Generic name	Date of approval	Classification
Combivir	Lamivudine and zidovudine	27 September 1997	NRTI
Emtriva	Emtricitabine, FTC	2 July 2003	NRTI
Epivir	Lamivudine, 3TC	17 November 1995	NRTI
Epzicom	Abacavir and lamivudine	2 August 2004	NRTI
Hivid	Zalcitabine, dideoxycytidine, ddC (no longer marketed)	19 June 1992	NRTI
Retrovir	Zidovudine, azidothymidine, AZT, ZDV	19 March 1987	NRTI
Trizivir	Abacavir, zidovudine, and lamivudine	14 November 2000	NRTI
Truvada	Tenofovir disoproxil fumarate and emtricitabine	2 August 2004	NRTI
Videx EC	Enteric coated didanosine, ddI EC	31 October 2000	NRTI
Videx	Didanosine, dideoxyinosine, ddI	9 October 1991	NRTI
Viread	Tenofovir disoproxil fumarate, TDF	26 October 2001	NRTI
Zerit	Stavudine, d4T	24 June 1994	NRTI
Ziagen	Abacavir sulfate, ABC	17 December 1998	NRTI
Edurant	Rilpivirine	20 May 2011	NNRTI
Intelence	Etravirine	18 January 2008	NNRTI
Rescriptor	Delavirdine, DLV	4 April 1997	NNRTI
Sustiva	Efavirenz, EFV	17 September 1998	NNRTI
Viramune	Nevirapine, NVP	21 June 1996	NNRTI
Viramune XR	Nevirapine, NVP	25 March 2011	NNRTI
Agenerase	Amprenavir, APV (no longer marketed)	15 April 1999	PI
Aptivus	Tipranavir, TPV	22 June 2005	PI
Crixivan	Indinavir, IDV	13 March 1996	PI
Fortovase	Saquinavir (no longer marketed)	7 November 1997	PI
Invirase	Saquinavir mesylate, SQV	6 December 1995	PI
Kaletra	Lopinavir and ritonavir, LPV/RTV	15 September 2000	PI
Lexiva	Fosamprenavir calcium, FOS-APV	20 October 2003	PI
Norvir	Ritonavir, RTV	1 March 1996	PI
Prezista	Darunavir	23 June 2006	PI
Reyataz	Atazanavir sulfate, ATV	20 June 2003	PI
Viracept	Nelfinavir mesylate, NFV	14 March 1997	PI
Fuzeon	Enfuvirtide, T-20	13 March 2003	FI
Selzentry	Maraviroc	6 August 2007	EI
Isentress	Raltegravir	12 October 2007	ISTI
Tivicay	Dolutegravir	13 August 2013	ISTI
Vitekta	Elvitegravir	24 September 2014	ISTI

EI, entry inhibitors; FI, fusion inhibitors; ISTI, integrase strand-transfer inhibitors; NNRTI, non-nucleoside reverse transcriptase inhibitors; NRTI, nucleoside reverse transcriptase inhibitors; PI, protease inhibitors.

Source: www.fda.gov/ForPatients/Illness/HIVAIDS/Treatment/ucm118915.htm.

Ethiopia accounting for the majority of cases (UNAIDS 2017). The UNAIDS 2016–2021 Universal Test and Treat (UTT) Strategy aims to reach the 90–90–90 treatment targets, encompassing (1) 90 percent anti-retroviral therapy coverage for all people with HIV; (2) having 90 percent of all HIV-positive people knowing their HIV status; and (3) having 90 percent of all people receiving anti-retroviral therapy achieving viral suppression. The World Health Organization (WHO) guidelines recommend a combination of two NRTIs and one NNRTI as initial first-line treatment and two NRTI and one PI for patients on second-line treatment therapy (WHO 2016). The idea behind combination anti-retroviral drug therapy is based on the synergistic or additive potential of two or more drugs with independent modes of action and different biochemical targets in order to improve therapeutic efficacy and prevent or delay the development of resistance of the virus to the individual components of the combination (Arts and Hazuda 2012).

Although there is a trend toward fewer infected people, the prevalence of HIV-positive people is still very high. The observed longevity for people living with HIV seems to be fueling the resurgence of new infections, with people no longer treating HIV as a serious issue. The chronic nature of HIV treatment has brought up new problems, especially to do with quality of life for patients. In sub-Saharan Africa, anti-retroviral drug-associated side-effects have been reported, including lipodystrophy, hypersensitivity, neurological disorders, neuropathy and cardiometabolic disorders, hepatotoxicity, hematological toxicity, rash, and renal effects (Abdissa et al. 2012; Angamo et al. 2017; Eluwa et al. 2012; Masenyetse et al. 2015; Mouton et al. 2016; Namme Luma et al. 2012; Ombeni and Kamuhabwa 2016). Some of the ADRs contribute to increasing proportions of patients becoming non-adherent to treatment. Non-adherence to ART is a recipe for development of virus resistance and the need to move patients to second- and third-line ART regimens, which are more expensive and are prone to the same problems as first-line regimens if not properly used.

Various genes have been implicated in variation of ART drug response and ADRs. Of significance is CYP2B6, which is a VIP enzyme in HIV pharmacogenetics because of its metabolism of two of the key drugs, efavirenz and nevirapine. Efavirenz is primarily metabolized by CYP2B6 to 8-hydroxy-EFV (8-OH-EFV), which accounts for >90 percent of its metabolism, while the rest is through other minor pathways (Ward et al. 2003). There is high interindividual variability in plasma EFV levels after drug administration at standard regimens. High EFV plasma concentration is associated with adverse effects such as neuropsychiatric manifestations.

CYP2B6 exhibits genetic polymorphisms, with over 40 alleles characterized to date, and more than a dozen SNPs that still have to be assigned to haplotypes (www.pharmvar.org/gene/CYP2B6). The most common functionally deficient alleles are *CYP2B6*6* [Q172H, K262R], which leads to lower expression in liver due to erroneous splicing, and *CYP2B6*18* [I328T], which does not express functional protein (Klein et al. 2005; Zanger and Klein 2013). Both alleles are highly prevalent in African populations and have been associated with high plasma efavirenz and nevirapine concentrations in African HIV/AIDS patients in Uganda, South Africa, Zimbabwe, Tanzania, Kenya, and Ghana (Mhandire et al. 2015; Mukonzo et al. 2014;

Ngaimisi et al. 2013; Nyakutira et al. 2008; Oluka et al. 2015; Sarfo et al. 2014; Swart et al. 2013). In South Africa, *CYP2B6*6* and **18* were also associated with higher efavirenz concentrations in hair, an alternative tool that could be used to determine long-term exposure in HIV/AIDS patients (Röhrich et al. 2016). High EFV levels are associated with increased incidence of efavirenz-associated neuropsychological toxicity (Gounden et al. 2010; Mukonzo et al. 2013). Therefore African population-based pharmacokinetic models based on *CYP2B6* genotypes may be used to predict efavirenz exposure. In addition, surveillance of central nervous system disorders incorporating *CYP2B6* genotypes and genotype-guided dosing have been proposed (Masimirembwa et al. 2016; Vo and Varghese Gupta 2016).

Although higher drug plasma concentrations are important for virological suppression, the associated side-effects lead to a poor quality of life among patients, and thus patients end up becoming non-adherent (Zanger and Klein 2013). Due to the high prevalence of *CYP2B6*6*, which is up to 50 percent in some African populations, patients present with EFV concentration above the safe level of 4 μg/ml. African and other population pharmacokinetic studies suggest that reducing the dose from the recommended 600 mg/day to as low as 300–400 mg/day would still be within therapeutic range, a strategy that may be beneficial for resource-limited countries in Africa (Masimirembwa et al. 2016; Mukonzo et al. 2014).

For nevirapine, the *CYP2B6*6* and **18* alleles have been associated with NVP bioavailability (Carr et al. 2014; Hicks et al. 2013; Maimbo et al. 2012). *CYP2B6*18* has been associated with nevirapine-induced Stevens–Johnson syndrome (SJS) and toxic epidermal necrolysis symptoms, suggesting that the adverse reactions were a result of increased circulating parent compound or possibly biotransformation of nevirapine. While all these findings suggest the need for genotype-guided dosing in African HIV/AIDS patients, more focused replication studies will be required to determine the clinical and economic relevance of pharmacogenomics in identifying the right doses for patients, thereby reducing non-adherence and viral resistance.

TB Management

Drugs used in first-line TB management include isoniazid (INH), rifampicin, ethambutol, and pyrazinamide. Adverse effects of anti-TB drugs include hepatotoxicity, gastrointestinal intolerance, and peripheral neuropathy. Pharmacogenomics of anti-TB drugs has focused largely on isoniazid and rifampicin, and aims to understand underlying mechanisms which cause these side-effects. However, it is important to note that genetic variation in transporters is an important factor in responses to anti-TB drugs. The solute carrier organic anion transporter family member 1B1 (*SLCO1B1*) presents a polymorphism, rs4149032, which is highly prevalent in South Africans and has been associated with reduced rifampicin concentrations (Chigutsa et al. 2011). This may have implications in development of drug resistance. The *SLCO1B1* polymorphism rs1104581 has been associated with increased bioavailability of rifabutin (an alternative rifamycin) in TB patients with HIV (Hennig et al. 2015). Recently, polymorphisms in SLCO1B1 and DME UGT1A1 were associated with susceptibility to anti-tuberculosis drug-induced hepatotoxicity (ATDH) in a case–control study in China (Sun et al. 2017). Therefore, polymorphisms in SLCO1B1 and other DMEs play an important role in pharmacogenetic variation, and further studies to determine dose adjustment in African populations should be conducted to reduce risk of toxic adverse effects (Chigutsa et al. 2011; Weiner et al. 2010).

Studies conducted in TB patients in African and Asian populations have showed that common polymorphisms in the gene encoding N-acetyltransferase enzyme (*NAT2*) were associated with toxic levels of isoniazid, increasing the risk of hepatotoxicity and suggesting a need for genotype-guided dosing stratification of isoniazid in chemotherapy for tuberculosis (Azuma et al. 2013; Jung et al. 2015; Toure et al. 2016). Polymorphisms in *NAT2* resulting in *NAT2*5*, **6*, **7*, and **14* alleles divide populations into slow and fast acetylators. NAT2 slow acetylators have increased exposure to INH compared to rapid acetylators, resulting in increased risk of INH-induced liver injury due to the slow clearance of acetyl hydrazine, a toxic metabolite of isoniazid in affected patients (Gupta et al. 2013; Wang et al. 2016).

Malaria Management

Malaria causes a significant burden on health systems in sub-Saharan African countries. Antimalarial drugs including artemether, lumefantrine, amodiaquine, primaquine, mefloquine, quinine, and sulfadoxine-pyrimethamine are substrates of CYP enzymes. Genetic polymorphisms in CYP genes *2C8*, *2D6*, *2B6*, and *3A4/3A5* have been implicated in treatment failure. Although the prevalence of these markers varies significantly across the region, the utility of these pharmacogenomics markers is not well established.

A historical example of how genetic variants affect malaria drugs is the glucose-6-phosphatase deficiency (G6PD) and primaquine. Despite concerns about toxicity of primaquine, it is effective in clearing mature gametocytes that persist after conventional treatment, thereby reducing transmission (Bretscher et al. 2017). Individuals with G6PD deficiency, an inherited disorder, are at risk of severe haemolytic anemia when given primaquine, and the effects present in a dose-dependent manner (Shekalaghe et al. 2010; Watson et al. 2017). G6PD deficiency affects 10–20 percent of the population in sub-Saharan Africa, where malaria is endemic (Howes et al. 2012; Von Seidlein et al. 2013). This is attributed to the prevalence of a range of mutations including the *G6PD A–*, *202G>A/376A>G*, and other uncharacterized polymorphisms (Shah et al. 2016).

The diagnosis and management of G6PD deficiency is important for malaria control and elimination, and mass drug administration programs (Von Seidlein et al. 2013). However, G6PD and malaria phenotypes are complex and risk misclassification due to population diversity and limited numbers of genetic markers (usually two) considered in current reports. Therefore, large-scale studies of malaria and G6PD enzymatic function across African populations require the identification and analysis of the full repertoire of G6PD genetic markers in order to provide meaningful prediction of malaria risk phenotypes and pharmacological response (Manjurano et al. 2015).

Identification of parasite genes underlying drug resistance and treatment failure is paramount to successful management of malaria. In uncomplicated malaria, artemisinin-based drug combinations are recommended, although resistance may be emerging (Rasmussen et al. 2017). Genomic-based methods have been used to identify drug resistance markers (Anderson et al. 2011). Monitoring and surveillance using genomic-based screening methods will be useful in understanding the genetic basis of antimalarial drugs, and can inform local treatment policies. Identification of resistance alleles can be used to delineate mechanisms of drug action and provides tools to determine disease evolution and spread (Apinjoh et al. 2017; Conrad et al. 2017).

Using Pharmacogenomics in Tailoring Therapy for Non-Communicable Disease Management in Africa

Although Africa has such a huge burden of infectious diseases (communicable), there is an exponential increase in chronic non-communicable diseases (NCDs) such as cardiovascular disease, diabetes, and cancer, which are projected to reach epidemic levels in the next decades (Nyirenda 2016). Despite advances in pharmacogenomics of therapeutics in NCD management, its applicability remains a challenge. In this section, we will highlight the pharmacogenomics of drugs used in cardiovascular diseases, diabetes, and sickle cell disease.

Pharmacogenomics of Warfarin

Warfarin is the most commonly used anticoagulant and exhibits a narrow therapeutic window. In the first few days of treatment, warfarin causes many adverse drug events (ADEs), with more than 20 percent of patients started on warfarin being hospitalized in the first six months (Dang et al. 2005; Mitchell et al. 2011; Perera et al. 2013). There is significant interindividual variability in warfarin response, and genetics accounts for up to 40 percent of this variance based primarily on polymorphisms in *CYP2C9*, *CYP4F2*, and *VKORC1* genes (Takeuchi et al. 2009). The effect of the main allelic variants *CYP2C9*2*, **3*, and the *VKORC1* promoter polymorphism (*−1639G>A*) were replicated in various Caucasian populations. Homozygosity for either *CYP2C9*2* or **3* or heterozygosity for only non-functional alleles is associated with less than 10 percent of normal CYP2C9 enzyme activity. However, these two variants are rare in Africans (Matimba et al. 2009; Mitchell et al. 2011; Schelleman et al. 2008). The homozygous state of the *VKORC1* polymorphism (genotype AA) is significantly represented in warfarin-sensitive patients, and patients with this genotype have been observed to require 50 percent lower warfarin doses to exert the same effects (Table 7.4). These variants have been incorporated into warfarin dosing algorithms and tested in various populations (Bader and Elewa 2016; Kubo et al. 2017).

Despite potential application of the dosing algorithm in improving guidance of the starting dose of warfarin in Caucasian populations, the pharmacogenetic dosing of warfarin is less applicable in African ancestry populations and the algorithm should be evaluated in each respective ethnic population (Kubo et al. 2017;

Table 7.4 Three ranges of expected maintenance of Coumadin (warfarin) daily doses based on *CYP2C9* and *VKORC1* genotypes

	CYP2C9					
VKORC1	*1/*1	*1/*2	*1/*3	*2/*2	*2/*3	*3/*3
GG	5–7 mg	5–7 mg	3–4 mg	3–4 mg	3–4 mg	0.5–2 mg
AG	5–7 mg	3–4 mg	3–4 mg	3–4 mg	0.5–2 mg	0.5–2 mg
AA	3–4 mg	3–4 mg	0.5–2 mg	0.5–2 mg	0.5–2 mg	0.5–2 mg

Reproduced from warfarin prescribing information insert. Available at: www.accessdata.fda.gov/drugsatfda_docs/label/2016/009218s116lbl.pdf, accessed February 20, 2018.

Limdi et al. 2015; Wadelius 2014). A GWAS study in African Americans discovered a genetic variant in the folate homeostasis gene, folylpolyglutamate synthase (*FPGS*), which was associated with lower warfarin doses in African Americans (Daneshjou et al. 2014). Another variant, rs12777823, located on the *CYP2C* gene cluster was associated with reduced warfarin dose in African Americans and could improve warfarin dose prediction using pharmacogenetic dosing algorithms. Despite these advances in application of pharmacogenomics in warfarin dosing, its utility in Africa is still unclear and more research encompassing other factors such as environment is required. Newly discovered genetic markers may be further analyzed and larger-scale replication studies conducted in African populations.

Pharmacogenomics of Antiplatelet Drugs

Clopidogrel is an antiplatelet agent administered for the treatment of patients undergoing percutaneous coronary interventions and is administered as an inactive prodrug that requires oxidation by the CYP19 system to an active metabolite, 2-oxo-clopidogrel. There is interethnic variation of *CYP2C19* alleles and metabolic phenotypes across world populations (Fricke-Galindo et al. 2016). The most common loss-of-function variant is *CYP2C19*2*, with a frequency of 25–30 percent (European and African descendants) and up to 70 percent of Asians carrying at least one loss-of-function polymorphism. Among Africans, the frequency ranges from 14 to 18 percent (Dandara et al. 2014). The loss of function of CYP2C19 impedes production of the active form of the drug, hence reducing its potency. The differences in the distribution of the loss-of-function variant result in differences in the way patients from these different populations respond to medication that is metabolized by CYP2C19. Thus, these observations have led the FDA to issue a warning on the label of clopidogrel. There are still conflicting reports on the clinical utility of CYP2C19 among patients for clopidogrel dosing, and there is limited evidence of pharmacogenomics testing for clopidogrel in Africa. More clinical studies are required.

Pharmacogenomics of Cholesterol Lowering Drugs

With the increasing incidence in non-communicable diseases (NCDs), statins, a class of lipid-lowering drugs that act by competitively inhibiting hydroxyl-methyl-glutaryl co-enzyme A reductase (HMGCR), have become a common feature of treatment. In addition to their main role, statins present with other pleiotropic effects such as anti-inflammation and anti-oxidative effects (Hu and Tomlinson 2013). There is no doubt that statins have outperformed other lipid-lowering drugs since their release (Gelissen and Brown 2012). However, statins present with some side-effects in up to 10 percent of patients, and these effects include muscle pain and weakness (myalgia or myopathy) (Canestaro et al. 2012; Soko et al. 2016).

Studies in Caucasian and Asian populations have consistently associated interindividual and interethnic variability in rosuvastatin pharmacokinetics with the polymorphisms *SLCO1B1 c.521T>C* (rs4149056 p. Val174Ala) and *ABCG2 c.421C>A* (rs2231142, p. Gln141Lys). However, a study by Soko and co-authors (unpublished, Pharmacogenetics Research Group, University of Cape Town) after screening nearly 1000 black Africans from nine ethnic African populations, reported *SLCO1B1 c.521C* variant frequencies of 0.0 percent (San) and *ABCG2 c.421A* variants ranging between 0.0 percent and 5.0 percent, yet their earlier study had shown that the pharmacokinetic parameters of statins in a black African population also showed variability, as is observed in Caucasian populations where the genetic basis is known. This led to the discovery of novel variants underlying this PK variability in African populations (Soko et al. 2016). These novel variants appeared to be African-specific and were identified in genes not previously thought to be important in statin response. Thus, although Caucasians and Africans present with similar interindividual differences in the pharmacokinetics of statins, it is important to note that the underlying genetic variation is not consistent. This has a bearing on pharmacogenomic testing in terms of what variants are put on testing panels for universal applicability.

Pharmacogenomics in the Management of Diabetes

In sub-Saharan Africa, diabetes and associated complications are posing an increased health and economic burden (Atun et al. 2017; Jaffar and Gill 2017). Diabetes management is complicated by the heterogeneous nature of the disease with respect to genetics, environmental factors, pathophysiology, and variation in therapeutic response to oral-glucose-lowering medicines: metformin, sulfonylureas, meglitinides, and thiazolidinediones. Guidelines for diabetes management are mostly based on studies conducted in other parts of the world and may not be optimal for sub-Saharan African populations due to genetic as well as environmental effects. Although extensive evidence exists for polymorphisms affecting pharmacokinetics and pharmacodynamics of metformin and sulfonylureas, few studies have been conducted in Africa. Candidate gene studies have reported the effect of polymorphisms in DMEs and transporters in efficacy of oral antidiabetic agents. Polymorphisms in drug transporters of the SLC superfamily have been associated with pharmacokinetic variation of metformin affecting drug tolerance and reduction in efficacy (Becker et al. 2009; Dujic et al. 2015).

In Africa, sulfonylurea drugs are prescribed as a second-line treatment and include glibenclamide glipizide, and glimepiride, which are extensively metabolized by the

CYP2C9 enzyme. *CYP2C9*2* and **3* alleles were associated with significantly greater reduction in glycated hemoglobin (HbA1c) levels in Chinese patients treated with sulfonylureas (Zhou et al. 2010). Furthermore, recent studies have indicated the interaction between *CYP2C9* and *POR* genes as important determinants of efficacy and severe adverse effects of sulfonylurea-induced hypoglycemia (Dujic et al. 2018).

GWAS studies have identified and replicated discovery of variants which can predict incident T2DM among individuals of various age ranges and diverse ethnic backgrounds. Some of these genetic variants were also associated with pharmacodynamics variation of sulfonylureas, causing treatment failure and decrease of HbA1c due to polymorphisms in *TCLF2* and *KCNQ1*, respectively (Holstein et al. 2011; Li et al. 2017). Other genes that affect pharmacodynamics of sulfonylureas include Sulfonylurea receptor 1 (SUR1, *ABCC8* gene) and K+ inward rectifier Kir6.2. Despite these advances, major barriers to the application of pharmacogenomics in diabetes care remain as more evidence and generation of actionable results is required (Floyd and Psaty 2016). As some genetic variants are absent or rare or not yet replicated, understanding population and phenotypic diversity in African populations is further warranted.

Pharmacogenomics of Sickle Cell Disease

Approximately 300,000 children are born with sickle cell disease (SCD) every year, and 75 percent are in sub-Saharan Africa (Modell and Darlison 2008). The geographical distribution of the sickle cell trait (carriers) mirrors that of malaria. In Africa, the highest prevalence of sickle cell trait occurs in countries such as Cameroon, Republic of Congo, Gabon, Ghana, Nigeria, and Uganda, at 10–45 percent of the population (Okwi et al. 2010). Sickle cell disease is a common inherited autosomal recessive condition, and is most prevalent in sub-Saharan Africa (Grosse et al. 2011; Makani et al. 2013). It is characterized by abnormality in the hemoglobin molecules which leads to sickle-shaped red blood cells. Complications of SCD include infections, attacks of severe pain, pulmonary hypertension, and increased risk of death. The pharmacological management of SCD includes vaccination, use of preventive antibiotics, analgesics to relieve pain, blood transfusion, and the drug hydroxyurea (HU) (Ndefo et al. 2008).

In Africa, *Plasmodium falciparum*-caused malaria is a significant cause of mortality in SCD patients (McAuley et al. 2010; Makani et al. 2010), and malaria infection has been reported to trigger cell crises in SCD patients. Therefore, malaria chemoprophylaxis is recommended for SCD patients living in high-risk areas of malaria. Chemoprophylaxis is complemented by blood transfusion (Oniyangi and Omari 2006). The most commonly used drugs for malaria treatment include chloroquine (Lewthwaite 1962), proguanil and pyrimethamine (Eke and Anochie 2003), and sulfadoxine-pyrimethamine (especially during the high-transmission season). Each of these drugs is associated with reduction in malaria prevalence but they have different effects among SCD patients (Aneni et al. 2013; Diop et al. 2011). Thus, management of malaria is a very complicated issue in populations that have SCD patients and sickle trait carriers.

Hydroxyurea is the only drug currently approved by the US FDA for the treatment of SCD and is effective in increasing total and fetal hemoglobin (HbF) and consequent reduction in the incidence of acute pain episodes, acute chest syndrome, and hospitalization (Charache et al. 1995). Despite its proven efficacy, HU-induced HbF shows wide interindividual variability; for example, it has been reported that HU upregulates endothelial surface expression of ICAM-1, a major receptor for *P. falciparum*-infected erythrocytes in the brain and which in some cases is thought to enhance parasite replication (Pino et al. 2006). Factors influencing HU response include age and baseline HbF, as well as genetics; children generally have higher HbF levels and better response to HU than adults (Ware et al. 2002).

Baseline HbF and HU-induced HbF are both inherited traits. HbF expression occurs through complex interactions at the beta-globin gene locus. HbF genes are genetically regulated, and the level of HbF and its distribution among sickle erythrocytes is highly variable. Therefore, it is imperative that more studies, especially of a pharmacogenomics nature, be carried out to evaluate genomic markers of optimal responses for these drugs, especially among African populations. Although several gene variants have been reported to be associated with HbF levels, only three genetic loci have been validated as strongly associated with higher HbF in SCD, primarily *Xmn1-HBG2* SNP, *HBS1L-MYB* intergenic region on chromosome 6q, and *BCL11A*, which contribute 20–50 percent variance in Caucasians (Lettre et al. 2008; Thein and Menzel 2009). In Africa, most studies are reported from Cameroon and Tanzania, confirming associations of SNPs in *BCL11A* and *HBS1L-MYB* and HbF levels with hospitalization (Mtatiro et al. 2014, 2015; Wonkam et al. 2014).

The underlying genetic effects of baseline HbF on SCD treatment outcomes means that the same markers can be applied in pharmacogenetic/genomic analysis (Green and Barral 2011). Pharmacogenetic analysis in children with SCA showed SNPs influencing baseline HbF, including five within *BCL11A* (Ware et al. 2011). A whole-exon sequencing genotype–phenotype association study identified a variant located in the *SALL2* gene which was associated with HU response. The *P840R* variant in the *SALL2* gene, rs61743453, was associated with a higher change in HbF in response to HU in both the discovery and validation cohorts. There are no studies to date that have applied pharmacogenomics of HU in African populations. However, the studies focusing on HbF levels and disease severity may provide a basis for conducting pharmacogenomics (Mtatiro et al. 2014, 2015; Wonkam et al. 2014).

Emerging evidence suggests that SNPs in genes encoding drug metabolism and transporters impact pharmacokinetics and pharmacodynamics of opioid drugs used for alleviating pain reported or observed among SCD patients. Single nucleotide polymorphisms in the *CYP2D6* gene may affect the pathway of codeine conversion to morphine (Brousseau et al. 2007; Shord et al. 2009; Yee et al. 2013). Morphine is frequently used to treat severe cases associated with SCD. Specifically, genes involving the opioid (e.g., *UGT2B7*, *OPRM1*, and *ABCB1* genes) and non-opioid (e.g., *COMT* gene) system can alter the efficacy of morphine (Darbari et al. 2008a, 2008b; Joly et al. 2012). Although it may be early days for pharmacogenomics in SCD management, follow-up of identified variants in other genes such as *OPRM1*, *HMOX-1*, *GCH1*, *VEGFA*, and *STAT6* could aid understanding of individual variability in the use of anti-inflammatory and opioid drugs in SCD treatment (Wonkam 2017). Due to the wide interindividual variability in the management of SCD, a multidisciplinary approach to research and larger-scale cohorts could provide invaluable resources for understanding disease and drug response mechanisms.

7.6 Role of African Genetic Diversity in Pharmacogenomics

There is growing evidence on the use of genetics in personalized medicine. Understanding underlying genetic mechanisms and their roles in clinical outcomes can provide guidance in pharmacological management and could provide a basis for tailoring individual therapy. In order to ascertain these effects, further studies are required in African populations. The diversity of the African genome has often been viewed as too complex to study. However, this diversity should be viewed as an opportunity to discover novel variants that can lead to further understanding of gene function and disease pathways and further support for drug discovery and development (Masimirembwa and Matimba 2012). We acknowledge the efforts in the development of pharmacogenomic resources and tools that are available for exploitation by African researchers for further development of pharmacogenomics. Although it may not be possible to apply individualized genotyping, findings can be useful for targeting therapies at the population level.

7.6.1 Understanding Gene Function and Disease Pathways

Although implementation of pharmacogenomics has been slow, there is an increase in generation of data and supporting evidence due to technological advances. It is now possible to sequence whole exomes or genomes using next-generation sequencing strategies, broadening pharmacogenetics to pharmacogenomics. Genome-wide association studies have played an important role in discovery of novel non-ADMET (where T is toxicology) gene candidates, highlighting interactions of drug response pathways and disease mechanisms in drug response. The challenge is that some of the genomic targets identified to be associated with response sometimes do not belong to specific pathways known to affect drug disposition. It is thus, in some cases, possible to identify targets significantly associated with drug response, yet their function involves host defenses, or other disease pathways.

African genomic diversity is an important precursor in functional pharmacogenomics toward understanding the relationships of specific variants and genes and pathways, with drug response using clinical, cell line, and animal models. Genomic signals of pharmacological relevance may be better elucidated in a more diverse background than in a homogeneous sample set, as in Caucasians. Pharmacogenomics in African populations supports probing the widest genetic diversity, which improves the chances of identifying novel and relevant mechanisms of disease and for targeting therapies to subpopulations (Matimba et al. 2016; Mpye et al. 2016).

7.6.2 Developing Targeted Therapeutics for African Populations

Variability in genes important for drug response has great implications for the safe and efficacious use of drugs and necessitates the inclusion of this genetic information into the discovery and development of new chemical entities for treatment of disease (Masimirembwa et al. 2014). Most, if not all, of the drugs in clinical settings were designed based on European (i.e., Caucasian) and Asian populations. Studies often extrapolate data from African American cohorts to describe and predict African populations, which in most cases is misleading due to the diversity of the African gene pool compared to that of African Americans. Additionally, differences in expression and distribution of genes have been reported among Africans in various regions of the continent (Dandara et al. 2004, 2014; Swart et al. 2012a, 2012b). As there is growing interest in conducting clinical trials in Africa in the future, pharmacogenomics knowledge from African populations can also inform and guide the design of clinical studies, which can result in population-/region-specific product labeling. However, one should not forget that, among Africans, the definition of drug does not end with conventional medicine, but also includes herbal medicine, which is used by a substantial number of people (Thomford et al. 2015).

Due to the high level of heterogeneity, a thorough characterization of the African genome is paramount as it is unsafe to extrapolate results from one African ancestral group to another (the evolutionary history of infectious diseases has been applied for targeted drug development for African populations) (Baker et al. 2017). For example, due to the high prevalence of G6PD deficiency as a result of genetic polymorphisms in malaria-endemic areas, some drugs such as primaquine, despite their effectiveness, require genetic testing and safety monitoring in these populations, and use of alternative drugs is preferable. Other examples include the need for population-specific dose adjustments for the anti-retroviral drug efavirenz and the anti-TB drug isoniazid, as described above.

7.6.3 Pharmacogenomics Databases and Other Resources for Research and Clinical Application

One of the major challenges in genomics research and its utilization is the mass accumulation of data. The application of bioinformatics to integrate and interrogate these data has made a sterling contribution. In order to standardize and make use of the same nomenclature, several databases are available that are regularly updated and curated by consortia of expertise in particular domains. For example, CPIC maintains databases and resources including recommendations for clinical use of pharmacogenomics knowledge. Following is a list of some of the relevant databases for pharmacogenomics data comparison:

- *PharmGKB:* www.pharmgkb.org; a comprehensive resource that curates knowledge about the impact of genetic variation on drug response for clinicians and researchers. PharmGKB coined the term *VIP genes* and curates variants with evidence of the effect of polymorphisms in gene expression, drug activity, and association with clinical phenotypes (Thorn et al. 2013).
- *CPIC:* www.pharmgkb.org/page/cpic; http://cpicpgx.ord; an international consortium of individual volunteers and dedicated staff who are interested in facilitating use of pharmacogenetic tests for patient care.

- *Electronic PharmacoGenomic Assistance* (ePGA): www.epga.gr; linked to PharmGKB, the ePGA provides personalized genotype-to-phenotype translation, linked to clinical guidelines, and can be used by clinicians.
- *DrugBank:* www.drugbank.ca; a unique bioinformatics and cheminformatics resource that combines detailed drug data with comprehensive drug target information.
- *PACdb:* www.pacdb.org; a pharmacogenetics cell line database for use as a central repository of pharmacology-related phenotypes that integrates genotypic, gene expression, and pharmacological data obtained from lymphoblastoid cell lines.
- *Human cytochrome P450 allele nomenclature:* www.pharmVar.org (formerly www.cypalleles.ki.se); a central repository for pharmacogene variation to facilitate allele (haplotype) designation and the interpretation of pharmacogenetic test results to guide precision.
- *CYP drug interaction table:* http://medicine.iupui.edu/clinpharm/ddis; a resource for CYP450 drug interactions for clinically relevant substrates.
- *FDA pharmacogenetics website:* www.fda.gov/drugs/science-research-drugs/table-pharmacogenomic-biomarkers-drug-labeling; tables of pharmacogenomic biomarkers in drug labeling.

Although extensive datasets exist from pharmacogenomics studies which include African Americans, the contribution of European ancestry resulting in admixture means that these data cannot be representative of African ancestry populations (Li et al. 2014). However, such data resources can be used as a basis for replication studies in African populations and the development of new hypotheses in disease and drug response mechanisms. Despite the existence of these well-curated databases, the pharmacogenetic data of African populations remains rudimentary and unexplored. In attempts to develop pharmacogenomic resources, more rigorous approaches are required, including population-wide and clinical cohort-based studies supported by biobanks and informatics experts (Masimirembwa and Matimba 2012; Matimba et al. 2008). Existing consortia such as the Human Heredity and Health in Africa and the African Pharmacogenomics Consortium will contribute to building capacity for pharmacogenomics and genomic medicine from the data resources being developed and the extensive training programs.

7.7 Conclusion

In this chapter an overview of pharmacogenomics was provided and the current status of knowledge and existing gaps in Africa were explored. Focusing on variation in genes coding for DMEs and transporters, the role of population genetic diversity in drug response was highlighted. Potential application of pharmacogenomics in Africa was discussed, giving examples of pharmacogene–drug combinations for commonly occurring diseases relevant in the region. However, we did not exhaustively examine every gene that has been shown to be of pharmacogenomic significance, and the element of traditional herbal medicine will be pursued in other publications. As pharmacogenomics knowledge remains sparse, there is a clear need for more large-scale studies in Africa. Understanding the genomics of Africans is a big step toward understanding the genetic determinants of disease susceptibility and drug response for all world populations due to the "African-origin" of all modern human beings. African genomes are likely to be a mirror to all other genomes of populations in the world. Thus, drug discovery pipelines making use of genetics should include African genomes to mitigate against drugs later presenting with side-effects or inefficacy in African populations. In conclusion, pharmacogenetics information on African populations is patchy. A lot of work still needs to be done to characterize African pharmacogenomic variation, especially taking advantage of the current advances in technologies and existence of continent-wide genomics consortia, as unraveling the pharmacogenomics of Africans is likely to be beneficial not only to Africans, but to most populations of the world.

Appendix

ABCB1:	ATP-binding cassette subfamily B member 1
ABCC8:	ATP-binding cassette transporter subfamily C member 8
ADE:	adverse drug event
ADME:	absorption, distribution, metabolism, excretion
ALL:	acute lymphocytic leukemia
ADR:	adverse drug reaction
ATDH:	anti-tuberculosis drug-induced hepatotoxicity
ART:	antiretroviral therapy
BCL11A:	B-cell lymphoma/leukemia 11A
CNV:	copy number variant
CPIC:	Clinical Pharmacogenetics Implementation Consortium
CYP:	cytochrome P450
EFV:	efavirenz
EM:	extensive metabolizer
FI:	fusion inhibitor
FDA:	Food and Drug Administration, USA
G6PD:	glucose-6-phosphate dehydrogenase
GCH1:	GTP cyclohydrolase 1
GSTM1:	glutathione S-transferase
GWAS:	genome-wise association
HbF:	fetal hemoglobin
HER2:	epdemral growth factor receptor 2
HIV:	human immunodeficiency virus
HLA:	human leucocyte antigen
HMOX1:	heme oxygenase 1
HU:	hydroxyurea
IM:	intermediate metabolizer
INH:	isoniazid
ISTI:	integrase strand-transfer inhibitor
KCNQ1:	potassium voltage-gated channel subfamily Q member 1

MDR: multi-drug resistant BC: ATP-binding cassette transporters
NAT2: N-acetyltransferase 2
NCD: non-communicable disease
NNRTI: non-nucleoside reverse transcriptase inhibitor
NRTI: nucleoside reverse transcriptase inhibitor
NUDT15: nudix hydrolase 15
OPRM1: opioid receptor mu 1
PCR: polymerase chain reaction
PD: pharmacodynamics
PI: protease inhibitor
PK: pharmacokinetics
PM: poor metabolizer
SALL2: sal-like protein 2
SERM: selective estrogen receptor modulator
SCD: sickle cell disease
SLC: solute carrier transporter
STAT6: signal transducer and activator of transcription 6)
SSA: sub-Saharan Africa
SNP: single nucleotide polymorphism
TB: tuberculosis
TCF7L2: transcription factor 7-like 2
TPMT: thiopurine methyltransferase or thiopurine S-methyltransferase
UGT: UDP-glucuronosyltransferaze
UM: ultrarapid metaboliser
UNAIDS: United Nations Programme on HIV/AIDS
UTT: Universal Test and Treat
VKORC1: vitamin K oxidoreductase complex subunit 1
VEGFA1: vascular endothelial growth factor A

References

Abdissa SG, Fekade D, Feleke Y, Seboxa T, and Diro E (2012). Adverse drug reactions associated with antiretroviral treatment among adult Ethiopian patients in a tertiary hospital. *Ethiop Med J* **50** (2): 107–113.

Ahmed S, Zhou Z, Zhou J, and Chen SQ (2016). Pharmacogenomics of drug metabolizing enzymes and transporters: relevance to precision medicine. *Genomics Proteomics Bioinformatics* **14**(5): 298–313.

Akinboro AO, Akinyemi SO, Olaitan PB, et al. (2014). Quality of life of Nigerians living with human immunodeficiency virus. *Pan Afr Med J* **18**: 234.

Aklillu E, Herrlin K, Gustafsson LL, Bertilsson L, and Ingelman-Sundberg M (2002). Evidence for environmental influence on CYP2D6-catalysed debrisoquine hydroxylation as demonstrated by phenotyping and genotyping of Ethiopians living in Ethiopia or in Sweden. *Pharmacogenetics* **12** (5): 375–383.

Anderson T, Nkhoma S, Ecker A, and Fidock D (2011). How can we identify parasite genes that underlie antimalarial drug resistance? *Pharmacogenomics* **12**(1): 59–85.

Aneni EC, Hamer DH, and Gill CJ (2013). Systematic review of current and emerging strategies for reducing morbidity from malaria in sickle cell disease. *Trop Med Int Health* **18**: 313–327.

Angamo MT, Curtain CM, Chalmers L, Yilma D, and Bereznicki L (2017). Predictors of adverse drug reaction-related hospitalization in Southwest Ethiopia: a prospective cross-sectional study. *PLoS One* 12(10): e0186631.

Ankathil R (2017). ABCB1 genetic variants in leukemias: current insights into treatment outcomes. *Pharmgenomics Pers Med* 10: 169–181.

Apinjoh TO, Mugri RN, Miotto O, et al. (2017). Molecular markers for artemisinin and partner drug resistance in natural *Plasmodium falciparum* populations following increased insecticide treated net coverage along the slope of mount Cameroon: cross-sectional study. *Infect Dis Poverty* 6(1): 136.

Arts EJ and Hazuda DJ (2012). HIV-1 antiretroviral drug therapy. *Cold Spring Harb Perspect Med* 2: a007161.

Atun R, Davies JI, Gale EAM, et al. (2017). Diabetes in sub-Saharan Africa: from clinical care to health policy. *Lancet Diabetes Endocrinol* 5(8): 622–667.

Azuma J, Ohno M, Kubota R, et al. (2013). Pharmacogenetics-based tuberculosis therapy research group. NAT2 genotype guided regimen reduces isoniazid induced liver injury and early treatment failure in the 6-month four-drug standard treatment of tuberculosis: a randomized controlled trial for pharmacogenetics-based therapy. *Eur J Clin Pharmacol* 69(5): 1091–1101.

Bader LA and Elewa H (2016). The impact of genetic and non-genetic factors on warfarin dose prediction in MENA region: a systematic review. *PLoS One* 11(12): e0168732.

Bains RK, Kovacevic M, Plaster CA, et al. (2013). Molecular diversity and population structure at the Cytochrome P450 3A5 gene in Africa. *BMC Genet* 14: 34.

Baker JL, Shriner D, Bentley AR, and Rotimi CN (2017). Pharmacogenomic implications of the evolutionary history of infectious diseases in Africa. *Pharmacogenomics J* 17(2): 112–120.

Bapiro TE, Hasler JA, Ridderstrom M, and Masimirembwa CM (2002). The molecular and enzyme kinetic basis for the diminished activity of the cytochrome P450 2D6.17 (CYP2D6.17) variant: potential implications for CYP2D6 phenotyping studies and the clinical use of CYP2D6 substrate drugs in some African populations. *Biochem Pharmacol* 64: 1387–1398.

Becker ML, Visser LE, Van Schaik RH, et al. (2009). Genetic variation in the multidrug and toxin extrusion 1 transporter protein influences the glucose-lowering effect of metformin in patients with diabetes: a preliminary study. *Diabetes* 58: 745–749.

Beutler E, Dern RJ, and Alving AS (1955). The hemolytic effect of primaquine: VI. An in vitro test for sensitivity of erythrocytes to primaquine. *J Lab Clin Med* 45: 40–50.

Bretscher MT, Griffin JT, Ghani AC, and Okell LC (2017). Modelling the benefits of long-acting or transmission-blocking drugs for reducing *Plasmodium falciparum* transmission by case management or by mass treatment. *Malar J* 16(1): 341.

Bringolf L, Pestalozzi B, Fink D, and Dedes K (2017). Exploring prognostic factors for HER2-positive metastatic breast cancer: a retrospective cohort study in a major Swiss hospital. *Swiss Med Wkly* 146: w14393.

Brousseau DC, Mccarver DG, Drendel AL, Divakaran K, and Panepinto JA (2007). The effect of CYP2D6 polymorphisms on the response to pain treatment for pediatric sickle cell pain crisis. *J Pediatr* 150: 623–626.

Byakika-Kibwika P, Lamorde M, Mayito J, et al. (2012). Significant pharmacokinetic interactions between artemether/lumefantrine and efavirenz or nevirapine in HIV-infected Ugandan adults. *J Antimicrob Chemother* 67: 2213–2221.

Canestaro WJ, Brooks DG, Chaplin D, et al. (2012). Statin pharmacogenomics: opportunities to improve patient outcomes and healthcare costs with genetic testing. *J Pers Med* 2: 158–174.

Carr DF, Chaponda M, Cornejo Castro EM, et al. (2014). CYP2B6 c.983T>C polymorphism is associated with nevirapine hypersensitivity in Malawian and Ugandan HIV populations. *J Antimicrob Chemother* 69: 3329–3334.

Carr DF, Bourgeois S, Chaponda M, et al. (2017). Genome-wide association study of nevirapine hypersensitivity in a sub-Saharan African HIV-infected population. *J Antimicrob Chemother* 72 (4): 1152–1162.

Caudle KE, Klein TE, Hoffman JM, et al. (2014). Incorporation of pharmacogenomics into routine clinical practice: the Clinical Pharmacogenetics Implementation Consortium (CPIC) guideline development process. *Curr Drug Metab* **15**(2): 209–217.

Chandra R (2017). The role of pharmacogenomics in precision medicine. *Med Lab Obs* **49**(9): 8–14.

Charache S, Terrin ML, Moore RD, et al. (1995). Effect of hydroxyurea on the frequency of painful crises in sickle cell anemia: investigators of the Multicenter Study of Hydroxyurea in Sickle Cell Anemia. *N Engl J Med* **332**: 1317–1322.

Chigutsa E, Visser ME, Swart EC, et al. (2011). The *SLCO1B1* rs4149032 polymorphism is highly prevalent in South Africans and is associated with reduced rifampin concentrations: dosing implications. *Antimicrob Agents Chemother* **55**(9): 4122–4127.

Choudhury A, Ramsay M, Hazelhurst S, et al. (2017). Whole-genome sequencing for an enhanced understanding of genetic variation among South Africans. *Nat Commun* **8**(1): 2062.

Collins SL, Carr DF, Pirmohamed M (2016). Advances in the pharmacogenomics of adverse drug reactions. *Drug Saf* **39**(1): 15–27.

Conrad MD, Mota D, Musiime A, et al. (2017). Comparative prevalence of *Plasmodium falciparum* resistance-associated genetic polymorphisms in parasites infecting humans and mosquitoes in Uganda. *Am J Trop Med Hyg* **97**(5): 1576–1580.

Crews KR, Gaedigk A, Dunnenberger HM, et al. (2014). Clinical Pharmacogenetics Implementation Consortium guidelines for cytochrome P450 2D6 genotype and codeine therapy: 2014 update. *Clin Pharmacol Ther* **95**(4): 376–382.

Crews KR, Caudle KE, Dunnenberger HM, Sadhasivam S, and Skaar TC (2015). Considerations for the utility of the CPIC guideline for CYP2D6 genotype and codeine therapy. *Clin Chem* **61**(**5**): 775–776.

Daka A, Dimovski A, Kapedanovska A, et al. (2015). Effects of single nucleotide polymorphisms and haplotypes of the *SLCO1B1* gene on the pharmacokinetic profile of atorvastatin in healthy Macedonian volunteers. *Pharmazie* **70**(7): 480–488.

Dandara C, Basvi PT, Bapiro TE, Sayi J, and Hasler JA (2004). Frequency of -163 C>A and 63 C>G single nucleotide polymorphism of cytochrome P450 1A2 in two African populations. *Clin Chem Lab Med* **42**: 939–941.

Dandara C, Lombard Z, Du Plooy I, et al. (2011). Genetic variants in *CYP* (-1A2, -2C9, -2C19, -3A4 and -3A5), *VKORC1* and *ABCB1* genes in a black South African population: a window into diversity. *Pharmacogenomics* **12**(12): 1663–1670.

Dandara C, Swart M, Mpeta B, Wonkam A, and Masimirembwa C (2014). Cytochrome P450 pharmacogenetics in African populations: implications for public health. *Expert Opin Drug Metab Toxicol* **10**: 769–785.

Dandara C, Suarez-Kurtz G, and Aklillu E (2016). Conference report: first Pharmacogenetics and Precision Medicine Conference in Africa (April 7–9, 2016, Cape Town, South Africa). *OMICS* **20**(9): 496–497.

Daneshjou R, Gamazon ER, Burkley B, et al. (2014). Genetic variant in folate homeostasis is associated with lower warfarin dose in African Americans. *Blood* **124**(14): 2298–2305.

Dang MT, Hambleton J, and Kayser SR (2005). The influence of ethnicity on warfarin dosage requirement. *Ann Pharmacother* **39**: 1008–1012.

Darbari DS, Minniti CP, Rana S, and Van Den Anker J (2008a). Pharmacogenetics of morphine: potential implications in sickle cell disease. *Am J Hematol* **83**: 233–236.

Darbari DS, Van Schaik RH, Capparelli EV, et al. (2008b). UGT2B7 promoter variant -840G>A contributes to the variability in hepatic clearance of morphine in patients with sickle cell disease. *Am J Hematol* **83**: 200–202.

den Dunnen JT, Dalgleish R, Maglott DR, et al. (2016). HGVS recommendations for the description of sequence variants: 2016 update. *Hum Mutat* **37**(6): 564–569.

Deng J, Vozmediano V, Rodriguez M, Cavallari LH, and Schmidt S (2017). Genotype-guided dosing of warfarin through modeling and simulation. *Eur J Pharm Sci* **109S**: S9–S14.

Dhoro M, Zvada S, Ngara B, et al. (2015). CYP2B6*6, CYP2B6*18, body weight and sex are predictors of efavirenz pharmacokinetics and treatment response: population pharmacokinetic modeling in an HIV/AIDS and TB cohort in Zimbabwe. *BMC Pharmacol Toxicol* **27**(16): 4.

Diop S, Soudre F, Seck M, et al. (2011). Sickle-cell disease and malaria: evaluation of seasonal intermittent preventive treatment with sulfadoxine-pyrimethamine in Senegalese patients: a randomized placebo-controlled trial. *Ann Hematol* **90**: 23–27.

Dooley KE, Flexner C, and Andrade AS (2008). Drug interactions involving combination antiretroviral therapy and other anti-infective agents: repercussions for resource-limited countries. *J Infect Dis* **198**: 948–961.

Dujic T, Zhou K, Donnelly LA, et al. (2015). Association of organic cation transporter 1 with intolerance to metformin in type 2 diabetes: a GoDARTS study. *Diabetes* **64**: 1786–1793.

Dujic T, Zhou K, Donnelly LA, et al. (2018). Interaction between variants in the *CYP2C9* and *POR* genes and the risk of sulfonylurea-induced hypoglycaemia: a GoDARTS study. *Diabetes Obes Metab* **20**(1): 211–214.

Eke FU and Anochie I (2003). Effects of pyrimethamine versus proguanil in malarial chemoprophylaxis in children with sickle cell disease: a randomized, placebo-controlled, open-label study. *Curr Ther Res Clin Exp* **64**: 616–625.

Eluwa GI, Badru T, Agu KA, et al. (2012). Adverse drug reactions to antiretroviral therapy (ARVs): incidence, type and risk factors in Nigeria. *BMC Clin Pharmacol* **12**: 7.

Evaluation of Genomic Applications in Practice and Prevention (EGAPP) Working Group (2013). Recommendations from the EGAPP Working Group: can testing of tumor tissue for mutations in EGFR pathway downstream effector genes in patients with metastatic colorectal cancer improve health outcomes by guiding decisions regarding anti-EGFR therapy? *Genet Med* **15**(7): 517–527.

Floyd JS and Psaty BM (2016). The application of genomics in diabetes: barriers to discovery and implementation. *Diabetes Care* **39**(11): 1858–1869.

Fohner A, Muzquiz LI, Austin MA, et al. (2013). Pharmacogenetics in American Indian populations: analysis of *CYP2D6*, *CYP3A4*, *CYP3A5*, and *CYP2C9* in the Confederated Salish and Kootenai Tribes. *Pharmacogenet Genomics* **23**: 403–414.

Fricke-Galindo I, Céspedes-Garro C, Rodrigues-Soares F, et al. (2016). Interethnic variation of *CYP2C19* alleles, "predicted" phenotypes and "measured" metabolic phenotypes across world populations. *Pharmacogenomics J* **16**(2): 113–123.

Gaedigk A, Blum M, Gaedigk R, Eichelbaum M, and Meyer UA (1991). Deletion of the entire cytochrome P450 *CYP2D6* gene as a cause of impaired drug metabolism in poor metabolizers of the debrisoquine/sparteine polymorphism. *Am J Hum Genet* **48**(5): 943–950.

Gaedigk A, Simon SD, Pearce RE, et al. (2008). The CYP2D6 activity score: translating genotype information into a qualitative measure of phenotype. *Clin Pharmacol Ther* **83**: 234–242.

Gaedigk A, Ingelman-Sundberg M, Miller NA, et al. (2017a). The Pharmacogene Variation (PharmVar) Consortium: incorporation of the human cytochrome P450 (CYP) allele nomenclature database. *Clin Pharmacol Ther.* DOI: 10.1002/cpt.910.

Gaedigk A, Sangkuhl K, Whirl-Carrillo M, Klein T, and Leeder JS (2017b). Prediction of CYP2D6 phenotype from genotype across world populations. *Genet Med* **19**(1): 69–76.

Ganesh SK, Arnett DK, Assimes TL, et al. (2013). Genetics and genomics for the prevention and treatment of cardiovascular disease: update – a scientific statement from the American Heart Association. *Circulation* **128**: 2813–2851.

Gelissen IC and Brown AJ (2012). Predicting response to statins by pharmacogenetic testing. *Pharmacogenomics* **13**: 1223–1225.

Gounden V, Van Niekerk C, Snyman T, and George JA (2010). Presence of the CYP2B6 516G> T polymorphism, increased plasma efavirenz concentrations and early neuropsychiatric side effects in South African HIV-infected patients. *AIDS Res Ther* **7**: 32.

Green NS and Barral S (2011). Genetic modifiers of HbF and response to hydroxyurea in sickle cell disease. *Pediatr Blood Cancer* **56**: 177–181.

Grosse SD, Odame I, Atrash HK, et al. (2011). Sickle cell disease in Africa: a neglected cause of early childhood mortality. *Am J Prev Med* **41**: S398–S405.

Guo Y, Shi L, Hong H, et al. (2013). Studies on abacavir-induced hypersensitivity reaction: a successful example of translation of pharmacogenetics to personalized medicine. *Sci China Life Sci* **56**(2): 119–124.

Gupta VH, Amarapurkar DN, Singh M, et al. (2013). Association of N-acetyltransferase 2 and cytochrome P450 2E1 gene polymorphisms with antituberculosis drug-induced hepatotoxicity in Western India. *J Gastroenterol Hepatol* **28**: 1368–1374.

Gurdasani D, Carstensen T, Tekola-Ayele F, et al. (2015). The African Genome Variation Project shapes medical genetics in Africa. *Nature* **517**(7534): 327–332.

Hansen NT, Brunak S, and Altman RB (2009). Generating genome-scale candidate gene lists for pharmacogenomics. *Clin Pharmacol Ther* **86**: 183–189.

He L, Vasiliou K and Nebert DW (2009). Analysis and update of the human solute carrier (SLC) gene superfamily. *Hum Genomics* **3**(2): 195–206.

Helland T, Henne N, Bifulco E, et al. (2017). Serum concentrations of active tamoxifen metabolites predict long-term survival in adjuvantly treated breast cancer patients. *Breast Cancer Res* **19**(1): 125.

Hennig S, Naiker S, Reddy T, et al. (2015). Effect of SLCO1B1 polymorphisms on rifabutin pharmacokinetics in African HIV-infected patients with tuberculosis. *Antimicrob Agents Chemother* **60**(1): 617–620.

Hicks JK, Swen JJ, Thorn CF, et al. (2013). Clinical Pharmacogenetics Implementation Consortium guideline for CYP2D6 and CYP2C19 genotypes and dosing of tricyclic antidepressants. *Clin Pharmacol Ther* **93**: 402–408.

Holstein A, Hahn M, Korner A, Stumvoll M, and Kovacs P (2011). TCF7L2 and therapeutic response to sulfonylureas in patients with type 2 diabetes. *BMC Med Genet* **12**: 30.

Howes RE, Piel FB, Patil AP, et al. (2012). G6PD deficiency prevalence and estimates of affected populations in malaria endemic countries: a geostatistical model-based map. *PLoS Med* **9**(11): e1001339.

Hu M and Tomlinson B (2013). Pharmacogenomics of lipid-lowering therapies. *Pharmacogenomics* **14**: 981–995.

Jaffar S and Gill G (2017). The crisis of diabetes in sub-Saharan Africa. *Lancet Diabetes Endocrinol* **5**(8): 574–575.

Johansson I and Ingelman-Sundberg M (2008). Current research in drug metabolism, drug transport and drug development. *Drug News Perspect* **21**(9): 518–528.

Joly P, Gagnieu MC, Bardel C, et al. (2012). Genotypic screening of the main opiate-related polymorphisms in a cohort of 139 sickle cell disease patients. *Am J Hematol* **87**: 534–536.

Jung JA, Kim TE, Lee H, et al. (2015). A proposal for an individualized pharmacogenetic-guided isoniazid dosage regimen for patients with tuberculosis. *Drug Des Devel Ther* **9**: 5433–5438. DOI: 10.2147/DDDT.S87131.

Kalman LV, Agúndez J, Appell ML, et al. (2016). Pharmacogenetic allele nomenclature: international workgroup recommendations for test result reporting. *Clin Pharmacol Ther* **99**(2): 172–185. DOI: 10.1002/cpt.280.

Kalow W (1956). Familial incidence of low pseudocholinesterase level. *Lancet* **268**: 576–577.

Kaye JB, Schultz LE, Steiner HE, et al. (2017). Warfarin pharmacogenomics in diverse populations. *Pharmacotherapy* **37**(9): 1150–1163.

Kharsany AB and Karim QA (2016). HIV infection and AIDS in sub-Saharan Africa: current status, challenges and opportunities. *Open AIDS J* **10**: 34–48.

Kitzmiller JP, Groen DK, Phelps MA, and Sadee W (2011). Pharmacogenomic testing: relevance in medical practice – why drugs work in some patients but not in others. *Cleve Clin J Med* **78**: 243–257.

Klein K, Lang T, Saussele T, et al. (2005). Genetic variability of CYP2B6 in populations of African and Asian origin: allele frequencies, novel functional variants, and possible implications for anti-HIV therapy with efavirenz. *Pharmacogenet Genomics* **15**(12): 861–873.

Kubo K, Ohara M, Tachikawa M, et al. (2017). Population differences in S-warfarin pharmacokinetics among African Americans, Asians and whites: their influence on pharmacogenetic dosing algorithms. *Pharmacogenomics J* **17**(6): 494–500.

Kudzi W, Addy BS, and Dzudzor B (2015). Knowledge of pharmacogenetics among healthcare professionals and faculty members of health training institutions in Ghana. *Ghana Med J* **49**(1): 50–56.

Lee J, Wang R, Yang Y, et al. (2015). The effect of ABCB1 C3435T polymorphism on cyclosporine dose requirements in kidney transplant recipients: a meta-analysis. *Basic Clin Pharmacol Toxicol* **117**(2): 117–125.

Lettre G, Sankaran VG, Bezerra MA, et al. (2008). DNA polymorphisms at the BCL11A, HBS1L-MYB, and beta-globin loci associate with fetal hemoglobin levels and pain crises in sickle cell disease. *PNAS* **105**: 11869–11874.

Lewthwaite CJ (1962). A trial of chemoprophylaxis in sickle cell anaemia: preliminary communication. *East Afr Med J* **39**: 196–199.

Li J, Lao X, Zhang C, et al. (2014). Increased genetic diversity of ADME genes in African Americans compared with their putative ancestral source populations and implications for pharmacogenomics. *BMC Genet* **15**: 52.

Li Q, Tang TT, Jiang F, et al. (2017). Polymorphisms of the KCNQ1 gene are associated with the therapeutic responses of sulfonylureas in Chinese patients with type 2 diabetes. *Acta Pharmacol Sin* **38**(1): 80–89.

Limdi NA, Brown TM, Yan Q, et al. (2015). Race influences warfarin dose changes associated with genetic factors. *Blood* **126**(4): 539–545.

Lynch T and Price A (2007). The effect of cytochrome P450 metabolism on drug response, interactions, and adverse effects. *Am Fam Physician* **76**(3): 391–396.

Maimbo M, Kiyotani K, Mushiroda T, Masimirembwa C, and Nakamura Y (2012). CYP2B6 genotype is a strong predictor of systemic exposure to efavirenz in HIV-infected Zimbabweans. *Eur J Clin Pharmacol* **68**: 267–271.

Makani J, Komba AN, Cox SE, et al. (2010). Malaria in patients with sickle cell anemia: burden, risk factors, and outcome at the outpatient clinic and during hospitalization. *Blood* **115**: 215–220.

Makani J, Ofori-Acquah SF, Nnodu O, Wonkam A, and Ohene-Frempong K (2013). Sickle cell disease: new opportunities and challenges in Africa. *Sci World J* **2013**: 193252.

Malik VS, Popkin BM, Bray GA, Despres JP, and Hu FB (2010). Sugar-sweetened beverages, obesity, type 2 diabetes mellitus, and cardiovascular disease risk. *Circulation* **121**: 1356–1364.

Manjurano A, Sepulveda N, Nadjm B, et al. (2015). African glucose-6-phosphate dehydrogenase alleles associated with protection from severe malaria in heterozygous females in Tanzania. *PLoS Genet* **11**(2): e1004960.

Masenyetse LJ, Manda SO, and Mwambi HG (2015). An assessment of adverse drug reactions among HIV positive patients receiving antiretroviral treatment in South Africa. *AIDS Res Ther* **12**: 6.

Masimirembwa C and Matimba A (2012). Pharmacogenomics in Africa: diversity as an opportunity for personalized healthcare in genomics applications for the developing world. In Nelson KE and Jones-Nelson B, eds., *Genomics Applications for the Developing World.* Springer, pp. 161–181.

Masimirembwa C, Persson I, Bertilsson L, Hasler J, and Ingelman-Sundberg M (1996). A novel mutant variant of the *CYP2D6* gene (*CYP2D6*17*) common in a black African population: association with diminished debrisoquine hydroxylase activity. *Br J Clin Pharmacol* **42**: 713–719.

Masimirembwa C, Dandara C, and Hasler J (2014). Population diversity and pharmacogenomics in Africa. In Padmanabhan S., ed., *Handbook of Pharmacogenomics and Stratified Medicine.* Academic Press, pp. 161–182.

Masimirembwa C, Dandara C, and Leutscher PD (2016). Rolling out efavirenz for HIV precision medicine in Africa: are we ready for pharmacovigilance and tackling neuropsychiatric adverse effects? *OMICS* **20**(10): 575–580.

Matimba A, Oluka MN, Ebeshi BU, et al. (2008). Establishment of a biobank and pharmacogenetics database of African populations. *Eur J Hum Genet* **16**(7): 780–783.

Matimba A, Del-Favero J, Van Broeckhoven C, and Masimirembwa C (2009). Novel variants of major drug-metabolising enzyme genes in diverse African populations and their predicted functional effects. *Hum Genomics* **3**(2): 169–190.

Matimba A, Dhoro M, and Dandara C (2016). Is there a role of pharmacogenomics in Africa? *Glob Health Epidemiol Genom* **1**. DOI: 10.1017/gheg.2016.4.

Maxwell RR and Cole PD (2017). Pharmacogenetic predictors of treatment-related toxicity among children with acute lymphoblastic leukemia. *Curr Hematol Malig Rep* 12(3): 176–186.

McAuley CF, Webb C, Makani J, et al. (2010). High mortality from *Plasmodium falciparum* malaria in children living with sickle cell anemia on the coast of Kenya. *Blood* 116: 1663–1668.

Meyer ZU, Schwabedissen HE, Albers M, et al. (2015). Function-impairing polymorphisms of the hepatic uptake transporter SLCO1B1 modify the therapeutic efficacy of statins in a population-based cohort. *Pharmacogenet Genomics* 25(1): 8–18.

Mhandire D, Lacerda M, Castel S, et al. (2015). Effects of *CYP2B6* and *CYP1A2* genetic variation on nevirapine plasma concentration and pharmacodynamics as measured by CD4 cell count in Zimbabwean HIV-infected patients. *OMICS* 19(9): 553–562.

Mitchell C, Gregersen N, and Krause A (2011). Novel *CYP2C9* and *VKORC1* gene variants associated with warfarin dosage variability in the South African black population. *Pharmacogenomics* 12: 953–963.

Moaddeb J and Haga SB (2013). Pharmacogenetic testing: current evidence of clinical utility. *Ther Adv Drug Saf* 4(4): 155–169.

Modell B and Darlison M (2008). Global epidemiology of haemoglobin disorders and derived service indicators. *Bull World Health Organ* 86: 480–487.

Moriyama T, Yang YL, Nishii R, et al. (2017). Novel variants in NUDT15 and thiopurine intolerance in children with acute lymphoblastic leukemia from diverse ancestry. *Blood* 130(10): 1209–1212.

Moss DM, Liptrott NJ, Siccardi M, and Owen A (2015). Interactions of antiretroviral drugs with the SLC22A1 (OCT1) drug transporter. *Front Pharmacol* 6: 78.

Moss JA (2013). HIV/AIDS review. *Radiol Technol* 84(3): 247–267.

Mouton JP, Njuguna C, Kramer N, et al. (2016). Adverse drug reactions causing admission to medical wards: a cross-sectional survey at 4 hospitals in South Africa. *Medicine (Baltimore)* 95(19): e3437.

Mpye K, Matimba A, Dzobo K, et al. (2016). Disease burden and the role of pharmacogenomics in African populations. *Glob Health Epidemiol Genom* 2. DOI: 10.1017/gheg.2016.21.

Mtatiro SN, Singh T, Rooks H, et al. (2014). Genome wide association study of fetal hemoglobin in sickle cell anemia in Tanzania. *PLoS One* 9: e111464.

Mtatiro SN, Mgaya J, Singh T, et al. (2015). Genetic association of fetal-hemoglobin levels in individuals with sickle cell disease in Tanzania maps to conserved regulatory elements within the MYB core enhancer. *BMC Med Genet* 16: 4.

Mukonzo JK, Okwera A, Nakasujja N, et al. (2013). Influence of efavirenz pharmacokinetics and pharmacogenetics on neuropsychological disorders in Ugandan HIV-positive patients with or without tuberculosis: a prospective cohort study. *BMC Infect Dis* 13: 261.

Mukonzo JK, Owen JS, Ogwal-Okeng J, et al. (2014). Pharmacogenetic-based efavirenz dose modification: suggestions for an African population and the different CYP2B6 genotypes. *PLoS One* 9(1): e86919.

Mutabazi-Mwesigire D, Seeley J, Martin F, and Katamba A (2014). Perceptions of quality of life among Ugandan patients living with HIV: a qualitative study. *BMC Public Health* 10(14): 343.

Namme Luma H, Doualla MS, Choukem SP, et al. (2012). Adverse drug reactions of highly active antiretroviral therapy (HAART) in HIV infected patients at the General Hospital, Douala, Cameroon: a cross sectional study. *Pan Afr Med J* 12: 87.

Ndefo UA, Maxwell AE, Nguyen H, and Chiobi TL (2008). Pharmacological management of sickle cell disease. *Pharm Ther* 33: 238–243.

Ngaimisi E, Habtewold A, Minzi O, et al. (2013). Importance of ethnicity, *CYP2B6* and *ABCB1* genotype for efavirenz pharmacokinetics and treatment outcomes: a parallel-group prospective cohort study in two sub-Saharan Africa populations. *PLoS One* 8: e67946.

Nyakutira C, Roshammar D, Chigutsa E, et al. (2008). High prevalence of the CYP2B6 516G->T(*6) variant and effect on the population pharmacokinetics of efavirenz in HIV/AIDS outpatients in Zimbabwe. *Eur J Clin Pharmacol* 64: 357–365.

Nyirenda MJ (2016). Non-communicable diseases in sub-Saharan Africa: understanding the drivers of the epidemic to inform intervention strategies. *Int Health* 8(3): 157–158.

Okwi AL, Byarugaba W, Ndugwa CM, et al. (2010). An up-date on the prevalence of sickle cell trait in Eastern and Western Uganda. *BMC Blood Disord* 10: 5.

Oluka MN, Okalebo FA, Guantai AN, Mcclelland RS, and Graham SM (2015). Cytochrome P450 2B6 genetic variants are associated with plasma nevirapine levels and clinical response in HIV-1 infected Kenyan women: a prospective cohort study. *AIDS Res Ther* 12: 10.
Ombeni W and Kamuhabwa AR (2016). Lipid profile in HIV-infected patients using first-line antiretroviral drugs. *J Int Assoc Provid AIDS Care* 15(2): 164–171.
Oniyangi O and Omari AA (2006). Malaria chemoprophylaxis in sickle cell disease. *Cochrane Database Syst Rev* 4: Cd003489.
Perera MA, Cavallari LH, Limdi NA, et al. (2013). Genetic variants associated with warfarin dose in African-American individuals: a genome-wide association study. *Lancet* 382(9894): 790–796.
Pino P, Taoufiq Z, Brun M, et al. (2006). Effects of hydroxyurea on malaria, parasite growth and adhesion in experimental models. *Parasite Immunol* 28: 675–680.
Pirmohamed M, Burnside G, Eriksson N, et al. (2013). A randomized trial of genotype-guided dosing of warfarin. *N Engl J Med* 369(24): 2294–2303.
Preissner SC, Hoffmann MF, Preissner R, et al. (2013). Polymorphic cytochrome P450 enzymes (CYPs) and their role in personalized therapy. *PLoS One* 8(12): e82562.
Price-Evans DA, Manley FA, and Mckusick VA (1960). Genetic control of isoniazid metabolism in man. *Br Med J* 2: 485–491.
Rajman I, Knapp L, Morgan T, and Masimirembwa C (2017). African genetic diversity: implications for Cytochrome P450 mediated drug metabolism and drug development. *EBioMedicine* 17: 67–74.
Rasmussen SA, Ceja FG, Conrad MD, et al. (2017). Changing antimalarial drug sensitivities in Uganda. *Antimicrob Agents Chemother* 61(12): e01516-17.
Relling MV and Klein TE (2011). CPIC: Clinical Pharmacogenetics Implementation Consortium of the Pharmacogenomics Research Network. *Clin Pharmacol Ther* 89(3): 464–467.
Relling MV, Gardner EE, Sandborn WJ, et al. (2013). Clinical Pharmacogenetics Implementation Consortium Guidelines for thiopurine methyltransferase genotype and thiopurine dosing: 2013 update. *Clin Pharmacol Ther.* 93: 324–325.
Rendic S and Guengerich FP (2010). Update information on drug metabolism systems: 2009, part II – summary of information on the effects of diseases and environmental factors on human cytochrome P450 (CYP) enzymes and transporters. *Curr Drug Metab* 11(1): 4–84.
Robarge JD, Li L, Desta Z, Nguyen A, and Flockhart DA (2007). The star-allele nomenclature: retooling for translational genomics. *Clin Pharmacol Ther* 82(3): 244–248.
Roden DM, Wilke RA, Kroemer HK, and Stein CM (2011). Pharmacogenomics: the genetics of variable drug responses. *Circulation* 123(15): 1661–1670.
Röhrich CR, Drögemöller BI, Ikediobi O, et al. (2016). *CYP2B6*6* and *CYP2B6*18* predict long-term efavirenz: exposure measured in hair samples in HIV-positive South African women. *AIDS Res Hum Retroviruses* 32(6): 529–538.
Ruiz-Iruela C, Padullés-Zamora N, Podzamczer-Palter D, et al. (2016). HLA-B*57:01 genotyping in the prevention of hypersensitivity to abacavir: 5 years of experience. *Pharmacogenet Genomics* 26(8): 390–396.
Saladores P, Mürdter T, Eccles D, et al. (2015). Tamoxifen metabolism predicts drug concentrations and outcome in premenopausal patients with early breast cancer. *Pharmacogenomics J* 15(1): 84–94.
Sarfo FS, Zhang Y, Egan D, et al. (2014). Pharmacogenetic associations with plasma efavirenz concentrations and clinical correlates in a retrospective cohort of Ghanaian HIV-infected patients. *J Antimicrob Chemother* 69(2): 491–499.
Schelleman H, Chen J, Chen Z, et al. (2008). Dosing algorithms to predict warfarin maintenance dose in Caucasians and African Americans. *Clin Pharmacol Ther* 84: 332–339.
Schmidt KT, Chau CH, Price DK, and Figg WD (2016). Precision oncology medicine: the clinical relevance of patient-specific biomarkers used to optimize cancer treatment. *J Clin Pharmacol* 56 (12): 1484–1499.
Shah SS, Rockett KA, Jallow M, et al. (2016). Heterogeneous alleles comprising G6PD deficiency trait in West Africa exert contrasting effects on two major clinical presentations of severe malaria. *Malar J* 15: 13.

Shekalaghe SA, Ter Braak R, Daou M, et al. (2010). In Tanzania, hemolysis after a single dose of primaquine coadministered with an artemisinin is not restricted to glucose-6-phosphate dehydrogenase-deficient (G6PD A-) individuals. *Antimicrob Agents Chemother* **54**(5): 1762–1768.

Shi C, Yan W, Wang G, et al. (2015). Pharmacogenetics-based versus conventional dosing of warfarin: a meta-analysis of randomized controlled trials. *PLoS One* **10**(12): e0144511.

Shord SS, Cavallari LH, Gao W, et al. (2009). The pharmacokinetics of codeine and its metabolites in Blacks with sickle cell disease. *Eur J Clin Pharmacol* **65**: 651–658.

Small CB, Margolis DA, Shaefer MS, and Ross LL (2017). HLA-B*57:01 allele prevalence in HIV-infected North American subjects and the impact of allele testing on the incidence of abacavir-associated hypersensitivity reaction in HLA-B*57:01-negative subjects. *BMC Infect Dis* **17**(1): 256.

Soko ND, Masimirembwa C, and Dandara C (2016). Pharmacogenomics of rosuvastatin: a glocal (global+local) African perspective and expert review on a statin drug. *OMICS* **20**(9): 498–509.

Sun Q, Liu HP, Zheng RJ, et al. (2017). Genetic polymorphisms of SLCO1B1, CYP2E1 and UGT1A1 and susceptibility to anti-tuberculosis drug-induced hepatotoxicity: a Chinese population-based prospective case–control study. *Clin Drug Investig* **37**(12): 1125–1136.

Sundelin E, Gormsen LC, Jensen JB, et al. (2017). Genetic polymorphisms in organic cation transporter 1 attenuates hepatic metformin exposure in humans. *Clin Pharmacol Ther* **102**(5): 841–848.

Swart M, Ren Y, Smith P, and Dandara C (2012a). ABCB1 4036A>G and 1236C>T polymorphisms affect plasma efavirenz levels in South African HIV/AIDS patients. *Front Genet* **3**: 236.

Swart M, Whitehorn H, Ren Y, et al. (2012b). PXR and CAR single nucleotide polymorphisms influence plasma efavirenz levels in South African HIV/AIDS patients. *BMC Med Genet* **13**: 112.

Swart M, Skelton M, Ren Y, et al. (2013). High predictive value of CYP2B6 SNPs for steady-state plasma efavirenz levels in South African HIV/AIDS patients. *Pharmacogenet Genomics* **23**: 415–427.

Swen JJ, Nijenhuis M, De Boer A, et al. (2011). Pharmacogenetics: from bench to byte – an update of guidelines. *Clin Pharmacol Ther* **89**: 662–673.

Takeuchi F, Mcginnis R, Bourgeois S, et al. (2009). A genome-wide association study confirms *VKORC1*, *CYP2C9*, and *CYP4F2* as principal genetic determinants of warfarin dose. *PLoS Genet* **5**(3): e1000433.

Tang T, Liu J, Zuo K, et al. (2015). Genotype-guided dosing of coumarin anticoagulants: a meta-analysis of randomized controlled trials. *J Cardiovasc Pharmacol Ther* **20**(4): 387–394.

Thein SL and Menzel S (2009). Discovering the genetics underlying foetal haemoglobin production in adults. *Br J Haematol* **145**: 455–467.

Thomford N, Dzobo K, Chopera D, et al. (2015). Pharmacogenomics implications of using herbal medicinal plants on African populations in health transition. *Pharmaceuticals* **8**: 637.

Thorn CF, Klein TE, and Altman RB (2013). PharmGKB: the Pharmacogenomics Knowledge Base. *Methods Mol Biol* **1015**: 311–320.

Toure MC, Niang A, Diop C, et al. (2016). Prevention of isoniazid toxicity by NAT2 genotyping in Senegalese tuberculosis patients. *Toxicol Rep* **3**: 826–831.

Twist GP, Gaedigk A, Miller NA, et al. (2016). Constellation: a tool for rapid, automated phenotype assignment of a highly polymorphic pharmacogene, CYP2D6, from whole-genome sequences. *NPJ Genom Med* **1**: 15007.

UNAIDS (2016). *2016–2021 Treatment Strategy*. Available at: www.unaids.org/sites/default/files/media_asset/20151027_UNAIDS_PCB37_15_18_EN_rev1.pdf.

UNAIDS (2017). *Data Book*. Available at: www.unaids.org/en/resources/documents/2017/2017_data_book.

Vasiliou V, Vasiliou K, and Nebert DW (2009). Human ATP-binding cassette (ABC) transporter family. *Hum Genomics* **3**(3): 281–290.

Vo TT and Varghese Gupta S (2016). Role of Cytochrome P450 2B6 pharmacogenomics in determining efavirenz-mediated central nervous system toxicity, treatment outcomes, and dosage adjustments in patients with human immunodeficiency virus infection. *Pharmacotherapy* **36** (12): 1245–1254.

Von Seidlein L, Auburn S, Espino F, et al. (2013). Review of key knowledge gaps in glucose-6-phosphate dehydrogenase deficiency detection with regard to the safe clinical deployment of 8-aminoquinoline treatment regimens: a workshop report. *Malar J* **12**: 112.

Wadelius M (2014). Warfarin pharmacogenetics: it matters if you're black or white. *Blood* **124**(14): 2171.

Wang P, Pradhan K, Zhong XB, and Ma X (2016). Isoniazid metabolism and hepatotoxicity. *Acta Pharm Sin B* **6**(5): 384–392.

Ward BA, Gorski JC, Jones DR, et al. (2003). The cytochrome P450 2B6 (CYP2B6) is the main catalyst of efavirenz primary and secondary metabolism: implication for HIV/AIDS therapy and utility of efavirenz as a substrate marker of CYP2B6 catalytic activity. *J Pharmacol Exp Ther* **306**(1): 287–300.

Ware RE, Eggleston B, Redding-Lallinger R, et al. (2002). Predictors of fetal hemoglobin response in children with sickle cell anemia receiving hydroxyurea therapy. *Blood* **99**: 10–14.

Ware RE, Despotovic JM, Mortier NA, et al. (2011). Pharmacokinetics, pharmacodynamics, and pharmacogenetics of hydroxyurea treatment for children with sickle cell anemia. *Blood* **118**: 4985–4991.

Watson J, Taylor WR, Menard D, Kheng S, and White NJ (2017). Modelling primaquine-induced haemolysis in G6PD deficiency. *Elife* **6**: e23061.

Weiner M, Peloquin C, Burman W, et al. (2010). Effects of tuberculosis, race, and human gene *SLCO1B1* polymorphisms on rifampin concentrations. *Antimicrob Agents Chemother* **54**(10): 4192–4200.

Weinshilboum RM and Wang L (2017). Pharmacogenomics: precision medicine and drug response. *Mayo Clin Proc* **92**(11): 1711–1722.

Wennerholm A, Johansson I, Hidestrand M, et al. (2001). Characterization of the CYP2D6*29 allele commonly present in a black Tanzanian population causing reduced catalytic activity. *Pharmacogenetics* **11**: 417–427.

Wennerholm A, Dandara C, Sayi J, et al. (2002). The African-specific *CYP2D617* allele encodes an enzyme with changed substrate specificity. *Clin Pharmacol Ther* **71**: 77–88.

WHO (2016). *Consolidated Guidelines on the Use of Antiretroviral Drugs for Treating and Preventing HIV Infection: Recommendations for a Public Health Approach*, 2nd edn. Available at: www.who.int/hiv/pub/arv/chapter4.pdf?ua=1 (accessed January 2, 2017).

Wonkam A (2017). Is there a role for pharmacogenetics in the treatment of sickle cell disease? *Pharmacogenomics* **18**(4): 321–325.

Wonkam A, Ngo Bitoungui VJ, Vorster AA, et al. (2014). Association of variants at *BCL11A* and *HBS1L-MYB* with hemoglobin F and hospitalization rates among sickle cell patients in Cameroon. *PLoS One* **9**: e92506.

Wright GE, Niehaus DJ, Drogemoller BI, et al. (2010). Elucidation of CYP2D6 genetic diversity in a unique African population: implications for the future application of pharmacogenetics in the Xhosa population. *Ann Hum Genet* **74**: 340–350.

Yang JJ and Bhojwani D (2013). Thiopurine S-methyltransferase pharmacogenetics in childhood acute lymphoblastic leukemia. *Methods Mol Biol* **999**: 273–284.

Yang JJ, Landier W, Yang W, et al. (2015). Inherited NUDT15 variant is a genetic determinant of mercaptopurine intolerance in children with acute lymphoblastic leukemia. *J Clin Oncol* **33**(11): 1235–1242.

Yee MM, Josephson C, Hill CE, et al. (2013). Cytochrome P450 2D6 polymorphisms and predicted opioid metabolism in African American children with sickle cell disease. *J Pediatr Hematol Oncol* **35**: e301–e305.

Zanger UM and Klein K (2013). Pharmacogenetics of cytochrome P450 2B6 (CYP2B6): advances on polymorphisms, mechanisms, and clinical relevance. *Front Genet* **4**: 24.

Zanger UM and Schwab M (2013). Cytochrome P450 enzymes in drug metabolism: regulation of gene expression, enzyme activities, and impact of genetic variation. *Pharmacol Ther* **138**(1): 103–141.

Zembutsu H (2015). Pharmacogenomics toward personalized tamoxifen therapy for breast cancer. *Pharmacogenomics* **16**: 287–296.

Zhou K, Donnelly L, Burch L, et al. (2010). Loss-of-function CYP2C9 variants improve therapeutic response to sulfonylureas in type 2 diabetes: a Go-DARTS study. *Clin Pharmacol Ther* **87**(1): 52–56.

8 Genomics of Cardiometabolic Disorders in Sub-Saharan Africa[1]

Sally N. Adebamowo, Fasil Tekola-Ayele, Adebowale A. Adeyemo, and Charles N. Rotimi

8.1 Introduction

Cardiometabolic disorders including hypertension, diabetes, dyslipidemia, obesity, coronary heart disease (CHD), and stroke have been established as major contributors to the global burden of disease, disability, and mortality. Although cardiometabolic disorders were considered to be primarily prevalent in high-income countries (HICs), current indications identify their increasing public health impact in all regions of the world, particularly in lower middle income countries (LMICs). Between 2013 and 2100, the populations of 35 countries, most of them LMICs, could more than triple (United Nations 2013). Among them, the populations of Burundi, Malawi, Mali, Niger, Nigeria, Somalia, Uganda, United Republic of Tanzania, and Zambia are projected to increase at least five-fold by 2100 (United Nations 2013). The fast rate of population growth, increase in life expectancy, urbanization, and the shift from communicable to non-communicable diseases in these countries suggests an alarming trend of increase in cardiometabolic diseases in Africa in the near future.

Risk factors for cardiometabolic disorders include excess weight gain, especially around the abdomen (abdominal obesity), low levels of high-density lipoprotein cholesterol (HDL-C), high levels of triglyceride, fasting blood sugar, and blood pressure. A clustering of at least three of these metabolic risk factors, metabolic syndrome (Expert Panel on Detection 2001), is associated with increased risk of stroke by 2–3-fold, CHD by 3–4-fold (Alberti et al. 2005), and diabetes by five-fold (Alberti et al. 2009). Other cardiometabolic risk factors include smoking (Goldstein et al. 2011; Shah and Cole 2010), physical inactivity (Lindström et al. 2003; McDonnell et al. 2013; Muhihi et al. 2012), urbanization (Mathenge et al. 2010; Riha et al. 2014; Sliwa et al. 2016), sleep apnea (Drager et al. 2013; Gonzaga et al. 2015; Li et al. 2014; Loke et al. 2012; Nannapaneni et al. 2013; Peppard et al. 2013; Redline et al. 2010), western dietary pattern (Aljefree and Ahmed 2015; Danaei et al. 2013), and genetics (Adeyemo et al. 2005b, 2010; Ariyaratnam et al. 2007; Laston et al. 2015; Marcadenti 2016; Rotimi et al. 2004; Sharma et al. 2013; Simino et al. 2012; Tekola-Ayele et al. 2014). Variation in lifestyle risk factors operates against a background of genetic susceptibility to increase the risk of an individual developing a

[1] An earlier version of this chapter has previously been published online, May 9, 2017 in *Public Health Genomics*, published by S. Karger AG: Adebamowo SN, Tekola-Ayele F, Adeyemo AA, and Rotimi CN (2017). Genomics of cardiometabolic disorders in sub-Saharan Africa. *Public Health Genomics*. DOI: 10.1159/000468535. Reproduced with permission.

cardiometabolic disorder. This complex interaction of multiple genetic and non-genetic factors makes understanding the pathogenesis of cardiometabolic traits challenging. Table 8.1 shows genome-wide association study (GWAS) publications that included data from sub-Saharan Africa (SSA).

A key advance in the drive to identify the genetic contributions to cardiometabolic diseases is the GWAS, an agnostic approach that aims to detect variant frequency differences in cases and controls by interrogating the entire genome. Initially developed as a method for testing association with hundreds of thousands to millions of genotyped single nucleotide polymorphisms (SNPs), its utility has been increased by *in silico* imputation using reference datasets and meta-analytic procedures that can combine association statistics across multiple GWASs. Large-scale GWASs have been remarkably successful in uncovering novel susceptibility loci for a wide range of complex human diseases, including obesity-related traits, coronary artery disease, hypertension, dyslipidemia, and diabetes (Ehret et al. 2011; Welter et al. 2014).

Africa is the most genetically diverse continent in the world, even though few studies have focused on characterizing the genome-wide genetic diversity of African populations (Jones 2014). Figure 8.1 shows the number of GWAS publications from SSA. The International HapMap (Baudot et al. 2009), 1000 Genomes (1000 Genomes Project Consortium 2010), and especially the African Genome Variation Projects (AGVP) (Gurdasani et al. 2014) have provided remarkable insights into the scope, historic, and health implications of the vast genetic variation that exists across African ancestry populations. The AGVP project, which generated both genotype and whole-genome sequence data from 20 African ethno-linguistic groups, revealed the previously unappreciated complex population structure, including regionally distinct patterns of admixture and evidence for substantial "Eurasian" and hunter-gatherer admixture across SSA. African populations are highly subdivided. The population structure across Africa is currently recognized as 11 ancestries that correspond to a combination of geographic and linguistic separation, compared to 12 ancestries detected in the rest of the world (Shriner et al. 2014). Genetic differentiation, measured by Fixation index (F_{ST}), between SSA ancestries can exceed that between pairs of non-African ancestries. The vastness and complexity of genetic variations across African populations have significant implications for human health. For example, genetic adaptations that took place across Africa, particularly against fatal pathogens and ecological forces, have resulted in elevated frequencies of alleles conferring survival advantages detectable in present-day African ancestry individuals on the continent and in the diaspora. Unfortunately, some of these alleles are maladaptive in modern-day environments and the discordance between ancestral genetic background and modern-day environmental exposures became pronounced in the African diaspora, contributing to the disproportionately high burden of some chronic diseases and health disparities in these groups (Rotimi et al. 2016). The aim of this review is to highlight how genomics is contributing to our understanding of the patho-biology of cardiometabolic disorders among African ancestry populations.

Table 8.1 GWAS publications including samples from sub-Saharan Africa, 2009 to October 2016

Reference	Disease/trait	Location(s)	Initial sample	Replication sample
Jallow et al. 2009	Malaria	Gambia	958 Gambian ancestry cases, 1382 Gambian ancestry controls	1087 Gambian ancestry cases, 2376 Gambian ancestry controls
Joubert et al. 2010	HIV (mother-to-child transmission)	Malawi	100 Malawian ancestry infant cases, 126 Malawian ancestry infant controls	
Kang et al. 2010	Anthropometric traits	Nigeria	743 African American individuals, 1188 Nigerian ancestry individuals	704 African American individuals from 306 families, 3471 African American and Caribbean individuals
Thye et al. 2010	Tuberculosis	Gambia, Ghana	2237 SSA ancestry cases, 3122 SSA ancestry controls	1462 SSA ancestry cases, 4604 SSA ancestry controls
Petrovski et al. 2011	HIV-1 susceptibility	Malawi	531 SSA ancestry cases, 848 SSA ancestry controls	
Ziliak et al. 2011	Response to platinum-based chemotherapy in head and neck cancers	Nigeria	90 East Asian ancestry lymphoblastoid cell lines	90 European ancestry lymphoblastoid cell lines, 59 Yoruba ancestry lymphoblastoid cell lines
Fox et al. 2011*	Blood pressure	Nigeria	7473 African American individuals	1188 SSA individuals, 10,694 African American individuals, 69,899 European ancestry individuals
Ramsuran et al. 2011	Platelet count	South Africa	115 African ancestry HIV-negative individuals	
Haiman et al. 2011	Prostate cancer	Ghana, Senegal	3425 African American cases, 3290 African American controls	1500 African American/ Caribbean cases, 1908 African American/ Caribbean controls, 344 SSA ancestry cases, 1361 SSA ancestry controls
Wheeler et al. 2011	Chemotherapeutic susceptibility	Nigeria	83 African American lymphoblastoid cell lines, 176 African ancestry lymphoblastoid cell lines	

Table 8.1 (cont.)

Reference	Disease/trait	Location(s)	Initial sample	Replication sample
Wheeler et al. 2013	Response to platinum-based chemotherapy (cisplatin)	Nigeria	176 SSA ancestry lymphoblastoid cell lines, 83 African ancestry lymphoblastoid cell lines, 175 East Asian ancestry lymphoblastoid cell lines, 174 European ancestry lymphoblastoid cell lines	
Carty et al. 2012	Height	Nigeria, USA	8149 African American female individuals	Up to 20,809 African American and African ancestry individuals
Lingappa et al. 2011	HIV-1 viral setpoint	Kenya, Uganda, Botswana, South Africa	496 African ancestry cases and 302 African ancestry exposed controls from serodiscordant couples	
O'Donnell et al. 2012	Capecitabine sensitivity	Nigeria	Up to 84 East Asian ancestry lymphoblastoid cell lines, up to 164 European ancestry lymphoblastoid cell lines, up to 173 African ancestry lymphoblastoid cell lines, up to 82 African American lymphoblastoid cell lines	
Thye et al. 2012	Tuberculosis	Ghana	1329 African ancestry cases, 1847 African ancestry controls	2024 African ancestry cases, 5154 African ancestry controls, 1025 Indonesian ancestry cases, 983 Indonesian ancestry controls, 4441 European ancestry cases, 5874 European ancestry controls

Table 8.1 (cont.)

Reference	Disease/trait	Location(s)	Initial sample	Replication sample
Tekola-Ayele et al. 2012	Podoconiosis	Ethiopia	194 African ancestry cases, 203 African ancestry controls	202 African ancestry family trios
Doumatey et al. 2012	C-reactive protein	Ghana, Nigeria	837 African American individuals	486 West African ancestry individuals
Timmann et al. 2012	Malaria	Gambia, Ghana	1325 African ancestry, 828 African ancestry controls	2229 African ancestry cases, 3526 African ancestry controls
Wheeler et al. 2013	Paclitaxel-induced neuropathy	Nigeria	77 European ancestry lymphoblastoid cell lines, 87 Yoruban ancestry lymphoblastoid cell lines, 83 African American lymphoblastoid cell lines	
Monda et al. 2013	Body mass index	Ghana, Nigeria	37,956 African American individuals, 1188 Nigerian ancestry individuals	27,661 African American/ Caribbean individuals, 4607 SSA ancestry individuals, 123,706 European ancestry individuals
Band et al. 2013	Malaria	Gambia, Kenya, Malawi	5425 African ancestry cases, 6891 African ancestry controls	
Franceschini et al. 2013*	Blood pressure	Seychelles, Nigeria, Ghana, South Africa	28,190 African American individuals, 1188 Nigerian ancestry individuals	69,395 European ancestry individuals, 5266 African American individuals, 6497 African ancestry individuals, 19,601 East Asian ancestry individuals
Chimusa et al. 2014	Tuberculosis	Gambia, Ghana, Malawi, South Africa	642 South African colored cases, 91 South African colored controls	
Willer et al. 2013*	Triglyceride	Seychelles, Uganda, Jamaica, USA	94,595 European ancestry individuals	4420 African ancestry individuals

Table 8.1 (cont.)

Reference	Disease/trait	Location(s)	Initial sample	Replication sample
Cook et al. 2014	Prostate cancer	Ghana	474 West African ancestry cases, 458 West African ancestry controls	5096 African American cases, 4972 African American controls
Al Olama et al. 2014	Prostate cancer	USA, Ghana, Barbados	34,379 European ancestry cases, 33,164 European ancestry controls, 5327 African ancestry cases, 5136 African ancestry controls, 2563 Japanese ancestry cases, 4391 Japanese ancestry controls, 1034 Latino cases, 1046 Latino controls	
Mtatiro et al. 2014	Fetal hemoglobin levels in sickle cell anemia	United Republic of Tanzania, UK	1213 Tanzanian ancestry cases	321 Caribbean and SSA cases
Aung et al. 2015	Exfoliation syndrome		1484 Japanese ancestry cases, 1188 Japanese ancestry controls	1530 East Asian ancestry cases, 6230 East Asian ancestry controls, 1064 Southern Indian ancestry cases, 3536 Southern Indian ancestry controls, 140 Hispanic cases, 233 Hispanic controls, 347 Middle Eastern ancestry cases, 1108 Middle Eastern ancestry controls, 3588 European ancestry cases, 9012 European ancestry controls, 102 South African Black cases, 379 South African Black controls, 50 Uighur cases, 49 Uighur controls, 80 cases, 180 controls

Table 8.1 (cont.)

Reference	Disease/trait	Location(s)	Initial sample	Replication sample
Curtis et al. 2015	Tuberculosis	Gambia, Ghana	5530 European ancestry cases, 5607 European ancestry controls	1085 Russian ancestry cases, 2865 Russian ancestry controls, 2277 SSA ancestry cases, 2360 SSA ancestry controls
Komatsu et al. 2015	Paclitaxel-induced cytotoxicity	Nigeria	57 Japanese ancestry lymphoblastoid cell lines, 59 Han Chinese ancestry lymphoblastoid cell lines	77 European ancestry lymphoblastoid cell lines, 87 Yoruba ancestry lymphoblastoid cell lines, 83 African ancestry lymphoblastoid cell lines
Band et al. 2015	Severe malaria	Malawi, Ghana, Tanzania, Mali, Cameroon, Burkina Faso, Gambia	5130 African ancestry cases, 5291 African ancestry controls	5535 African ancestry cases, 9948 African ancestry controls
Kim et al. 2016	Plasma homocysteine	Nigeria	853 African American cognitively normal individuals, 45 African American dementia cases, 934 Yoruba ancestry cognitively normal individuals, 26 Yoruba ancestry dementia cases	

*Cardiometabolic disorders; location = country of recruitment.
Source: GWAS catalog, accessed November 2016.

8.2 Hypertension

In 2008, approximately 40 percent of adults aged 25 and above were diagnosed with hypertension worldwide; the number of people with the condition rose from 600 million in 1980 to one billion in 2008 (World Health Organization 2010). The increasing

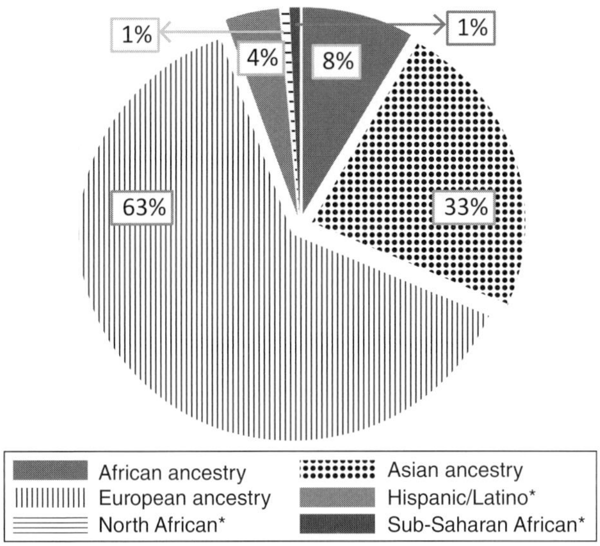

Figure 8.1 GWAS publications by ancestry of discovery sample. * used as a population label. *n* = 5794 GWAS publications after exclusions of unspecified, Oceania, Native Americans. Early studies did not consistently label ancestry. Also, some discovery studies used multiple ancestries. Therefore, some ancestry categories may be underestimated. African ancestry includes African Americans and Afro-Americans. The data used to create this figure were obtained from the GWAS catalog, accessed October 2016.

prevalence of hypertension is attributed to population growth, aging, and behavioral risk factors, such as unhealthy diet, harmful use of alcohol, salt sensitivity, lack of physical activity, excess weight, exposure to persistent stress, and Westernization (Armando et al. 2015; Kaufman et al. 1996; World Health Day 2013). Overall, HICs have a lower prevalence of hypertension (35 percent), compared to LMIC (~40 percent). With prevalence of 46 percent, Africa has the highest prevalence of hypertension (World Health Organization 2008, 2010). Figure 8.2 shows the prevalence of raised blood pressure in SSA, by country income group.

Admixture mapping analysis suggested that African ancestry is associated with increased risk of hypertension (Zhu et al. 2005). Some studies reported heritability estimates of systolic blood pressure ranging from 34 percent to 68 percent and diastolic blood pressure ranging from 29 percent to 65 percent in African populations (Adeyemo et al. 2002; Gu et al. 1998; Rotimi et al. 1999). Early GWAS studies investigating risk variant associations with hypertension and/or blood pressure traits in Europeans were not very successful (Diabetes Genetics Initiative of Broad Institute of Harvard and MIT, Lund University, and Novartis Institutes of BioMedical Research 2007; Wellcome Trust Case Control C 2007). More recently, large-scale GWASs predominantly involving European ancestry study subjects have identified numerous loci associated with blood pressure traits (Ehret et al. 2011; Levy et al. 2010). A recent meta-analysis of gene expression data identified 34 differentially expressed genes for blood pressure, and discovered genes that may explain 5–9 percent of blood pressure

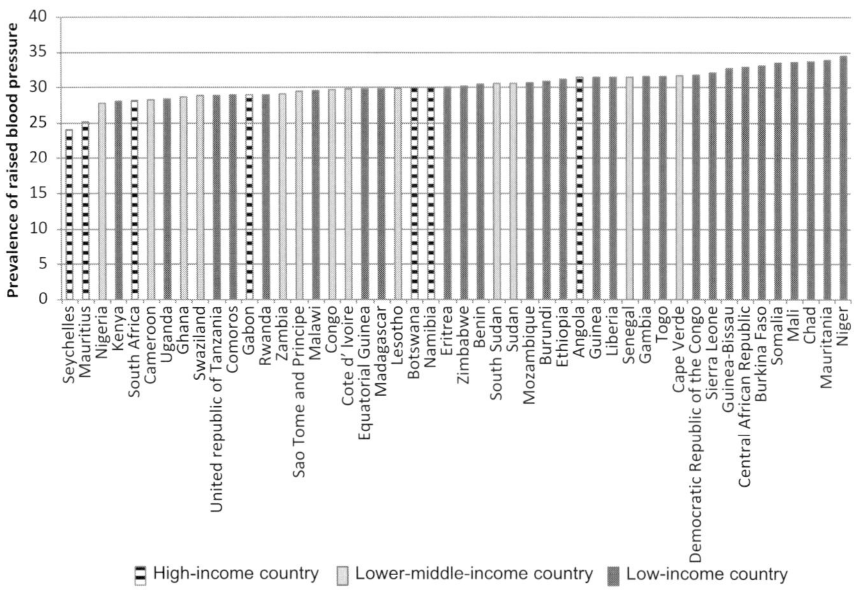

Figure 8.2 Prevalence of raised blood pressure among adults in sub-Saharan Africa by country income groups in 2014. Raised blood pressure = systolic 140 mmHg or diastolic 90 mmHg. The data used to create this figure were obtained from the Global Health Observatory Data Repository, accessed August 2016. Age-standardized data for males and females were used. The country income group was based on the World Bank 2014 classification.

variability (Huan et al. 2015). The first GWAS for hypertension and blood pressure was conducted in 1017 African Americans enrolled from the Washington, DC metropolitan region of the United States in the Howard University Family Study. This study identified multiple SNPs reaching genome-wide significance for systolic BP in or near genes *CACNA1H*, *IPO7*, *PMS1*, *SLC24A4*, and *YWHAZ*, with a replication of some of the SNPs in a sample of West Africans. Two of these genes, *SLC24A4* (a sodium/potassium/calcium exchanger) and *CACNA1H* (a voltage-dependent calcium channel), are potential candidate genes for blood pressure regulation, and the latter is a drug target for a class of calcium channel blockers (Adeyemo et al. 2009). A larger GWAS for blood pressure using the Candidate Gene Association Resource (CARe) consortium consisting of 8591 African Americans identified strong signals for blood pressure near *GPR98*, *ARRDC3*, *C21orf91*, *SLC25A42*, and *HLA-B* genes (Fox et al. 2011). Although the association in both of these studies among African Americans has not been successfully replicated in independent African American samples (Kidambi et al. 2012), some genes associated with blood pressure in other studies, including *SH2B3*, *STK39*, *CDH13*, *TBX3-TBX5*, and *CSK-ULK3*, have been replicated in studies involving African Americans (Adeyemo et al. 2009; Fox et al. 2011; Kidambi et al. 2012).

In a large population sample from rural Nigeria, a genome-wide scan for blood pressure in families found evidence of significant linkage on chromosomes 6 and 7 (Adeyemo et al. 2005a). Further linkage disequilibrium mapping identified at least seven loci, including an intronic SNP in the *PARK2* gene (Tayo et al. 2009), in the region previously identified by admixture mapping for hypertension in populations of African ancestry (Zhu et al. 2005). Genetic variation in the PARK2 gene has also been shown to be significantly associated with blood pressure in Korean populations (Jin et al. 2011).

Alternative methods of identifying the genetic underpinnings of hypertension are being explored, including pathway analysis of GWAS results (Torkamani et al. 2008) and targeting of candidate genes (Sõber et al. 2009). A study in Africa that applied a pathway-focused approach to genomic association scans for hypertension found associations between multiple variants in several genes in the adrenergic alpha-1 receptor (ADRA1) pathway and hypertension in Yoruba Nigerians (Reder et al. 2012). A meta-analysis of genome-wide linkage scans for blood pressure variation in Nigerians and African Americans reported association in two loci, 2p14–p13.1 and 7p21.3–p15.3, the second locus being attributed to the Nigerian sample and suggesting a unique locus for blood pressure variation in people of African ancestry (Rice et al. 2006).

Most genes near blood pressure GWAS loci are not known to be mechanistically associated with blood pressure regulation (Ehret et al. 2011). Therefore, further studies are needed to determine whether the genes implicated in GWAS demonstrate functional relations to blood pressure physiology and to uncover the molecular actions and interactions of genetic and environmental factors involved in blood pressure regulation (Huan et al. 2015). In a study using a different approach, members of the Framingham Heart Study were screened for variants in genes causing rare diseases characterized by blood pressure change. This study found rare variants in three genes, *SLC12A3* (*NCCT*), *SLC12A1* (*NKCC2*), and *KCNJ1* (*ROMK*), which alter renal salt handling, were associated with blood pressure variation in the general population (Ji et al. 2013).

8.3 Diabetes

The rapidly increasing prevalence of type 2 diabetes (T2D) is causing a huge public health burden in Africa. The number of people in SSA with diabetes is projected to rise from 7.2 million in the year 2000 to 18.7 million in 2030, representing an alarming 161 percent increase (Wild et al. 2004). The projected increase in the prevalence of diabetes in SSA is much higher than the projected global average of 114 percent (Wild et al. 2004). Figure 8.3 shows the prevalence of diabetes among adults in SSA.

Type 2 diabetes results from complex interplays between genetic and non-genetic factors, including lifestyle, sociocultural, ecological, and demographic characteristics (Jackson 2008). Genomic studies that employed the GWAS approach have identified more than 100 genetic loci associated with T2D in predominantly European-ancestry study subjects (Welter et al. 2014). In continental Africans, the only published study to date that applied a genome-wide approach in T2D research is the Africa America Diabetes Mellitus (AADM) study that enrolled participants recruited from Ghana, Kenya, and Nigeria (Rotimi et al. 2001). Multipoint linkage analysis in the AADM study revealed suggestive evidence of linkage on chromosomes 12, 19, and 20, the

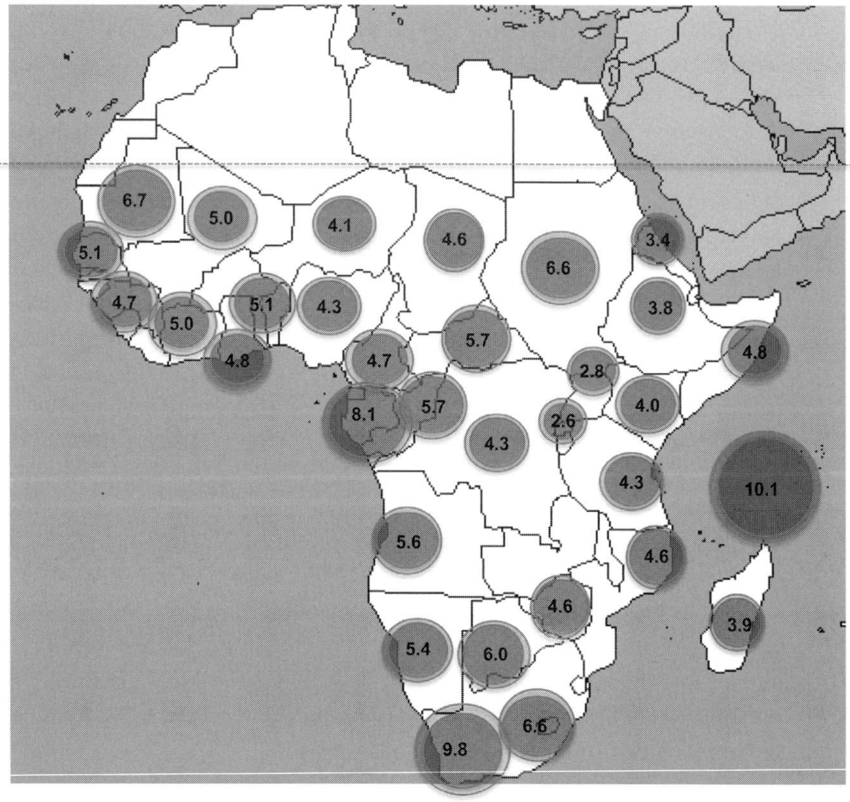

Figure 8.3 Prevalence of diabetes among adults in SSA in 2014. This figure was created using crude prevalence estimates for both sexes. The data were obtained from the Global Health Observatory.

strongest evidence for linkage being on chromosome 20q13.3, a locus that had also been reported to be linked to T2D in non-Africans (Rotimi et al. 2004). No T2D GWAS involving continental African populations has been published to date (Table 8.2). However, our group recently published the first large-scale transferability of T2D loci in continental Africans. In a replication study of 1775 unrelated Africans genotyped on the Affymetrix Axiom PanAFR SNP array, 38 percent of 106 known T2D GWAS loci showed transferability and the majority of the loci that showed replication were fine-mapped (Adeyemo et al. 2015). An extensive review of published genome-wide and candidate gene studies of T2D in SSA can be located elsewhere (Tekola-Ayele et al. 2014).

The gene transcription factor 7-like 2 (*TCF7L2*), particularly the intronic SNP rs7903146, has been one of the strongest and most widely reproduced associations with T2D across diverse ethnic groups (Cauchi et al. 2007). This *TCF7L2* association

Table 8.2 Genes associated with cardiometabolic disease traits identified in SSA versus other populations

	African	African American	European	Asian
Hypertension	*ADH7, ALDH1A2*	*SLC12A9, IGSF5, ADH7, KLHL29, ALDH1A2, ZBTB38, LOC105377436, OPRM1, TIAM1, UBE3AP2, LOC105377369-CCDC34P1, IGSF5*	*AN06, ATP2B1, PLCE1, BAT2, ZFAT, TMEM133, BAT5, XRCC4, CACNB2, CBLN2, FLJ32810, EBF1, CDH13, SLC12A9, NPR3, GNAS, GPR39, EDN3, C5orf23, CHRM3, ZP4, RYR2, HFE, MACROD2, MSRA, MYO6, MOV10, ZNF831, RANBP3L, FES, FURIN, SLC39A8*	*UMOD, FSTL4, FGF5, NR, CACNA1D, MOV10, CASZ1, CYP21A2, MED13L, SLC4A7, GUCY1A3, SOX6, ATP2B1, ALDH2*
Diabetes	*INS-IGF2, TCF7L2, TCF2, FTO, SLC308A*	*SLC44A3, F3, RBM43, RND3, GALNTL4, CPA6, LOC729013, BARX2, TMEM45B, HLA-B, GALNT14, CAPN13, HECW1, INTS8, ASCL2, INS-IGF2, TCF7L2, TH, KCNQ1, ANXA2, HMGA2, TMEM175, SLC12A9, KLHL29, SLC30A1, RNA5SP110, RNA5SP111*	*ADAMTS9, ADCY5, ANK1, AP3S2, ARL15, DCD, BCL11A, C10orf35, C14orf70, C2CD4A, DGKB, C2CD4B, C6orf173, CDC123, CDC123, DUSP9, ETV1, FAF1, CAMK1D, CDKAL, CDKAL1, FTO, GIPR, GLIS3, GRB14, HMG20A, CDKN2A, HHEX, IDE, HMGA2, HNF1A, HNF4A, CDKN2B, LGR5, TSPAN8, CENTD2, ARAP1, CHCHD9, HNF1B, TCF2, IGF2BP2, IL20RA, IRS1, JAZF1, KCNJ11, KCNK16, KCNQ1, KLF14, KLHDC5, LGR5, TP53INP1, TSPAN8, LPP, LYPLAL1, MC4R, NOTCH2, ADAM30, PLEKHA1, PPARG, MPHOSPH9, MTNR1B, PCBD2, PRC1, PROX1, PLEKHA1, PTEN, RNF6, RREB1, SSR1, SLC30A8, SPRY2, ST64GAL1, SYN2, PPARG, TCF19, POU5F1, TCF7L2, THADA, TLE1, TLE4, TMEM154,*	*ANK1, AP3S2, CMIP, GCC1, GPSM1, PTPRD, DUSP9, GLIS3, KCNK16, MAEA, C2CD4B, C2CD4A, CDKAL1, CDKN2A, CDKN2B, FITM2, R3HDML, HNF4A, HMG20A, HNF1B, HNF4A, IGF2BP2, KCNQ1, ZFAND3, MGC21675, MIR129, LEP, NXN, PEPD, PALM2, PAX4, AKAP2, GCC1, PSMD6, SGCG, SACS, RHOU, WWOX, SLC16A13, SLC30A8, SRR, ST6GAL1, SYK, TCF7L2, UBE2E2, VPS26A, WISP1*

Table 8.2 (cont.)

	African	African American	European	Asian
			VEGFA, WFS1, ZBED3, ZFAND6, ZMIZ1, CRHR2, UBE2E2, TMEM18, TMEM75	
Dyslipidemia			*ABCA1, ABCG8, ANGPTL4, APOB, APOA1, XKR6, APOC3, APOA4, APOA5, APOC2, APOE, APOC4, APOC1, CELSR2, PSRC1, SORT1, CETP, GALNT2, FADS1, FADS2, FADS3, GCKR, HMGCR, HNF1A, HNF4A, LCAT, LDLR, LIPC, LIPG, LPL, MAFB, MVK, MMAB, MLXIPL, PCSK9, NCAN, CILP2, PBX4, PSRC1, CELSR2, PLTP, TRIB1, TTC39B, AMAC1L2*	
Obesity			*ADCY3, POMC, ADCY9, ADCY9, ARG1, TNNI3K, UBE2E3, WWOX, ZZZ3, TCF4, SLC8A1, TFAP2B, TMEM18, SCNN1A, SDCCAG8, SEC16B, SH2B1, ASAH1, BC041448, BDNF, C5orf37, CADM2, CAMK1G, CCDC77, CMYA5, ETV5, FANCL, FAT1, MTNR1A, FARS2, FLJ39743, GPRC5B, HNF4G, HOXB5, HS6ST3, INHBB, KC6, KCNMA1, LEPR, LHFPL3, LINGO2, RABEP2, SH2B1, RPL27A, RPTOR, LOC144233, MAF, MAP2K5, MC4R, MRPS33P4, MSRA, TNKS, NCAM2, NEGR1, PRL, PRMT6, PTER, RGS6, RMST, NINJ1,*	*MDFIC, PROX1, FTO*

Table 8.2 (cont.)

	African	African American	European	Asian
			NPC1, NRXN3, OLFM4, PACS1, PCDH9, OTUD1, KIAA1217, PITPNB, PRKCH	
CAD/CHD		*CELSR2, PSRC1, SORT1, CRYGN, WDR86-AS1, LINC00333, LOC105360289, LOC101926941-RPS12P10*	*ABO, ADAMTS7, ANKS1A, CXCL12, FN, ANRIL, CDKN2A, CDKN2B, FHL5, FNDC1, C12orf43, HNF1A, GPC5, HHIPL1, ILR6, KIAA1462, LDLR, COL4A1, COL4A2, LIPA, LPA, LRFN2, HCG27, HLA-C, MIA3, MRAS, MRPS6, ZNF259, APOA1, APOC3, APOA4, APOA5, SNF8, GIP, UBE2Z, ATP5G1, WDR12, SLC22A3, LPAL2, LPA, VEGFA, ZC3HC1, RASD1, SMCR3, PEMT, SORT1, TCF21, PHACTR1, PSRC1, SH2B3, SMAD3, NT5C2, CNNM2, CYP17A1, MTAP, PCSK9, MTHFD1L, PHACTR1, PPAP2B*	*CDKN2A, RPL6-PTPN11, TTC32, WDR35, GUCY1A3, ATP2B1, C12orf51, TCF21, CUX2, CDKN2B, ALDH2, MYL2, FLT1, ACAD10, PHACTR1*
Stroke		*ALDH1A2, GAPDHP71, C21orf81, AGBL1, TSG1, WDFY4, UBASH3B, CDC5L, CHD3, ELTD1, CLDN17, GATA3, FLRT2, TRNAK27, GLULP3, LIPC, HDAC9, HPS4, IL1RN, IL1F10, IL36RN, LIP1, LOC100507163, MSX2, NKX2-5, ABCC13, TGFBI,*	*AIM1, ALDH2, HDAC9, IMPA2, NR, FOXF2, FOXQ1, PITX2, SPSB4, ZFHX3*	

Table 8.2 (cont.)

	African	African American	European	Asian
		SPINK2, SUMO2P6, PTPRG, MICAL2, RNU6-36, MIR135A2, AQP9		

Source: GWAS catalog, accessed February 2017.

was first reported by the deCODE group of investigators (Grant et al. 2006), followed by successful fine-mapping and evidence for recent selection in West African populations (621 cases and 448 controls) enrolled in the AADM study (Grant et al. 2006; Helgason et al. 2007). This study fine-mapped the association between the *TCF7L2* rs7903146 T allele and T2D, leveraging on the weaker linkage disequilibrium in African genomes. This study was among the earliest empirical demonstrations of trans-ethnic fine-mapping of a disease locus (Helgason et al. 2007). The frequency of rs7901695 C allele ranges from 0.01 to 0.49 in global populations, with African-ancestry populations having up to 40-fold higher frequency (Shriner et al. 2010).

Currently ongoing studies are applying genome-wide approaches such as whole-genome genotyping and sequencing in larger samples of subjects recruited from Africa to enhance our understanding of the genetic basis of T2D unique to Africans and shared among diverse population groups. It is anticipated that large-scale genomic projects including those supported by the Human Heredity and Health in Africa (H3Africa) initiative (Badalà et al. 2008) co-funded by the US National Institutes of Health and the UK Wellcome Trust will generate data that shed new light into the genetic architecture of T2D. The H3Africa T2D project is designed to study the burden, etiology, and spectra of diabetes in Africa. The study is projected to enroll and examine 12,000 cases of T2D and 12,000 ethnically matched controls from 12 centers across nine countries in SSA (www.h3africa.org).

8.4 Dyslipidemia

Dyslipidemia is characterized by a spectrum of qualitative lipid abnormalities reflecting perturbations in the structure, metabolism, and biological activities of both atherogenic lipoproteins and anti-atherogenic HDL-C. These abnormalities include an elevation of lipoproteins containing apolipoprotein B (apoB), elevated triglycerides, increased levels of low-density lipoprotein (LDL), and low levels of HDL-C (Kaur 2014). Data from recent studies showed that the prevalence of low HDL was ~13 percent, high LDL was ~23 percent, and high triglyceride was 7–21 percent in SSA (Cheserek et al. 2014; Doupa et al. 2014; Sabir et al. 2013); in the United States the prevalence of low HDL was 17 percent (Carroll et al. 2012), high LDL was 32 percent, and high triglyceride was 25 percent (Mozaffarian et al. 2015). Blood lipid levels are influenced by multiple environmental and genetic factors. Heritability estimates for blood lipids are high, including ~40–60 percent for HDL-C, ~40–50 percent for LDL, and ~35–48 percent for triglycerides (Weiss et al. 2006), suggesting that genetic factors play a significant role in determining the variation of blood lipid levels.

Most of our knowledge of lipoprotein genetics stems from monogenic familial disorders of lipoprotein metabolism with extreme phenotypes (Rotondo and Davidson 2010) – for example, familial hypercholesterolemia (FH), an inherited disorder of lipoprotein metabolism, which is characterized by highly elevated serum levels of LDL (Soutar and Naoumova 2007). Familial hypercholesterolemia is the most common dominantly inherited disorder in humans, affecting 1 in about 250 people (Nordestgaard et al. 2013; Watts et al. 2014). Most people with the disease have mutations in the LDL receptor (Leigh et al. 2008), apoB (Innerarity et al. 1987), or *PCSK9* genes (Abifadel et al. 2003). However, in most studies, despite next-generation sequencing, no mutation can be identified in up to 20 percent of patients with a clinical diagnosis of definite heterozygous FH (Vandrovcova et al. 2013). In a small sample of FH patients in central and southern Tunisia, two novel mutations in the LDL gene were identified, CYS127S and D245N (Slimane et al. 2002).

A combined analysis of the AADM cohort and other population datasets revealed that the relationship between HDL-C and kidney function varies by ancestry, with an observed inverse relationship in African ancestry populations (West Africans and African Americans), and positive relationship in East Asians (Han Chinese) and European Americans (Bentley et al. 2012). The observed inverse relationship between HDL and kidney function in African ancestry populations was strongest in individuals with the nephropathy risk genotype of SNP rs73885319 in the *APOL1* gene. Higher HDL-C was significantly associated with worse kidney function in subjects with the risk genotype but not in those without this genotype. It is believed that the *APOL1* risk variant rose to high frequency in West African populations to provide protection against a deadly form of African sleeping sickness (Bentley et al. 2012; Genovese et al. 2010).

Loci associated with dyslipidemia have been identified in several populations. Genome-wide association studies involving up to 20,000 individuals of European ancestry have identified over 30 genetic loci contributing to variation in serum and lipid concentrations (Chasman et al. 2009; Kathiresan et al. 2008a; Kathiresan et al. 2008b; Kooner et al. 2008). Meta-analyses of GWASs among 100,000 individuals of European descent ascertained in the United States, Europe, or Australia showed 95 significantly associated loci with lipid traits, which in aggregate may explain 10–12 percent of the total variance (Teslovich et al. 2010). In individuals of African descent, sequence variants in individual genes shown to be consistently related to blood lipid phenotypes include *PCSK9* (Cohen et al. 2005), *C679X* (Hooper et al. 2007), and apolipoprotein E (Masemola et al. 2007) with LDL; *LPL*, *CETP* (Ellman et al. 2015), and apolipoprotein E (Masemola et al. 2007) with HDL-C; and apolipoprotein A5 with triglycerides (Pennacchio et al. 2002).

8.5 Obesity

Obesity is a major global health challenge. The number of overweight and obese individuals increased from 857 million in 1980 to 2.1 billion in 2013 (Murray et al. 2014). The rate of obesity is increasing globally, and in 2013 the prevalence of obesity was higher in women than men in developed and developing countries (Murray et al. 2014). In Africa, there are marked variations in the rate of obesity across countries, with distinct regional patterns. Within SSA in 2013, the age-standardized prevalence of obesity ranged from 4.4 percent (4.2–4.6) in the eastern to 11.7 percent (10.9–12.4) in the southern regions for men; and from 8.8 percent (8.4–9.1) in the eastern to 37.0 percent (35.9–38.1) in the southern regions for women; the rates were higher than 25 percent and 45 percent for men and women respectively in some northern African countries (Murray et al. 2014). Increased economic growth and sedentary lifestyles may explain the regional variations of obesity in Africa. Furthermore, increased levels of physical inactivity and consumption of calorie-dense foods may explain the upward trend of obesity in SSA (Akarolo-Anthony and Adebamowo 2014; Akarolo-Anthony et al. 2013, 2014; Minicuci et al. 2014; Njelekela et al. 2011; Pradeilles et al. 2015; Rguibi and Belahsen 2004).

The association between variants in the fat mass and obesity associated (*FTO*) gene and BMI is well established in populations of European descent, and GWASs have identified several strongly correlated SNPs in intron 1 of *FTO* associated with increased BMI and increased risk of obesity, with subsequent replication in other studies (Frayling et al. 2007; Thorleifsson et al. 2009). Although the *FTO* gene is the most replicated gene for obesity in humans, its association with obesity and BMI has not been consistently replicated in all populations (Hennig et al. 2009; Li et al. 2008; Liu et al. 2008) and the mechanistic basis of this association remains unclear. A recent study provided some evidence that the association of *FTO* with obesity does not involve the *FTO* gene but a repressor binding site that resides within an intron in the *FTO* gene. Among Africans, the *FTO* rs17817449 was associated with higher BMI among black South Africans (p = 0.022) (Lombard et al. 2012), but no association was observed among Gambians (Hennig et al. 2009). In another study, SNPs in the *FTO* gene previously reported to be associated with obesity in Europeans, including rs9939609, rs1121980, rs17817449, and rs7204609, were tested among populations of African ancestry (Adeyemo et al. 2010). The association with obesity was replicated for two SNPs, rs1121980 and rs7204609, at $p < 0.05$ in West Africans but not in African Americans (Adeyemo et al. 2010). The *FTO* region harbors the strongest genetic association with obesity, yet the mechanistic basis of this association remains elusive.

Several other polymorphisms have been found to be associated with obesity or BMI in some African populations, but were not subsequently replicated. The leptin receptor (*LEPR*) gene rs1137101 was significantly associated with BMI or obesity risk in some Tunisian populations, but not among South Africans (Ben Ali et al. 2009; Boumaiza et al. 2012b; Lombard et al. 2012). Likewise, although the *ENPP1* rs1044498 was associated with obesity in Moroccans, no significant association was observed among South Africans (Matsha et al. 2010). However, the *RETN* rs1862513 and rs3745367 have been found to be positively associated with obesity risk among several African populations (Boumaiza et al. 2012a; El-shal et al. 2013).

A recent systematic review on genomics of obesity in Africa, based mostly on studies that used the candidate gene approach, reported that the majority of polymorphisms investigated by single studies in genes such as *ACE*, *ADIPOQ*, *ADRB2*, *AGRP*, *AR*, *CAPN10*, *CD36*, *C7orf31*, *DRD4*, *FTO*, *MC3R*, *MC4R*, *SGIP1*, and *LEP* were found to be associated with various measures of obesity (BMI, waist circumference, and percentage fat mass) in different ethnic groups (Yako et al. 2015). Among these loci, *FTO*, *LEP*, *MC3R*, *MC4R*, and *SGIP1* were associated with lower measures of obesity indices. Some of the SNPs investigated in the same population group and located in one gene had opposite effects on obesity indices. Of the 17 genes reported to be significantly associated with obesity in Africa, only *FTO* and *MC4R* polymorphisms have been previously validated by GWAS that showed a directionally consistent association with obesity in non-African populations (Yako et al. 2015). Further replication studies and large-scale genomic studies are needed to validate and ascertain the role of polymorphisms and the underlying genetic mechanisms of obesity in African populations.

8.6 Coronary Artery Disease/Coronary Heart Disease

Coronary artery disease (CAD) is a heart disorder that results from interaction between multiple environmental and genetic factors. Coronary heart disease, a result of CAD, was the leading cause of death globally among men and women in both 1990 and 2013 (Shepard et al. 2015). In 2013, CHD was responsible for nearly half of all deaths from cardiovascular diseases, causing as many deaths as chronic obstructive pulmonary disease (COPD), diabetes mellitus, cirrhosis, lung cancer, and liver cancer combined (Shepard et al. 2015). Although SSA experiences the world's lowest CHD death rates (Moran et al. 2015), the number of deaths due to CHD increased by 87 percent from 1990 to 2013 in SSA, likely due to aging and growth of the SSA population (Mensah et al. 2015).

Several GWASs among individuals of European descent have reported a strong association between common variants on chromosome 9p21 and risk of CAD, and this association has been replicated in populations from North Africa (Humphries et al. 2010; McPherson et al. 2007; Samani et al. 2007). The chromosome 9p21 locus is also associated with other cardiovascular disease phenotypes, including abdominal aortic aneurysm and intracranial aneurysm (Helgadottir et al. 2008). A study of the haplotype diversity and allelic history of the 9p21.3 region among unrelated individuals from 51 populations of non-European origin (Human Genome Diversity Project) found that variants rs10116277 and rs2383207 located in the intron of cyclin-dependent kinase inhibitor 2B antisense RNA 1 (*CDKN2B-AS1/ANRIL*) gene were by far the most common in populations from SSA (Silander et al. 2009).

A meta-analysis of 14 CAD GWASs, including thousands of cases and controls, validated the association of ten previously reported loci and identified 13 new variants associated with heart disease (Schunkert et al. 2011). Few candidate gene studies have been conducted on CAD in North Africa. Results from recent studies show that there is an independent association between rs699 located in exon 2 of the angiotensinogen (*AGT*) gene and CAD among Tunisians (Abboud et al. 2010) and Egyptians (Abd El-Aziz et al. 2012). The *AGT* rs699 SNP is a well-characterized genetic polymorphism of the renin-angiotensin system (RAS) and is associated with known risk factors of heart disease (Tsai et al. 2009; Yang et al. 2010). Other RAS polymorphisms associated with CAD in Africans include rs4762 in the *AGT* gene, rs5186 in the ADP-ribosyltransferase 1 (*ATR1*) gene, and Glu298Asp polymorphism in exon 7 of the endothelial nitric oxide synthase 3 (*eNOS3*) gene (Abboud et al. 2010; Abd El-Aziz et al. 2012; Motawi et al. 2011). It is possible that RAS polymorphism and hypertension play a role in CAD (Cooper et al. 2000; Rotimi et al. 1997).

Chronic kidney disease (CKD) is an independent risk factor for the development of CAD (Sarnak et al. 2003). Recently the apolipoprotein L1 (*APOL1*) gene was identified as a risk locus for CKD in African Americans, and these findings were confirmed in Nigerian Yoruba CKD patients (Genovese et al. 2010; Tayo et al. 2013). Identified *APOL1* genetic variants G1 and G2 are both very frequent on African-derived chromosomes and strongly associated with focal segmental glomerulosclerosis and HIV-associated nephropathy (Genovese et al. 2010; Kopp et al. 2011). G1 and G2 alleles appear to be most frequent in West Africa, with the highest allele frequencies reported in Nigeria and Ghana (Kopp et al. 2011; Tzur et al. 2012). It is hypothesized that the *APOL1* renal disease risk variants evolved in SSA thousands of years ago to confer protection against the trypanosome parasite that causes African sleeping sickness (Genovese et al. 2010). Furthermore, some studies have shown an association between *APOL1* genotype and cardiovascular disease among African Americans (Freedman et al. 2015; Mukamal et al. 2016). While one study showed that *APOL1* nephropathy variants were associated with lower levels of subclinical atherosclerosis in African Americans with T2D (Freedman et al. 2015), another showed that *APOL1* genotype was associated with subclinical atherosclerosis and incident myocardial infarction, and concluded that *APOL1* trypanolytic variants may account for a substantial proportion of the excess risk of chronic disease in African Americans (Mukamal et al. 2016). Large-scale studies to examine the association between *APOL1* and CAD among Africans are warranted.

8.7 Stroke

Stroke is a focal neurological deficit resulting from disruption of the cerebral blood supply. About 80 percent of strokes are ischemic, resulting from an obstruction of blood flow, whereas 15–20 percent are hemorrhagic, due to bleeding into the brain (Sharma and Meschia 2013). Worldwide in 2010, the prevalence of stroke was 33 million, with 16.9 million people having a first stroke. The majority of global stroke burden is in LMICs (Mensah et al. 2015). In these countries, 2.8 million individuals died of ischemic stroke and 3 million of hemorrhagic stroke; 39.4 million disability-adjusted life years (DALYs) were lost because of ischemic stroke and 62.8 million because of hemorrhagic stroke (Mensah et al. 2015). Data from Africa suggests that about 66 percent of strokes are ischemic while 34 percent are hemorrhagic (O'Donnell et al. 2010). The number of deaths from stroke nearly doubled from 1990 to 2013, but overall age-adjusted stroke mortality rates decreased by 1 percent in SSA (increased by 9 percent for ischemic stroke and decreased by 10 percent for hemorrhagic stroke) (Mensah et al. 2015).

Evidence from studies of family history suggests that genetic predisposition is important (Sharma et al. 2013). These studies have consistently found that a family history of stroke is more common in individuals with ischemic stroke compared with controls (Jerrard-Dunne et al. 2003). In the Framingham Heart Study, a documented parental ischemic stroke by the age of 65 years was associated with a three-fold increase in ischemic stroke risk in offspring, even after adjustment for other known stroke risk factors (Seshadri et al. 2010). Studies evaluating the familial risk of hemorrhagic stroke demonstrate that in first-degree relatives a subject's odds of having a stroke increase by as much as two- to six-fold (Woo et al. 2002). Limited data from twin studies suggest stroke is more common in monozygotic compared with dizygotic twins (Flossmann et al. 2004).

Some monogenic disorders can cause stroke. Sickle cell disease is one of the most common severe monogenic disorders worldwide (Weatherall et al. 2005). A hereditary blood disorder, it is an established risk factor for stroke. The global distribution of this disease is indicative of two factors: selection for carriers through their survival advantage in malaria-endemic regions; and subsequent migration (Rees et al. 2010). The prevalence of sickle cell disease is highest in SSA. Although the scarcity of diagnostic facilities means that precise data are not available, a recent estimate suggests that more than 230,000 affected children are born in this region every year (0.74 percent of the births in SSA), which is about 80 percent of the global total (Modell and Darlison 2008). There is a first-stroke incidence of 1.02 per 100 patient-years between the ages of two and five years, and 11 percent of patients with sickle cell disease have had a stroke by the age of 20 years (Ohene-Frempong et al. 1998). Other monogenic disorders that can cause stroke include cerebral autosomal dominant arteriopathy with subcortical infarcts and leucoencephalopathy (CADASIL) (Sharma et al. 2013), autosomal dominant retinal vasculopathy, and homocystinemia (Lanfranconi and Markus 2010; Sharma and Meschia 2013).

Linkage studies for stroke have been successfully conducted, with the identification of *STRK1* gene in the Icelandic population as phosphodiesterase 4D (*PDE4D*) (Gretarsdottir et al. 2003); however, this association could not be replicated in other European populations (Bevan et al. 2008). Nonetheless, the rs918592 SNP variant of *PDE4D* was found to be significantly associated with stroke in a small sample of young African American women. A further linkage study in Iceland found a 13q12 locus associated with both stroke and myocardial infarction. Within this region, a candidate gene, lipoxygenase-activating protein (*ALOX5AP*), was proposed as the underlying gene (Helgadottir et al. 2004).

Mutations in the e*NOS3* were found to be associated with ischemic stroke risk in some, including African Americans and French (Elbaz et al. 2000; Grewal et al. 2007) but not all populations (MacLeod et al. 1999; Moe et al. 2008). *IL6* SNP rs2069832 (Cole et al. 2008) and *CSN3* (Cole et al. 2012) were also significantly associated with ischemic stroke in African Americans. Certain SNPs, including *ANXA2* rs11853426, *TEK* rs489347, and *TGFBR3* rs284875 variants, have been associated with increased stroke risk in sickle cell disease (Flanagan et al. 2011).

Genes with significant polymorphisms related to hemorrhagic stroke include the *APOE* ε2, *APOE* ε4 methylenetetrahydrofloate reductase (*MTHFR*), *IL6*, tumor necrosis factor (TNF)-α, angiotensin converting enzyme (ACE), factor VII, factor XIII, platelet activating factor, and β-tubulin (Biffi et al. 2011; Brouwers et al. 2012; Yang et al. 2012).

Various GWASs have been carried out, identifying a few associated loci including a risk variant on 4q25, possibly on *PITX2*; and another variant associated with atrial fibrillation in the *ZFHX3* gene at 16q22 (Gretarsdottir et al. 2008; Gudbjartsson et al. 2009). A recent whole-exome sequencing (WES) study identified two modifier mutations, *GOLGB1* (Y1212C) and *ENPP1* (K173Q), associated with protection from stroke in a cohort of children with sickle cell anemia (Flanagan et al. 2013). Larger stroke GWASs are underway, including the NIH-funded Stroke Investigative Research and Education Network (SIREN) project, which proposed to explore genomic factors in stroke in 6000 native West Africans (3000 case–control pairs) in comparison with 1000 African Americans (80 percent of whom are of West African ancestral origin) and 12,000 Americans of European ancestry in the REasons for Geographic and Racial Differences in Stroke (REGARDS) study.

8.8 Future Directions

The global burden of cardiometabolic diseases is increasing at an alarming rate. The brunt of this epidemic will have disproportionate public health and economic impacts in SSA because the public health infrastructure and policy framework are not adequately prepared to curb disease occurrence, diagnose early, and manage chronic diseases and associated comorbidities. Given the role of complex interactions of genetic and environmental factors in cardiometabolic diseases, genetic research in African populations can unravel the underlying pathogenesis of cardiometabolic diseases. The genetic diversity and narrow LD structure of African genomes is advantageous for fine-mapping disease loci and discovering novel genetic loci associated with cardiometabolic diseases. To drive these efforts forward, it will be important to (1) expand recruitment of large population-based cohorts from diverse African population groups; (2) perform deep phenotyping of research subjects; (3) link genomics research with health care programs and epidemiological surveys; (4) integrate multiomics studies in tissues more relevant to specific diseases; and (5) continuously improve genotyping arrays to accommodate the genetic diversity of African genomes.

References

1000 Genomes Project Consortium (2010). A map of human genome variation from population-scale sequencing. *Nature* **467**(7319): 1061–1073.

Abboud N, Ghazouani L, Kaabi B, et al. (2010). Evaluation of the contribution of renin angiotensin system polymorphisms to the risk of coronary artery disease among Tunisians. *Genet Test Mol Biomarkers* **14**(5): 661–666.

Abd El-Aziz TA, Hussein YM, Mohamed RH, et al. (2012). Renin-angiotensin system genes polymorphism in Egyptians with premature coronary artery disease. *Gene* **498**(2): 270–275.

Abifadel M, Varret M, Rabes JP, et al. (2003). Mutations in PCSK9 cause autosomal dominant hypercholesterolemia. *Nat Genet* **34**(2): 154–156.

Adeyemo AA, Omotade OO, Rotimi CN, et al. (2002). Heritability of blood pressure in Nigerian families. *J Hypertens* **20**(5): 859–863.

Adeyemo A, Luke A, Wu X, et al. (2005a). Genetic effects on blood pressure localized to chromosomes 6 and 7. *J Hypertens* **23**(7): 1367–1373.

Adeyemo AA, Johnson T, Acheampong J, et al. (2005b). A genome wide quantitative trait linkage analysis for serum lipids in type 2 diabetes in an African population. *Atherosclerosis* **181**(2): 389–397.

Adeyemo A, Gerry N, Chen G, et al. (2009). A genome-wide association study of hypertension and blood pressure in African Americans. *PLoS Genet* **5**(7): 1–11.

Adeyemo A, Chen G, Zhou J, et al. (2010). FTO genetic variation and association with obesity in West Africans and African Americans. *Diabetes* **59**(6): 1549–1554.

Adeyemo AA, Tekola-Ayele F, Doumatey AP, et al. (2015). Evaluation of genome wide association study associated type 2 diabetes susceptibility loci in sub Saharan Africans. *Frontiers in Genetics* **6**: 2–9.

Akarolo-Anthony SN and Adebamowo CA (2014). Prevalence and correlates of leisure-time physical activity among Nigerians. *BMC Public Health* **14**: 529.

Akarolo-Anthony SN, Odubore FO, Yilme S, et al. (2013). Pattern of dietary carbohydrate intake among urbanized adult Nigerians. *Int J Food Sci Nutr* **64**(3): 292–299.

Akarolo-Anthony SN, Willett WC, Spiegelman D, et al. (2014). Obesity epidemic has emerged among Nigerians. *BMC Public Health* **14**(1): 455.

Alberti KGMM, Zimmet P, and Shaw J (2005). The metabolic syndrome: a new worldwide definition. *Lancet* **366**(9491): 1059–1062.

Alberti KG, Eckel RH, Grundy SM, et al. (2009). Harmonizing the metabolic syndrome: a joint interim statement of the International Diabetes Federation Task Force on Epidemiology and Prevention; National Heart, Lung, and Blood Institute; American Heart Association; World Heart Federation; International. *Circulation* **120**(16): 1640–1645.

Aljefree N and Ahmed F (2015). Association between dietary pattern and risk of cardiovascular disease among adults in the Middle East and North Africa Region: a systematic review. *Food Nutr Res* **59**. DOI: 10.3402/fnr.v59.27486.

Al Olama AA, Kote-Jarai Z, Berndt SI, et al. (2014). A meta-analysis of 87,040 individuals identifies 23 new susceptibility loci for prostate cancer. *Nat Genet* **46**: 1103–1109.

Ariyaratnam R, Casas JP, Whittaker J, et al. (2007). Genetics of ischaemic stroke among persons of non-European descent: a meta-analysis of eight genes involving approximately 32,500 individuals. *PLoS Medicine* **4**(4): 728–736.

Armando I, Villar VAM, and Jose PA (2015). Genomics and pharmacogenomics of salt-sensitive hypertension. *Curr Hyperten Rev* **11**(1): 49–56.

Aung T, Ozaki M, Mizoguchi T, et al. (2015). A common variant mapping to CACNA1A is associated with susceptibility to exfoliation syndrome. *Nat Genet* **47**(4): 387–392.

Badalà F, Nouri-mahdavi K, and Raoof DA (2008). Enabling the genomic revolution in Africa: H3Africa is developing capacity for health-related genomics research in Africa. *The Computer* **144**(5): 724–732.

Band G, Le QS, Jostins L, et al. (2013). Imputation-based meta-analysis of severe malaria in three African populations. *PLoS Genetics* 9(5): e1003509.

Band G, Rockett KA, Spencer CC, and Kwiatkowski DP (2015). A novel locus of resistance to severe malaria in a region of ancient balancing selection. *Nature* 526(7572): 253–257.

Baudot A, Real FX, Izarzugaza JM, and Valencia A (2009). From cancer genomes to cancer models: bridging the gaps. *EMBO Reports* 10(4): 359–366.

Ben Ali S, Kallel A, Sediri Y, et al. (2009). LEPR p.Q223R polymorphism influences plasma leptin levels and body mass index in Tunisian obese patients. *Arch Med Res* 40(3): 186–190.

Bentley AR, Doumatey AP, Chen G, et al. (2012). Variation in APOL1 contributes to ancestry-level differences in HDLc-kidney function association. *Int J Nephrol.* DOI: 10.1155/2012/748984.

Bevan S, Dichgans M, Gschwendtner A, et al. (2008). Variation in the *PDE4D* gene and ischemic stroke risk: a systematic review and meta-analysis on 5200 cases and 6600 controls. *Stroke* 39 (7): 1966–1971.

Biffi A, Anderson CD, Jagiella JM, et al. (2011). APOE genotype predicts extent of bleeding and outcome in lobar intracerebral hemorrhage. *Lancet Neurol* 10(8): 702–709.

Boumaiza I, Omezzine A, Rejeb J, et al. (2012a). Association between four resistin polymorphisms, obesity, and metabolic syndrome parameters in Tunisian volunteers. *Genet Test Mol Biomarkers* 16(12): 1356–1362.

Boumaiza I, Omezzine A, Rejeb J, et al. (2012b). Relationship between leptin G2548A and leptin receptor Q223R gene polymorphisms and obesity and metabolic syndrome risk in Tunisian volunteers. *Genet Test Mol Biomarkers* 16(7): 726–733.

Brouwers HB, Biffi A, Ayres AM, et al. (2012). Apolipoprotein E genotype predicts hematoma expansion in lobar intracerebral hemorrhage. *Stroke* 43(6): 1490–1495.

Carroll MD, Kit BK, and Lacher DA (2012). Total and high-density lipoprotein cholesterol in adults: National Health and Nutrition Examination Survey, 2009–2010. *NCHS Data Brief* 92(92): 1–8.

Carty CL, Johnson NA, Hutter CM, et al. (2012). Genome-wide association study of body height in African Americans: the Women's Health Initiative SNP Health Association Resource (SHARe). *Hum Mol Genet* 21(3): 711–720.

Cauchi S, El Achhab Y, Choquet H, et al. (2007). TCF7L2 is reproducibly associated with type 2 diabetes in various ethnic groups: a global meta-analysis. *J Mol Med* 85(7): 777–782.

Chasman DI, Pare G, Mora S, et al. (2009). Forty-three loci associated with plasma lipoprotein size, concentration, and cholesterol content in genome-wide analysis. *PLoS Genetics* 5(11). DOI: 10.1371/journal.pgen.1000730.

Cheserek MJ, Wu GR, Shen LY, et al. (2014). Disparities in the prevalence of metabolic syndrome and its components among university employees by age, gender and occupation. *J Clin Diagn Res* 8 (2): 65–69.

Chimusa ER, Zaitlen N, Daya M, et al. (2014). Genome-wide association study of ancestry-specific TB risk in the South African Coloured population. *Hum Mol Genet* 23(3): 796–809.

Cohen J, Pertsemlidis A, Kotowski IK, et al. (2005). Low LDL cholesterol in individuals of African descent resulting from frequent nonsense mutations in PCSK9. *Nat Genet* 37(2): 161–165.

Cole JW, Brown DW, Giles WH, et al. (2008). Ischemic stroke risk, smoking, and the genetics of inflammation in a biracial population: the stroke prevention in young women study. *Thrombosis J* 6: 11.

Cole JW, Stine OC, Liu X, et al. (2012). Rare variants in ischemic stroke: an exome pilot study. *PLoS One* 7(4). DOI: 10.1371/journal.pone.0035591.

Cook MB, Wang Z, Yeboah ED, et al. (2014). A genome-wide association study of prostate cancer in West African men. *Hum Genet* 133(5): 509–521.

Cooper RS, Guo X, Rotimi CN, et al. (2000). Heritability of angiotensin-converting enzyme and angiotensinogen: a comparison of US blacks and Nigerians. *Hypertens (Dallas, Tex: 1979)* 35(5): 1141–1147.

Curtis J, Luo Y, Zenner HL, et al. (2015). Susceptibility to tuberculosis is associated with variants in the ASAP1 gene encoding a regulator of dendritic cell migration. *Nat Genet* **47**(5): 523–527.

Danaei G, Singh GM, Paciorek CJ, et al. (2013). The global cardiovascular risk transition: associations of four metabolic risk factors with national income, urbanization, and Western diet in 1980 and 2008. *Circulation* **127**(14): 1493–1508.

Diabetes Genetics Initiative of Broad Institute of Harvard and MIT, Lund University, and Novartis Institutes of BioMedical Research, Saxena R, Voight BF, et al. (2007). Genome-wide association analysis identifies loci for type 2 diabetes and triglyceride levels. *Science* **316**(5829): 1331–1336.

Doumatey AP, Chen G, Tekola Ayele F, et al. (2012). C-reactive protein (CRP) promoter polymorphisms influence circulating CRP levels in a genome-wide association study of African Americans. *Hum Mol Genet* **21**(13): 3063–3072.

Doupa D, Seck SM, Dia CA, et al. (2014). Dyslipidemia, obesity and other cardiovascular risk factors in the adult population in Senegal. *Pan Afr Med J* **19**: 1–7.

Drager LF, Togeiro SM, Polotsky VY, et al. (2013). Obstructive sleep apnea: a cardiometabolic risk in obesity and the metabolic syndrome. *J Am Coll Cardiol* **62**(7): 569–576.

Ehret GB, Munroe PB, Rice KM, et al. (2011). Genetic variants in novel pathways influence blood pressure and cardiovascular disease risk. *Nature* **478**(7367): 103–109.

Elbaz A, Poirier O, Moulin T, et al. (2000). Association between the Glu298Asp polymorphism in the endothelial constitutive nitric oxide synthase gene and brain infarction: the GENIC investigators. *Stroke* **31**(7): 1634–1639.

Ellman N, Keswell D, Collins M, et al. (2015). Ethnic differences in the association between lipid metabolism genes and lipid levels in black and white South African women. *Atherosclerosis* **240** (2): 311–317.

El-shal AS, Pasha HF, and Rashad NM (2013). Association of resistin gene polymorphisms with insulin resistance in Egyptian obese patients. *Gene* **515**(1): 233–238.

Expert Panel on Detection, Evaluation and Treatment of High Blood Cholesterol in Adults (2001). Executive summary of the third report of the national cholesterol education program Expert Panel on Detection, Evaluation and Treatment of High Blood Cholesterol in Adults (Adult Treatment Panel III). *JAMA* **285**(19): 2486–2497.

Flanagan JM, Frohlich DM, Howard TA, et al. (2011). Genetic predictors for stroke in children with sickle cell anemia. *Blood* **117**(24): 6681–6684.

Flanagan JM, Sheehan V, Linder H, et al. (2013). Genetic mapping and exome sequencing identify two mutations associated with stroke protection in pediatric patients with sickle cell anemia. *Blood* **121**(16): 3237–3245.

Flossmann E, Schulz UGR, and Rothwell PM (2004). Systematic review of methods and results of studies of the genetic epidemiology of ischemic stroke. *Stroke* **35**(1): 212–227.

Fox ER, Young JH, Li Y, et al. (2011). Association of genetic variation with systolic and diastolic blood pressure among African Americans: the candidate gene association resource study. *Hum Mol Genet* **20**(11): 2273–2284.

Franceschini N, Fox E, Zhang Z, et al. (2013). Genome-wide association analysis of blood-pressure traits in African-ancestry individuals reveals common associated genes in African and non-African populations. *Am J Hum Genet* **93**(3): 545–554.

Frayling TM, Timpson NJ, Weedon MN, et al. (2007). A common variant in the *FTO* gene is associated with body mass index and predisposes to childhood and adult obesity. *Science* **316** (5826): 889–894.

Freedman BI, Langefeld CD, Lu L, et al. (2015). *APOL1* associations with nephropathy, atherosclerosis, and all-cause mortality in African Americans with type 2 diabetes. *Kidney Int* **87**(1): 176–181.

Genovese G, Friedman DJ, Ross MD, et al. (2010). Association of trypanolytic ApoL1 variants with kidney disease in African-Americans. *Science* **329**(5993): 841–845.

Goldstein LB, Bushnell CD, Adams RJ, et al. (2011). Guidelines for the primary prevention of stroke: a guideline for healthcare professionals from the American Heart Association/American Stroke Association. *Stroke* 42(2): 517–584.

Gonzaga C, Bertolami A, Bertolami M, et al. (2015). Obstructive sleep apnea, hypertension and cardiovascular diseases. *J Hum Hypertens* 29(12): 705–712.

Grant SF, Thorleifsson G, Reynisdottir I, et al. (2006). Variant of transcription factor 7-like 2 (TCF7L2) gene confers risk of type 2 diabetes. *Nat Genet* 38(3): 320–323.

Gretarsdottir S, Thorleifsson G, Reynisdottir ST, et al. (2003). The gene encoding phosphodiesterase 4D confers risk of ischemic stroke. *Nat Genet* 35(2): 131–138.

Gretarsdottir S, Thorleifsson G, Manolescu A, et al. (2008). Risk variants for atrial fibrillation on chromosome 4q25 associate with ischemic stroke. *Ann Neurol* 64(4): 402–409.

Grewal RP, Dutra AV, Liao YC, et al. (2007). The intron 4c allele of the NOS3 gene is associated with ischemic stroke in African Americans. *BMC Med Genet* 8: 76.

Gu C, Borecki I, Gagnon J, et al. (1998). Familial resemblance for resting blood pressure with particular reference to racial differences: preliminary analyses from the HERITAGE Family Study. *Hum Biol* 70(1): 77–90.

Gudbjartsson DF, Holm H, Gretarsdottir S, et al. (2009). A sequence variant in ZFHX3 on 16q22 associates with atrial fibrillation and ischemic stroke. *Nat Genet* 41(8): 876–878.

Gurdasani D, Carstensen T, Tekola-Ayele F, et al. (2014). The African Genome Variation Project shapes medical genetics in Africa. *Nature* 517(7534): 327–332.

Haiman CA, Chen GK, Blot WJ, et al. (2011). Genome-wide association study of prostate cancer in men of African ancestry identifies a susceptibility locus at 17q21. *Nat Genet* 43(6): 570–573.

Helgadottir A, Manolescu A, Thorleifsson G, et al. (2004). The gene encoding 5-lipoxygenase activating protein confers risk of myocardial infarction and stroke. *Nat Genet* 36(3): 233–239.

Helgadottir A, Thorleifsson G, Magnusson KP, et al. (2008). The same sequence variant on 9p21 associates with myocardial infarction, abdominal aortic aneurysm and intracranial aneurysm. *Nat Genet* 40(2): 217–224.

Helgason A, Palsson S, Thorleifsson G, et al. (2007). Refining the impact of TCF7L2 gene variants on type 2 diabetes and adaptive evolution. *Nat Genet* 39(2): 218–225.

Hennig BJ, Fulford AJ, Sirugo G, et al. (2009). FTO gene variation and measures of body mass in an African population. *BMC Med Genet* 10: 21.

Hooper AJ, Marais AD, Tanyanyiwa DM, and Burnett JR (2007). The C679X mutation in *PCSK9* is present and lowers blood cholesterol in a Southern African population. *Atherosclerosis* 193(2): 445–448.

Huan T, Esko T, Peters MJ, et al. (2015). A meta-analysis of gene expression signatures of blood pressure and hypertension. *PLoS Genet* 11(3): e1005035.

Humphries SE, Drenos F, Ken-Dror G, and Talmud PJ (2010). Coronary heart disease risk prediction in the era of genome-wide association studies: current status and what the future holds. *Circulation* 121(20): 2235–2248.

Innerarity TL, Weisgraber KH, Arnold KS, et al. (1987). Familial defective apolipoprotein B-100: low density lipoproteins with abnormal receptor binding. *PNAS* 84(19): 6919–6923.

Jackson FLC (2008). Ethnogenetic layering (EL): an alternative to the traditional race model in human variation and health disparity studies. *Ann Hum Biol* 35(2): 121–144.

Jallow M, Teo YY, Small KS, et al. (2009). Genome-wide and fine-resolution association analysis of malaria in West Africa. *Nat Genet* 41(6): 657–665.

Jerrard-Dunne P, Cloud G, Hassan A, and Markus HS (2003). Evaluating the genetic component of ischemic stroke subtypes: a family history study. *Stroke* 34(6): 1364–1369.

Ji W, Foo JN, O'Roak BJ, et al. (2013). Rare independent mutations in renal salt handling genes contribute to blood pressure variation. *Nat Genet* 40(5), 592–599.

Jin HS, Hong KW, Kim BY, et al. (2011). Replicated association between genetic variation in the *PARK2* gene and blood pressure. *Clinica Chimica Acta* **412**(17–18): 1673–1677.

Jones B (2014). Population genetics: the African Genome Variation Project. *Nat Rev Genet* **16**: 68–69.

Joubert BR, Lange EM, Franceschini N, et al. (2010). A whole genome association study of mother-to-child transmission of HIV in Malawi. *Genome Med* **2**(3): 17.

Kang SJ, Chiang CW, Palmer CD, et al. (2010). Genome-wide association of anthropometric traits in African- and African-derived populations. *Hum Mol Genet* **19**(13): 2725–2738.

Kathiresan S, Willer C, Peloso GM, et al. (2008a). Common variants at 30 loci contribute to polygenic dyslipidemia. *Nat Genet* **41**(1): 56–65.

Kathiresan S, Melander O, Guiducci C, et al. (2008b). Six new loci associated with blood low-density lipoprotein cholesterol, high-density lipoprotein cholesterol or triglycerides in humans. *Nat Genet* **40**(2): 189–197.

Kaufman JS, Owoaje EE, James SA, et al. (1996). Determinants of hypertension in West Africa: contribution of anthropometric and dietary factors to urban–rural and socioeconomic gradients. *Am J Epidemiol* **143**(12): 1203–1218.

Kaur J (2014). A comprehensive review on metabolic syndrome. *Cardiol Res Pract.* DOI: 10.1155/2014/943162.

Kidambi S, Ghosh S, Kotchen JM, et al. (2012). Non-replication study of a genome-wide association study for hypertension and blood pressure in African Americans. *BMC Med Genet* **13** (1): 27.

Kim S, Nho K, Ramanan VK, et al. (2016). Genetic influences on plasma homocysteine levels in African Americans and Yoruba Nigerians. *J Alzheimers Dis* **49**(4): 991–1003.

Komatsu M, Wheeler HE, Chung S, et al. (2015). Pharmacoethnicity in paclitaxel-induced sensory peripheral neuropathy. *Clin Cancer Res* **21**(19): 4337–4346.

Kooner JS, Chambers JC, Aguilar-Salinas CA, et al. (2008). Genome-wide scan identifies variation in *MLXIPL* associated with plasma triglycerides. *Nat Genet* **40**(2): 149–151.

Kopp JB, Nelson GW, Sampath K, et al. (2011). *APOL1* genetic variants in focal segmental glomerulosclerosis and HIV-associated nephropathy. *JASN* **22**(11): 2129–2137.

Lanfranconi S and Markus HS (2010). *COL4A1* mutations as a monogenic cause of cerebral small vessel disease: a systematic review. *Stroke* **41**(8): e513–e518.

Laston SL, Voruganti VS, Haack K, et al. (2015). Genetics of kidney disease and related cardiometabolic phenotypes in Zuni Indians: the Zuni Kidney Project. *Front Genet* **5**: 1–12.

Leigh SE, Foster AH, Whittall RA, et al. (2008). Update and analysis of the University College London low density lipoprotein receptor familial hypercholesterolemia database. *Ann Hum Genet* **72**(4): 485–498.

Levy D, Ehret GB, Rice K, et al. (2010). Genome-wide association study of blood pressure and hypertension. *Population (English Edition)* **41**(6): 677–687.

Li H, Wu Y, Loos RJ, et al. (2008). Variants in the fat mass- and obesity-associated (*FTO*) gene are not associated with obesity in a Chinese Han population. *Diabetes* **57**(1): 264–268.

Li M, Hou WS, Zhang XW, and Tang ZY (2014). Obstructive sleep apnea and risk of stroke: a meta-analysis of prospective studies. *Int J Cardiol* **172**(2): 466–469.

Lindstrom J, Louheranta A, Mannelin M, et al. (2003). The Finnish Diabetes Prevention Study (DPS): Lifestyle intervention and 3-year results on diet and physical activity. *Diabetes Care* **26**(12): 3230–3236.

Lingappa JR, Petrovski S, Kahle E, et al. (2011). Genomewide association study for determinants of HIV-1 acquisition and viral set point in HIV-1 serodiscordant couples with quantified virus exposure. *PloS One* **6**(12): e28632.

Liu YJ, Liu XG, Wang L, et al. (2008). Genome-wide association scans identified CTNNBL1 as a novel gene for obesity. *Hum Mol Genet* **17**(12): 1803–1813.

Loke YK, Brown JW, Kwok CS, et al. (2012). Association of obstructive sleep apnea with risk of serious cardiovascular events: a systematic review and meta-analysis. *Circ Cardiovasc Qual Outcomes* **5**(5): 720–728.

Lombard Z, Crowther NJ, van der Merwe L, et al. (2012). Appetite regulation genes are associated with body mass index in black South African adolescents: a genetic association study. *BMJ Open* **2**(3): e000873–e000873.

MacLeod MJ, Dahiyat MT, Cumming A, et al. (1999). No association between Glu/Asp polymorphism of *NOS3* gene and ischemic stroke. *Neurology* **53**(2): 418–420.

Marcadenti A (2016). Diet, cardiometabolic factors and type-2 diabetes mellitus: the role of genetics. *Curr Diabetes Rev* **12**(4): 322–330.

Masemola ML, Alberts M, and Urdal P (2007). Apolipoprotein E genotypes and their relation to lipid levels in a rural South African population. *Scand J Public Health* **69**(Suppl. 69): 60–65.

Mathenge W, Foster A, and Kuper H (2010). Urbanization, ethnicity and cardiovascular risk in a population in transition in Nakuru, Kenya: a population-based survey. *BMC Public Health* **10**: 569.

Matsha T, Fanampe, Yako Y, et al. (2010). Association of the *ENPP1* rs997509 polymorphism with obesity in South African mixed ancestry learners. *East Afr Med J* **87**(8): 323–329.

McDonnell MN, Hillier SL, Hooker SP, et al. (2013). Physical activity frequency and risk of incident stroke in a national US study of blacks and whites. *Stroke* **44**(9): 2519–2524.

McPherson R, Pertsemlidis A, Kavaslar N, et al. (2007). A common allele on chromosome 9 associated with coronary heart disease. *Science* **316**(5830): 1488–1491.

Mensah GA, Roth GA, Sampson UK, et al. (2015). Mortality from cardiovascular diseases in sub-Saharan Africa, 1990–2013: a systematic analysis of data from the Global Burden of Disease Study 2013. *Cardiovasc J Afr* **26**(2 Suppl. 1): S6–S10.

Minicuci N, Biritwum RB, Mensah G, et al. (2014). Sociodemographic and socioeconomic patterns of chronic non-communicable disease among the older adult population in Ghana. *Global Health Action* **7**(1). DOI: 10.3402/gha.v7.21292.

Modell B and Darlison M (2008). Global epidemiology of haemoglobin disorders and derived service indicators. *Bull WHO* **86**(6): 480–487.

Moe KT, Woon FP, De Silva DA, et al. (2008). Association of acute ischemic stroke with the *MTHFR* C677T polymorphism but not with *NOS3* gene polymorphisms in a Singapore population. *Eur J Neurol* **15**(12): 1309–1314.

Monda KL, Chen GK, Taylor KC, et al. (2013). A meta-analysis identifies new loci associated with body mass index in individuals of African ancestry. *Nat Genet* **45**(6): 690–966.

Moran AE, Forouzanfar MH, Roth GA, et al. (2015). The Global Burden of Disease 2010 Study. *Lancet*, **129**(14), 1493–1501.

Motawi T, Shaker O, Taha M, et al. (2011). Endothelial nitric oxide synthase and angiotensinogen gene polymorphism in coronary artery diseases in Egypt. *Angiology* **62**(2): 191–197.

Mozaffarian D, Benjamin EJ, Go AS, et al. (2015). Heart disease and stroke statistics: 2015 update – a report from the American Heart Association. *Circulation* **131**(4): e29–e322.

Mtatiro SN, Singh T, Rooks H, et al. (2014). Genome wide association study of fetal hemoglobin in sickle cell anemia in Tanzania. *PloS One* **9**(11): e111464.

Muhihi A, Njelekela M, Mpembeni R, et al. (2012). Physical activity and cardiovascular disease risk factors among young and middle-aged men in urban Mwanza, Tanzania. *Pan Afr Med J* **11**: 11.

Mukamal KJ, Tremaglio J, Friedman DJ, et al. (2016). *APOL1* genotype, kidney and cardiovascular disease, and death in older adults. *Arterioscler Thromb Vasc Biol* **36**(2): 398–403.

Murray CJL, Ortblad KF, Guinovart C, et al. (2014). Global, regional, and national incidence and mortality for HIV, tuberculosis, and malaria during 1990–2013: a systematic analysis for the Global Burden of Disease Study 2013. *Lancet* **384**(9947): 1005–1070.

Nannapaneni S, Ramar K, and Surani S (2013). Effect of obstructive sleep apnea on type 2 diabetes mellitus: a comprehensive literature review. *World J Diabetes* **4**(6): 238–244.

Njelekela MA, Liu E, Mpembeni R, et al. (2011). Socio-economic status, urbanization, and cardiometabolic risk factors among middle-aged adults in Tanzania. *East Afr J Public Health* **8**(3): 216–223.

Nordestgaard BG, Chapman MJ, Humphries SE, et al. (2013). Familial hypercholesterolaemia is underdiagnosed and undertreated in the general population: guidance for clinicians to prevent coronary heart disease. *Eur Heart J* **34**(45): 3478–3490.

O'Donnell MJ, Xavier D, Liu L, et al. (2010). Risk factors for ischaemic and intracerebral haemorrhagic stroke in 22 countries (the INTERSTROKE study): a case–control study. *Lancet* **376**(9735): 112–123.

O'Donnell PH, Stark AL, Gamazon ER, et al. (2012). Identification of novel germline polymorphisms governing capecitabine sensitivity. *Cancer* **118**(16): 4063–4073.

Ohene-Frempong K, Weiner SJ, Sleeper LA, et al. (1998). Cerebrovascular accidents in sickle cell disease: rates and risk factors. *Blood* **91**(1): 288–294.

Pennacchio LA, Olivier M, Hubacek JA, et al. (2002). Two independent apolipoprotein A5 haplotypes influence human plasma triglyceride levels. *Hum Mol Genet* **11**(24): 3031–3038.

Peppard PE, Young T, Barnet JH, et al. (2013). Increased prevalence of sleep-disordered breathing in adults. *Am J Epidemiol* **177**(9): 1006–1014.

Petrovski S, Fellay J, Shianna KV, et al. (2011). Common human genetic variants and HIV-1 susceptibility: a genome-wide survey in a homogeneous African population. *AIDS* **25**(4): 513–518.

Pradeilles R, Griffiths PL, Norris SA, Feeley AB, and Rousham EK (2015). Socio-economic influences on anthropometric status in urban South African adolescents: sex differences in the birth to twenty plus cohort. *Public Health Nutr* **18**(16): 2998–3012.

Ramsuran V, Kulkarni H, He W, et al. (2011). Duffy-null-associated low neutrophil counts influence HIV-1 susceptibility in high-risk South African black women. *Clin Infect Dis* **52**(10):1248–1256.

Reder NP, Tayo BO, Salako B, et al. (2012). Adrenergic alpha-1 pathway is associated with hypertension among Nigerians in a pathway-focused analysis. *PLoS ONE* **7**(5): 3–10.

Redline S, Yenokyan G, Gottlieb DJ, et al. (2010). Obstructive sleep apnea–hypopnea and incident stroke: the Sleep Heart Health Study. *Am J Respir Crit Care Med* **182**(2): 269–277.

Rees DC, Williams TN, and Gladwin MT (2010). Sickle-cell disease. *Lancet* **376**(9757): 2018–2031.

Rguibi M and Belahsen R (2004). Overweight and obesity among urban Sahraoui women of South Morocco. *Ethn Dis* **14**(4): 542–547.

Rice EW, Rich WK, Johnson CH, and Lye DJ (2006). The role of flushing dental water lines for the removal of microbial contaminants. *Public Health Reports* **121**(3): 270–274.

Riha J, Karabarinde A, Ssenyomo G, et al. (2014). Urbanicity and lifestyle risk factors for cardiometabolic diseases in rural Uganda: a cross-sectional study. *PLoS Med* **11**(7). DOI: 10.1371/journal.pmed.1001683.

Rotimi C, Cooper R, Ogunbiyi O, et al. (1997). Hypertension, serum angiotensinogen, and molecular variants of the angiotensinogen gene among Nigerians. *Circulation* **95**(10): 2348–2350.

Rotimi CN, Cooper RS, Cao G, et al. (1999). Maximum-likelihood generalized heritability estimate for blood pressure in Nigerian families. *Hypertension* **33**(3): 874–878.

Rotimi CN, Dunston GM, Berg K, et al. (2001). In search of susceptibility genes for type 2 diabetes in West Africa: the design and results of the first phase of the AADM study. *Ann Epidemiol* **11**(1): 51–58.

Rotimi CN, Chen G, Adeyemo AA, et al. (2004). A genome-wide search for type 2 diabetes susceptibility genes in West Africans: the Africa America Diabetes Mellitus (AADM) Study. *Diabetes* **53** (3): 838–841.

Rotimi CN, Tekola-Ayele F, Baker JL, et al. (2016). The African diaspora: history, adaptation and health. *Curr Opin Genet Dev* **41**: 77–84.

Rotondo D and Davidson J (2010). Genetics and molecular biology: identification of dyslipidemia genes. *Curr Opin Lipidol* **21**(6): 548–589.

Sabir AA, Isezuo SA, Ohwovoriole AE, et al. (2013). Rural–urban difference in plasma lipid levels and prevalence of dyslipidemia in Hausa-Fulani of north-western Nigeria. *Ethn Dis* **23**(3): 374–378.

Samani NJ, Erdmann J, Hall AS, et al. (2007). Genomewide association analysis of coronary artery disease. *N Engl J Med* **357**(5): 443–453.

Sarnak MJ, Levey AS, Schoolwerth AC, et al. (2003). Kidney disease as a risk factor for development of cardiovascular disease: a statement from the American Heart Association Councils on Kidney in Cardiovascular Disease, High Blood Pressure Research, Clinical Cardiology, and Epidemiology and Prevention. *Hypertension* **42**(5): 1050–1065.

Schunkert H, Konig IR, Kathiresan S, et al. (2011). Large-scale association analyses identifies 13 new susceptibility loci for coronary artery disease. *Nat Genet* **43**(4): 333–338.

Seshadri S, Beiser A, Pikula A, et al. (2010). Parental occurrence of stroke and risk of stroke in their children: the Framingham Study. *Circulation* **121**(11): 1304–1312.

Shah RS and Cole JW (2010). Smoking and stroke: the more you smoke the more you stroke. *Expert Rev Cardiovasc The* **8**(7): 917–932.

Sharma P and Meschia JF (2013). Stroke genetics. *Stroke Genetics* **20**(2): 1–320.

Sharma P, Yadav S, and Meschia JF (2013). Genetics of ischaemic stroke. *J Neurol Neurosurg Psychiatr* **84**(12): 1302–1308.

Shepard D, VanderZanden A, Moran A, et al. (2015). Ischemic heart disease worldwide, 1990 to 2013: estimates from the Global Burden of Disease Study 2013. *Circulation* **8**(4): 455–456.

Shriner D, Adeyemo A, Chen G, Rotimi CN, et al. (2010). Practical considerations for imputation of untyped markers in admixed populations. *Genet Epidemiol* **34**(3): 258–265.

Shriner D, Tekola-Ayele F, Adeyemo A, and Rotimi CN (2014). Genome-wide genotype and sequence-based reconstruction of the 140,000 year history of modern human ancestry. *Scientific Reports* **4**: 6055.

Silander K, Tang H, Myles S, et al. (2009). Worldwide patterns of haplotype diversity at 9p21.3, a locus associated with type 2 diabetes and coronary heart disease. *Genome Med* **1**(5): 51.

Simino J, Rao DC, and Freedman BI (2012). Novel findings and future directions on the genetics of hypertension. *Curr Opin Nephrol Hypertens* **21**(5): 500–507.

Slimane MN, Lestavel S, Clavey V, et al. (2002). CYS127S (FH-Kairouan) and D245N (FH-Tozeur) mutations in the LDL receptor gene in Tunisian families with familial hypercholesterolaemia. *J Med Genet* **39**(11): e74.

Sliwa K, Acquah L, Gersh BJ, and Mocumbi AO (2016). Impact of socioeconomic status, ethnicity, and urbanization on risk factor profiles of cardiovascular disease in Africa. *Circulation* **133**(12): 1199–1208.

Sõber S, Org E, Kepp K, et al. (2009). Targeting 160 candidate genes for blood pressure regulation with a genome-wide genotyping array. *PLoS One* **4**(6). DOI: 10.1371/journal.pone.0006034.

Soutar AK and Naoumova RP (2007). Mechanisms of disease: genetic causes of familial hypercholesterolemia. *Nat Clin Pract Cardiovasc Med* **4**: 214–225.

Tayo BO, Luke A, Zhu X, Adeyemo A, and Cooper RS (2009). Association of regions on chromosomes 6 and 7 with blood pressure in Nigerian families. *Circulation* **2**(1): 38–45.

Tayo BO, Kramer H, Salako BL, et al. (2013). Genetic variation in *APOL1* and *MYH9* genes is associated with chronic kidney disease among Nigerians. *Int Urol Nephrol* **45**(2): 485–494.

Tekola-Ayele F, Adeyemo A, Finan C, et al. (2012). HLA class II locus and susceptibility to podoconiosis. *New Eng J Med* **366**(13): 1200–1208.

Tekola-Ayele F, Adeyemo A, and Rotimi CN (2014). Genetic epidemiology of type 2 diabetes and cardiovascular diseases in Africa. *Progress Cardiovasc Dis* **56**(3): 251–260.

Teslovich T, Musunuru KS, Smith AV, et al. (2010). Biological, clinical and population relevance of 95 loci for blood lipids. *Nature* **466**(7307): 707–713.

Thorleifsson G, Walters GB, Gudbjartsson DF, et al. (2009). Genome-wide association yields new sequence variants at seven loci that associate with measures of obesity. *Nat Genet* **41**(1): 18–24.

Thye T, Vannberg FO, Wong SH, et al. (2010). Genome-wide association analyses identifies a susceptibility locus for tuberculosis on chromosome 18q11.2. *Nat Genet* **42**(9): 739–741.

Thye T, Owusu-Dabo E, Vannberg FO, et al. (2012). Common variants at 11p13 are associated with susceptibility to tuberculosis. *Nat Genet* **44**(3): 257–259.

Timmann C, Thye T, Vens M, et al. (2012). Genome-wide association study indicates two novel resistance loci for severe malaria. *Nature* **489**(7416): 443–446.

Torkamani A, Topol EJ, and Schork NJ (2008). Pathway analysis of seven common diseases assessed by genome-wide association. *Genomics* **92**(5): 265–272.

Tsai CT, Hwang JJ, Lai LP, et al. (2009). Interaction of gender, hypertension, and the angiotensinogen gene haplotypes on the risk of coronary artery disease in a large angiographic cohort. *Atherosclerosis* **203**(1): 249–256.

Tzur S, Rosset S, Skorecki K, et al. (2012). *APOL1* allelic variants are associated with lower age of dialysis initiation and thereby increased dialysis vintage in African and Hispanic Americans with non-diabetic end-stage kidney disease. *Nephrol Dial Transplant* **27**(4): 1498–1505.

United Nations (2013). World Population Prospects: The 2012 Revision. Highlights and Advance Tables. *Popul Dev Rev* **36**: 775–801.

Vandrovcova J, Thomas ER, Atanur SS, et al. (2013). The use of next-generation sequencing in clinical diagnosis of familial hypercholesterolemia. *Genet Med* **15**: 1–10.

Watts GF, Gidding S, Wierzbicki AS, et al. (2014). Integrated guidance on the care of familial hypercholesterolemia from the International FH Foundation. *J Clin Lipidol* **8**(2): 148–172.

Weatherall D, Hofman K, Rodgers G, et al. (2005). A case for developing North–South partnerships for research in sickle cell disease. *Blood* **105**(3): 921–923.

Weiss LA, Pan L, Abney M, and Ober C (2006). The sex-specific genetic architecture of quantitative traits in humans. *Nat Genet* **38**(2): 218–222.

Wellcome Trust Case Control C (2007). Genome-wide association study of 14,000 cases of seven common diseases and 3,000 shared controls. *Nature* **447**(7145): 661–678.

Welter D, MacArthur J, Morales J, et al. (2014). The NHGRI GWAS Catalog, a curated resource of SNP-trait associations. *Nucleic Acids Res* **42**(D1): 1001–1006.

Wheeler HE, Gorsic LK, Welsh M, et al. (2011). Genome-wide local ancestry approach identifies genes and variants associated with chemotherapeutic susceptibility in African Americans. *PloS One* **6**(7): e21920.

Wheeler HE, Gamazon ER, Stark AL, et al. (2013). Genome-wide meta-analysis identifies variants associated with platinating agent susceptibility across populations. *Pharmacogenomics J* **13**(1): 35–43.

Wild S, Roglic G, Green A, Sicree R, and King H (2004). Global prevalence of diabetes: estimates for the year 2000 and projection for 2030. *Diabetes Care* **27**(5): 1047–1053.

Willer CJ, Schmidt EM, Sengupta S, et al. (2013). Discovery and refinement of loci associated with lipid levels. *Nat Genet* **45**(11): 1274–1283.

Woo D, Sauerbeck LR, Kissela BM, et al. (2002). Genetic and environmental risk factors for intracerebral hemorrhage: preliminary results of a population-based study. *Stroke* **33**(5): 1190–1195.

World Health Day (2013). A global brief on hypertension. www.who.int/cardiovascular_diseases/publications/global_brief_hypertension/en.

World Health Organization (2008). Global Health Observatory Data Repository. Available at: http://apps.who.int/gho/data/?theme=main.

World Health Organization (2010). *Global Status Report on Noncommunicable Diseases 2010*. New York: World Health Organization.

Yako YY, Echouffo-Tcheugui JB, Balti EV, et al. (2015). Genetic association studies of obesity in Africa: a systematic review. *Obesity Reviews* **16**(3): 259–272.

Yang JK, Zhou JB, Xin Z, et al. (2010). Interactions among related genes of renin-angiotensin system associated with type 2 diabetes. *Diabetes Care* **33**(10): 2271–2273.

Yang Q, Liu B, and Zhang L (2012). Genetics of intracerebral hemorrhage: insights from candidate gene approaches. *Neurology India* **60**(1): 3.

Zhu X, Luke A, Cooper RS, et al. (2005). Admixture mapping for hypertension loci with genome-scan markers. *Nat Genet* **37**(2): 177–181.

Ziliak D, O'Donnell PH, Im HK, et al. (2011). Germline polymorphisms discovered via a cell-based, genome-wide approach predict platinum response in head and neck cancers. *Transl Res* **157**(5): 265–272.

9 Breast Cancer in African Populations

Konduru S. Sastry and Lotfi Chouchane

9.1 Introduction

Breast cancer is the second most common form of malignant neoplasia in the world, with an estimated 1.67 million new cancer cases diagnosed in 2012, accounting for 25 percent of all cancers (Ferlay et al. 2013). It is the most frequently diagnosed type of cancer in women in both the developed and less developed regions. The biological and pathological diversity of this disease as influenced by distinct phenotypes, differences in prognosis, and mixed responses to treatment limit the progress to reduce the burden of this disease (Kestler et al. 2011). Thus, the incidence of breast cancer is rising in many parts of the world, especially in developing countries. Breast cancer does not manifest equally in various regions of the world. Its incidence is higher in North America and Northern and Western Europe compared to Southern Europe, and lowest in Africa and Asia (Ferlay et al. 2013).

Africa is the second largest and second most populous continent, with an estimated population of 1.033 billion people in 2013 (World Populations Review 2015). Africa is home to 54 recognized sovereign states and countries, nine territories, and two de facto independent states. Breast cancer is the most common cancer in women from African populations, with a mortality of 10–26 per 100,000 women per year (Ferlay et al. 2013). Although the incidence remains lower in African countries, it is rising steadily, and the exact burden of this disease is not known. Thus, most of these statistics from several African countries are likely to be under-representations of the true numbers because of financial and infrastructural limitations. This is particularly true of communities that still lack access to proper health care (Hayanga and Newman 2007). Lack of research in several African countries substantially hinders the progress in understanding the various aspects of the disease in these populations.

Here, we review published scientific literature on breast cancer in African countries and analyze available data for epidemiological, clinical, and pathological features of the disease. In addition, we also analyze genetic risk factors, specific molecular features of breast cancer, and unique features of inflammatory breast cancer in African populations.

9.2 Incidence and Mortality of Breast Cancer in African Populations

According to GLOBOCAN 2012, the age-standardized incidence rate (ASR) for breast cancer varies between countries in Africa (Figure 9.1), and the rate is growing in

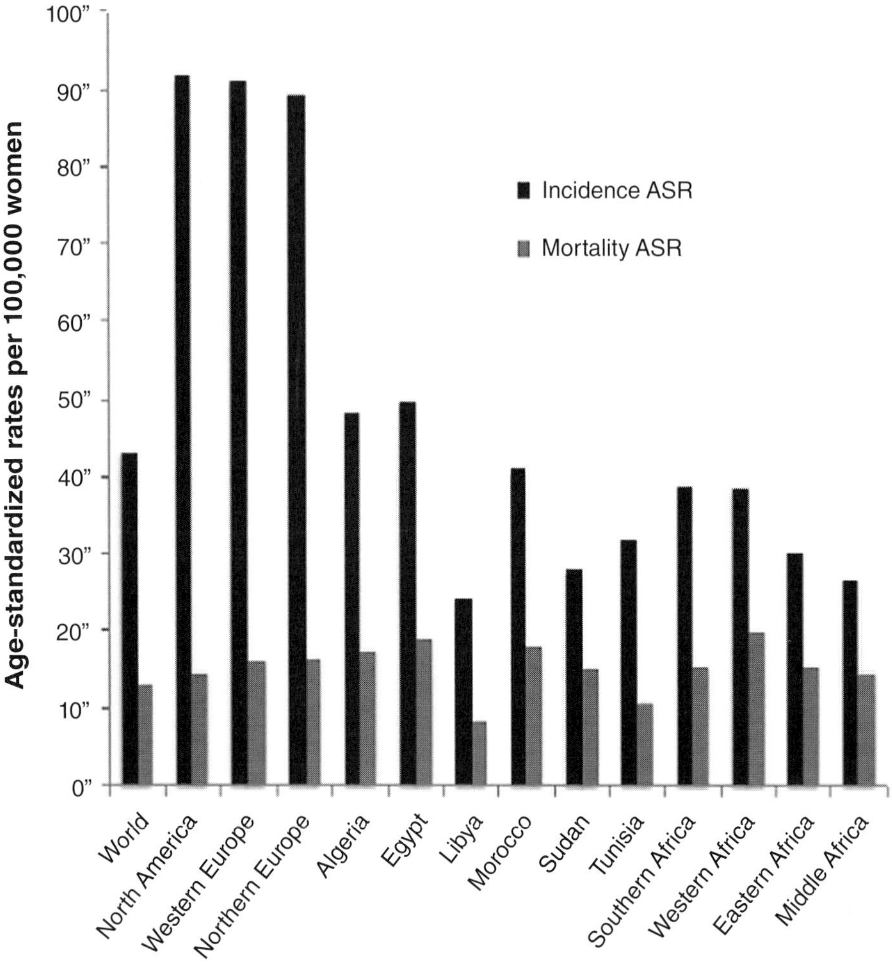

Figure 9.1 Age-standardized breast cancer incidence and mortality rates per 100,000 women according to GLOBOCAN 2012.

many countries in this area (Chouchane et al. 2013). For example, in Tunisia 17 cases per 100,000 women were reported in 1994, increasing to 31.8 in 2012 (Chouchane et al. 2013; Ferlay et al. 2010, 2013). Similarly, in Libya, according to Sabratha Cancer Registry, the incidence increased from 40 per 100,000 in 2002 to 50 per 100,000 in 2006 (Boder et al. 2011). In Algeria, the disease burden increased from 28.6 in 2002 to 48.5 in 2012 (Ferlay et al. 2010, 2013). In Nigeria, breast cancer incidence has shown a marked increase from 13.7 per 100,000 in the 1970s to 50.4 in 2012 (Boder et al. 2011). A similar trend of increased incidence of breast cancer was documented in other African populations (Chokunonga et al. 2013; Parkin et al. 2008; Vanderpuye et al. 2017).

The exact incidence rate of breast cancer in several African populations is not known because of lack of comprehensive data. In a defined geographic region in

Cairo, Egypt, initial screening in 4116 women aged 35–64 has detected eight breast cancer cases per 1000 women, and two additional cases upon rescreening (Boulos et al. 2005). Similarly, in Tunisia, mammography screening detected 4.4–4.6 cases per 1000 women (Zaanouni et al. 2009). These findings suggest that breast cancer may be more common than reported in African countries.

Despite the low incidence rates of breast cancer in Africans compared to Western populations, the ratio of mortality to incidence is higher in Africa (Vanderpuye et al. 2017). The mortality ASRs of breast cancer in Africa range from 10 to 26 per 100,000 women. The mortality rates in some African countries such as Nigeria, Egypt, and Ethiopia are among the highest in the world.

9.3 Clinicopathological Features

9.3.1 Age at Presentation

Several reports from various African countries indicate that the average age at presentation of breast cancer in African women is a decade earlier than in Western countries. In a comparative clinical study, we demonstrated that South Mediterranean breast cancer patients are ten years younger than French patients (Chalabi et al. 2008). A study in African populations reported that the peak age incidence of breast cancer was 40–49 years, accounting for 36 percent, at least ten years younger than in Western countries, where the average age of onset is 60 years (Nggada et al. 2008). Most breast cancer cases in Africa are found in the fourth and fifth decades of life (Forae et al. 2014). In Tunisian women, the median age at diagnosis was 48 years and the mean age was 50.2 years. A total of 57.1 percent of patients were younger than 50 years old (Missaoui et al. 2012). Similarly, in women from Morocco and Egypt, the mean age of breast cancer diagnosis is 45–50 years and 51 years, respectively (Salhia et al. 2011; Slaoui et al. 2014). Studies have reported that African American women and black women residing in the UK have a similar age of presentation of breast cancer to native African women (Forae et al. 2014). Similar patterns of younger age at presentation have been reported in other African populations (Edge et al. 2014; Othieno-Abinya et al. 2002).

9.3.2 Tumor Stage and Type

Many African women with palpable breast cancer do not seek medical help, leading to late presentation and advanced disease at first diagnosis. In many publications, it is reported that breast cancer is more aggressive in Africa than in Western countries (Corbex et al. 2014). In a study comprising 203 breast cancer cases in Egypt, the tumors were >2 cm in size with either intermediate grade (82.2 percent) or high grade (17.1 percent); none were low grade (Salhia et al. 2011). In a similar study in Tunisian women with breast cancer, 81 percent of individuals displayed tumor sizes >2 cm, and in 34 percent of cases the tumor size was >5 cm. Stage II and Stage III or IV cancers represented 48 percent and 41 percent, respectively. Only 11 percent of tumors were detected at an early stage (Missaoui et al. 2011). A study in Nigerians revealed that Stages III and IV constituted 76.2 percent of all breast cancer cases (Forae et al. 2014). A similar pattern of tumor size, grade, and staging has been reported in other African countries (Bird et al. 2008; Elgaili et al. 2010; Van Bogaert 2013).

Based on histopathology, invasive ductal carcinoma was the most prevalent type reported in African patients, accounting for 70–90 percent of all breast cancers. Ductal carcinoma *in situ* represents fewer than 5 percent of cases (Forae et al. 2014). Medullary carcinoma is more common in African populations than in European cases (Boder et al. 2011). The majority of cancers were detected to have lymph node involvement. The proportion of lymph node involvement was 74 percent in Libya (Boder et al. 2011), 70 percent in Egypt (Salhia et al. 2011), and 69 percent in Morocco (Ismaili et al. 2014).

9.3.3 Hormone Receptor Subtypes

The distribution of molecular subtypes of breast cancer varies between African populations. In a study of Egyptian women, the immunohistochemical analysis revealed that the majority of tumors were Luminal-A (44.3 percent), 24.6 percent were Luminal-B, and less than 12 percent were Her2+/ER–, unclassified or Basal-like (Salhia et al. 2011). In Morocco, Luminal-A subtype was predominant, with a prevalence of 53.6 percent. Luminal-B, Her2+, Basal-like, and unclassified subtypes represented 16.4 percent, 12.6 percent, 12.6 percent, and 4.9 percent, respectively (Slaoui et al. 2014). In a retrospective study from Tunisia, triple-negative tumors accounted for 23 percent, followed by Luminal-A 51 percent, Luminal-B 13 percent, and Her2+ 13 percent (Fourati et al. 2012). In the Senegal/Nigerian populations, 27 percent of tumors were Luminal-A, while Luminal-B accounted for 2 percent and HER2+/ER negative represented 15 percent. Triple-negative and unclassified tumors were 27 percent and 28 percent, respectively (Huo et al. 2009). Triple-negative tumors are aggressive and have poor survival (Boyle 2012). Several other studies reported a high proportion (17–28 percent) of triple-negative breast cancer in African populations (Amir et al. 2000; El-Hawary et al. 2012; Fourati et al. 2012; Rais et al. 2012) as well as in African American women (Morris and Mitchell 2008).

9.3.4 Inflammatory Breast Cancer

Inflammatory breast cancer (IBC) is rare, but the most lethal form of breast cancer, characterized by erythema, edema, peau d'orange, diffuse enlargement, and ulceration, resulting from an increased number of dermal tumor lymphatic emboli that block the drainage of lymph fluids (Dawood et al. 2011; Singletary and Cristofanilli 2008). Compared to other types of breast cancer, IBC affects women at younger age, progresses rapidly, and tends to be more advanced at first diagnosis, with a poor prognosis (Boussen et al. 2008; Maalej et al. 1999; Soliman et al. 2009).

Inflammatory breast cancer is the most aggressive form of breast cancer and constitutes 1–2 percent of all breast cancers in the United States (Anderson et al. 2005–2006). However, there is an apparent striking geographic pattern of IBC, with a higher percentage of cases reported in African populations, i.e., 5–11 percent of all breast cancers, as best documented in Tunisia, Egypt, and Morocco (Dawood et al. 2011; Molckovsky et al. 2009; Singletary and Cristofanilli 2008, Soliman et al. 2011). Although historically very high proportions of IBC were reported in Tunisia (59 percent) and Egypt (65 percent), more recent epidemiologic analysis of all the new cases of breast cancer diagnosed in Tunisia found that the incidence of IBC cases represents 6.2 percent of the total cases (Boussen et al. 2008), still substantially higher than US populations. In line with these observations, a recent population-based study confirmed previous reports that IBC in the Gharbiah region of Egypt is higher than in the USA and constitutes up to 11 percent (Soliman et al. 2009). Thus, it seems that the prevalence of IBC in African populations is high compared to developed countries.

Risk factors for IBC are not clear, but some established risk factors including ethnicity, obesity, socioeconomic level, and young age may explain differences in epidemiological aspects of African, European, and US populations (Schlichting et al. 2012a, 2012b). Recent studies report that the incidence of IBC is declining in some African countries as a result of stringent diagnostic criteria, improvement in socio-economic conditions, early diagnosis, and increased awareness (Boussen et al. 2008, Labidi et al. 2008).

Significant molecular and pathological differences have been reported between IBC in African and Western populations, which might be linked to the more aggressive clinical behavior of IBC in African women. A high level of oncogene protein RhoCGTPase overexpression was found in only 14 percent (one of seven) IBC tumors from the USA compared with 87 percent (40 of 46) in Egyptian IBC patients (Lo et al. 2008). Another study revealed that IBC tumors from Egyptian patients exhibited higher levels of RhoC overexpression compared to patients from Tunisia and Morocco (Soliman et al. 2011). The elevated expression of RhoC plays a pivotal role in the metastatic behavior of IBC by modulating the expression of several genes involved in angiogenesis, cell cycle progression, lymphangiogenesis, and cell adhesion and invasion (Wu et al. 2004). In addition, IBC tumors from the Egyptian patients had a higher number of tumor emboli than those of US patients. The quantity of dermal tumor emboli has been shown to be associated with poor prognosis and positive lymph node involvement for IBC patients. A high frequency of a mouse mammary tumor virus (MMTV)-like sequence was also noticed in Tunisian patients (74 percent) compared with patients from the USA (36 percent), Italy (38 percent), and Australia (42 percent) (Levine et al. 2004). Together, these molecular and clinicopathological differences might suggest that there is a greater potential for metastasis in IBC patients from Africa than from the USA.

In summary, breast cancer in African populations affects patients who are a decade younger, with advanced stages at presentation, generally with poorer clinicopathologic variables such as larger tumor size, lymph node positivity, poorly differentiated histopathology, and maintenance of higher ER and PR negativity, suggesting aggressive behavior. Significant disparities are evident with respect to molecular expression pattern and subtype distribution between African and Western patients.

9.4 Screening

Breast cancer burden and mortality can be substantially reduced through efforts directed at screening and early detection. A recent study estimated that mammographic screening led to a reduction in the risk of breast cancer mortality by about 15 percent (Gotzsche and Jorgensen 2013). Despite several awareness campaigns conducted in some African populations, national-level organized mammography screening programs still do not exist in these countries (Abdulrahman and Rahman 2012; El Saghir et al. 2007). Only recently have a few promising screening initiatives emerged in some countries, with the objectives to detect small tumors and subclinical breast cancers. Mammography is generally not recommended for women below 35 years. Since substantial numbers of breast cancer patients in African countries are in the age group 35–40, it is advised to screen by mammography starting at age of 40 instead of 50. Due to the predominance of advanced breast cancer and the high cost of structured screening, a different strategy such as "prevention of advanced disease at presentation" has been recommended (El Saghir and Abulkhair 2010). The goal would be the detection of small tumors before they grow further or involve the skin or chest wall, or metastasize to other organs.

Several barriers appear to influence women's participation in screening. Insufficient levels of education about breast cancer, poverty, and difficulty in accessing health care facilities prohibit women from participating in breast screening (Abdulrahman and Rahman 2012; Anderson et al. 2011; Galukande and Kiguli-Malwadde 2010; Okobia et al. 2006). Embarrassment and shyness to uncover breasts to a physician, family relationships, fatalism, and consultation of traditional healers are some of the culture-specific barriers (Azaiza and Cohen 2006; El Saghir and Abulkhair 2010; Ermiah et al. 2012; Galukande and Kiguli-Malwadde 2010; Najjar and Easson 2010). Several women expressed the belief that a cancer diagnosis can reflect negatively on the family of an affected individual, particularly females. In many cases, this concern led individuals to keep their cancer diagnoses a secret or avoid seeking medical attention when symptoms of cancer were suspected (Chouchane et al. 2013). Myths and misconceptions about the disease include beliefs that cancer is contagious, that undergoing an operation can cause cancer to spread, and that cancer is a death sentence (Morse et al. 2014). Improvement in awareness and health care facilities and implementation of mammography screening could help with early detection of the disease.

9.5 Hereditary Risk Factors

It is well established that genetic factors play a substantial role in epidemiology and pathogenesis of both familial and sporadic breast cancer. About 5–10 percent of breast cancer cases are reported to be hereditary, resulting directly from genetic changes inherited from a parent. Recent genome-wide association studies (GWASs) have identified and confirmed various novel breast cancer susceptibility variants, including four rare high-penetrance genes (*BRCA1*, *BRCA2*, *TP53*, and *PTEN*), four rare moderate-penetrance genes (*CHEK2*, *ATM*, *BRIP1*, and *PALB2*), and around 20 common low-penetrance variants in 19 genes or loci that contribute to a woman's risk of breast cancer (Easton et al. 2007; Hunter et al. 2007). Germline mutations in *BRCA1* and *BRCA2* contribute to a significant number of familial and hereditary breast cancers (Zhang et al. 2011). The proportion of high-risk families with breast cancer cases due to mutations in these tumor suppressor genes varies widely among populations.

9.5.1 *BRCA1* and *BRCA2* Mutations in African Populations

The *BRCA1* gene is mapped to chromosome 17q21 and consists of 24 exons. It encodes a protein of 220 kDa, composed of 1863 amino acids (Chen et al. 1996). *BRCA2* gene is located on the long arm of chromosome 13. It contains 27 exons and encodes a 380 kDa protein (Bertwistle et al. 1997). Both *BRCA1* and *BRCA2* are tumor suppressor proteins and help in maintaining genomic stability through their role in DNA damage signaling and repair (Joukov et al. 2006; Patel et al. 1998). *BRCA1* and *BRCA2* are linked to 70 percent of breast cancer in high-risk families, and account for <10 percent of all breast cancer cases, probably because of their low allele frequencies in the general population (Honrado et al. 2004). The analysis of the *BRCA1/2* genes, as well as other genes conferring increased risk of hereditary breast cancer (e.g., *TP53*, *PTEN*, *STK11*, *CDH1*), is poorly studied in African populations. A few reports, including from our group, reported *BRCA1/2* mutations associated with hereditary breast cancers in a small number of families from African countries. Reports suggest that the proportion of *BRCA1* and *BRCA2* mutations are higher in African populations. The frequency of *BRCA1* mutations in Moroccan women with family histories of breast cancer or ovarian cancer ranged between 12.8 percent and 31.6 percent (Laraqui et al. 2013; Tazzite et al. 2012). Two separate studies performed in Tunisian women reported a prevalence of 15.6 percent and 37.5 percent of *BRCA1* mutations (Mahfoudh et al. 2012a; Troudi et al. 2008). Similarly, a much higher frequency has been observed in Algerian patients: 9.8 percent (5/51) of early-onset sporadic and 36.4 percent (4/11) of familial cases were found to be associated with *BRCA1* mutations (Uhrhammer et al. 2008). Similar high frequencies of *BRCA* mutations were reported in other African populations (Fackenthal et al. 2012).

The prevalence of *BRCA1* mutations is 1.5–2 times higher than that of *BRCA2* as documented in Tunisia. Our previous study screened 16 high-risk breast cancer Tunisian families comprising 24 patients, and identified four deleterious mutations in *BRCA1*, of which c.212+2insG was found to be novel and Tunisian-specific (Mahfoudh et al. 2012a). Our study also revealed that the c.798_799delTT was the most commonly observed mutation and suggests that it is a Tunisian and northern African founder mutation. The c.5266dupC mutation in exon 20 was found to be the most frequent *BRCA1* gene mutation among Egyptian women, followed by c.181T>G in exon 5 and c.68 69delAG in exon 2 (Ibrahim et al. 2010). In the Moroccan population, eight distinct *BRCA1* mutations (c.181T>G, c.68 69delAG, c.798 799delTT, c.1016dupA, c.2805delA, c.3279delC, c.5062 5064delGTT, and c.5095C>T) were observed (Laraqui et al. 2013; Tazzite et al. 2012). In addition, 32 *BRCA1* variants including distinct polymorphisms and unclassified variants were identified. The most frequently detected unclassified variant was c.5117G>C (p.Gly1706Ala) (Laraqui et al. 2013). In 49 out of 59 breast cancer patients in Sudan, a total of 33 *BRCA1* variants were detected. Of these, c.1088A>G:p.Asn363Ser and c.5090G>A:p. Cys1697Tyr mutations were found to be novel (Biunno et al. 2014). A study performed in Algerian women (Uhrhammer et al. 2008) identified five deleterious *BRCA1* mutations among 51 early-onset sporadic cases and four mutations among 11 families (c.46 74del29, c.83 84delTG, c.202+1G>A, and c.798 799delTT). Screening of 70 Nigerian breast cancer patients with early onset of the disease revealed that 23 percent of them had *BRCA* mutations (Gao et al. 2000). Two novel protein-truncating *BRCA1* mutations (Q1090X and 1742insG) and one protein-truncating *BRCA2* (3034del4 or 3036del4) mutation were identified (Gao et al. 2000). Disease-causing mutations were identified in 20 percent (18/90) of South African breast/ovarian cancer families (Reeves et al. 2004). Four Ashkenazi Jewish families were identified with the 185delAG mutation, whereas two Afrikaner and one Ashkenazi Jewish family were found to harbor the 5382insC mutation. Five of the families (5.56 percent), all of whom were Afrikaners, were found to carry the novel E881X mutation. A frameshift truncating *BRCA1* mutation, 926ins10, was found in a breast/ovarian cancer family from Ivory Coast (Stoppa-Lyonnet et al. 1997). These results suggest that *BRCA* mutations might contribute to breast cancer in African populations; however, much larger cohort sizes of patients would be needed to establish the most accurate frequency and spectrum of *BRCA* mutations and their role in disease contribution.

9.6 Association of Genetic Polymorphism with Breast Cancer in African Populations

Only a small minority (~5 percent) of patients with breast cancer develops the disease as a result of inheritance of germline mutations in dominant, highly penetrant susceptibility genes such as *BRCA1* and *BRCA2*. However, single nucleotide polymorphisms (SNPs) in genes involved in the complex mechanisms of carcinogenesis may confer low-penetrant susceptibility to breast cancer in a significant proportion of the remaining patients. Such gene polymorphisms, in combination with nutritional and environmental factors, can affect breast cancer development.

A genetic variant might be associated with disease risk in some populations but not in others. Several factors can contribute to this: the variant might not be causal, the variant allele might be rare in some populations, genetic and environmental backgrounds can modify the risk, and there might be distributions of breast cancer subtypes across populations (Steven 2011). Most of the studies that address the association between breast cancer risk and genetic factors were performed in the populations of the United States, Europe, and Asia; only a few such studies were carried out in African populations. Different genetic associations between polymorphisms and breast cancer risk in various ethnic groups have been found for many genetic variants. For example, statistically significant associations were found between *GSTT1* (null/present), *SULT1A1* (R213H), and *XRCC1* (Arg399Gln) and breast cancer risk in the overall population, but only non-significant associations were found in Asians (Peng et al. 2011).

In a few studies in African populations, various SNPs have been associated with breast cancer risk. We observed a significant association between an *IL-8* (-251) allele and increased risk of breast cancer, as well as the presence of lymph node metastasis, high SBR grading, and poor prognosis with reduced overall survival in a Tunisian population (Snoussi et al. 2006). Our study also noted an association between *IL-8* (-251) AA genotype and aggressive forms of breast carcinoma as defined by the large tumor size, high grade, and lymph node metastasis. In another recent study, we identified an association of ApoA1-75 G/A and +83 C/T genotypes with breast cancer susceptibility and lymph node metastases, respectively, in a Tunisian population. Apolipoprotein A1, a major apoprotein of high-density lipoproteins, plays an important role in tumor invasion and metastasis (Hamrita et al. 2011).

Previous GWASs identified a link between breast cancer risk and polymorphisms in *FGFR2*, *TNRC9*, *MAP3K1*, *LSP1*, 2q35, and 8q240 genes. We performed a GWAS to assess whether these polymorphisms confer breast cancer risk in Tunisian women (Shan et al. 2012). Our study revealed that only five out of nine GWAS breast cancer loci were found to be significantly associated with breast cancer in Tunisians, highlighting the existence of different genetic susceptibility patterns in African populations than those reported for other populations. These differences could stem from disparities in genetic background, including differences in allele frequencies and LD pattern, and gene–environment interactions.

Variation of *VEGF-A* gene might influence breast cancer susceptibility, as SNPs in this gene have been associated with breast cancer risk. A case–control study in Moroccan women identified that women carrying -1154AG + AA, -2578AC + AA, -460CT and CT + CC VEGF genotypes had a reduced risk of developing breast cancer (Rahoui et al. 2014). The SNP rs763110 (-844 T/C) of the *FASL* gene is located within a putative binding motif of CAAT/enhancer-binding protein β transcription factor. We observed FASL-844 CC genotype and FASL-844 C allele were significantly associated with SBR 1–2 tumor grade (Mahfoudh et al. 2012b). In patients with diagnosis age $\leq$50 years, FASL-844 CC genotype and C allele showed significant associations with T(1)–T(2) clinical tumor size and SBR grade 1–2. p53 plays a key role in maintaining genomic integrity and preventing tumorigenesis. The G13964C intronic variant in the p53 gene was significantly associated with familial breast cancer risk in Tunisian women (Trifa et al. 2010).

Functional polymorphisms in HER2 and estrogen receptor alpha can potentially affect breast cancer risk, as well as cancer phenotype and outcome. A significant association was found between breast cancer risk and the short tandem repeat (STR)

[H(AC)I4] in HER2 and the ESR1 2014G allele of estrogen receptor alpha in Tunisian women (Kallel et al. 2009, 2010). Similarly, a significantly higher risk of breast cancer was observed among carriers of homozygous ESR1 325CC and heterozygous Her-2/neu 655 Ile/Val in Sudanese women (Siddig et al. 2008). Leptin and leptin receptor have been implicated in processes leading to breast cancer initiation and progression. The LEPR Gln223Arg polymorphism in the extracellular domain of the LEPR receptor gene has been shown to be associated with a modestly increased risk of premenopausal breast cancer in Nigerian women (Okobia et al. 2008).

The insulin-like growth factor (IGF) signaling pathway is thought to play a major role in the etiology of breast cancer. Significant associations with breast cancer were observed across approximately 50 kb of DNA sequence encompassing three exons in the 3′ end of *IGFBP2* and three exons in the 3′ end of *IGFBP5* genes on chromosome 2q35 in African American and Nigerian women (Garner et al. 2008). While *CYP1A1 M1* polymorphism (CYP1A1M1 [T/C]) conferred reduced risk of breast cancer, homozygosity for the *CYP1A1 M3* (C/C) was associated with increased risk in Nigerian women (Okobia et al. 2005).

In summary, studies conducted in various African populations suggest that while the association of some SNPs with breast cancer overlap with those of other populations, several other SNPs seem to be specific to the indigenous population, highlighting the need for replication in other African populations. New GWASs in African women should be conducted to reveal new causal variants that are needed to fully uncover the genetic basis for breast cancer susceptibility in African populations. This promises to assist in the identification of novel breast cancer-related genes as well as in the development of new diagnostic tests and therapeutics aimed at early detection and for predicting an individual's likelihood of developing breast cancer. With the emphasis on developing simpler SNP assays that can simultaneously evaluate larger numbers of SNPs, it would help in the development of a product that would consolidate several predictive cancer-related polymorphisms and mutations into a single diagnostic test (Engle et al. 2006).

9.6.1 Consanguinity

The incidence of consanguinity is high in North African countries. The rate of consanguineous marriages in Sudan, Egypt, Morocco, Algeria, and Tunisia are between 15 and 50 percent, of which more than half are predominantly first-cousin marriages (Bener et al. 2010; Chouchane et al. 2013; Elalaoui et al. 2013; Romdhane et al. 2014). The epidemiological studies that have examined adult breast cancer risk and consanguinity have reported conflicting results. Consanguinity, which increases the chances of homozygosity, is linked to an increased overall risk of cancer and elevated risk of breast cancer (Gilani and Kamal 2004). This increased risk was explained by Khoury et al.: By creating a homozygous state for a recessive tumor gene, consanguinity may provide a congenital setting for multistage carcinogenesis and, theoretically, increased cancer risk (Khoury et al. 1987). On the contrary, some studies have shown that consanguinity has no effect, or may even have a mild protective effect on breast cancer risk, probably because the long-term practice of consanguinity may decrease the frequency of deleterious genes or eliminate them from a population. A lower risk of breast cancer was reported among those whose parents were consanguineous, compared to those whose parents were non-consanguineous (Mahfoudh et al. 2012a). Similarly, another study suggested that parental consanguinity in Moroccan women might not be associated with an altered risk of breast cancer (Romdhane et al. 2014).

9.7 Expression Profiling Studies in African Populations

Until recently, tumor-related factors such as lymph node status, tumor size, histologic grade, proliferative index, and age were considered as prognostic factors, while hormone receptor and Her2/neu status served as both prognostic and predictive factors. Although these parameters are useful in predicting the outcome, they do not fully take into account the biologic complexity of an individual's tumor. For example, morphologically identical tumors can result in divergent clinical outcomes, which can be attributed to molecular class differences that exist among the histologically similar types. A detailed knowledge about these molecular classes and their specific identification could lead to the development of a diagnostically more advanced and prognostically more beneficial reporting system that could offer much more precise prediction scores.

Recent high-throughput proteomic and DNA array experiments have been used to accurately diagnose and molecularly classify tumors, assess their propensity to metastasize, and predict response to combination chemotherapy compared to traditional histologic parameters. In a seminal study by Perou et al. it was shown that breast cancer could be divided into four intrinsic types, Luminal-A, Luminal-B, Basal-like, and HER2 overexpressing, according to their gene expression profiles; these authors also demonstrated the ability to predict clinical outcomes independent of other prognostic factors (Perou et al. 1999, 2000). A previous study compared the gene expression profile of breast cancer patients who relapsed with metastasis within five years to those who did not, and identified specific gene expression signatures that most reliably identified good and poor prognostic tumors (van 't Veer et al. 2002).

Gene expression profiling has also been used to design genomic tests that could provide a better prediction of clinical outcomes relative to traditional approaches. Commercial tests based on different gene expression signatures are currently available, such as Oncotype DX from Genomic Health and MammaPrint from Agenda. OncotypeDx assay measures the expressions of 21 genes (16 cancer-related and five references) and is designed to predict the likelihood of breast cancer recurrence. MammaPrint measures the expressions of 70 genes and is used to predict patients into "good" or "poor" risk groups (Paik et al. 2004; van 't Veer et al. 2002).

More recently, a new class of small non-coding RNAs, microRNAs (miRNAs), able to regulate gene expression has been linked to several human diseases, including breast cancer. A pilot study performed by Roth et al. provided the first evidence that tumor-associated circulating miRNAs are elevated in the blood of breast cancer patients and are associated with tumor progression (Roth et al. 2010). They found that miR-10b, miR-34a, and miR-155 could discriminate M1-patients from healthy controls. Growing evidence demonstrates that miRNAs can correlate with well-defined clinico-pathological features and disease outcomes.

We compared the transcriptomic and clinical data of breast cancer patients between South Mediterranean countries such as Tunisia, Lebanon, and Morocco, and European (French) women (Chalabi et al. 2008). Several differentially expressed genes were identified and three of them, *MMP9*, *VEGF*, and *PHB1*, were related to invasion and angiogenesis. Disparities have been observed between these two groups: MMP9 was upregulated in South Mediterranean patients, whereas VEGF was upregulated in European tumors. While downregulation of MMP9 and PHB1 correlated with ER positivity, the downregulation of VEGF was associated with SBR grade in South Mediterranean tumors. The much lower prevalence of ER+ tumors might explain the underexpression of VEGF in South Mediterranean tumors because VEGF gene transcription requires a variant of estrogen response element.

An increased expression of cytokeratins KRT8 and KRT18 and TFAP2C may contribute to higher prevalence of Luminal-B subtypes noticed in South Mediterranean women. There was also an upregulation of CEACAM5 in South Mediterranean tumors in relation to the age at diagnosis. Since higher CEA levels are usually associated with metastases, South Mediterranean tumors may be more aggressive than European tumors. These observations could be linked to the fact that South Mediterranean tumors display bigger tumor diameter, more frequent SBR grade III, and greater positive node invasion. The major differences observed between these two populations were highly correlated with social factors such as breastfeeding and a high number of pregnancies in South Mediterranean populations. For example, increased expression of BRCA1 in relation to breastfeeding, and GJA1 (Connexin 43) downregulation in association with number of pregnancies in South Mediterranean tumors was observed.

A study performed in Egyptian breast cancer patients tried to identify the miRNA expression level of eight miRNAs and eight metastatic-related genes in 40 breast cancer samples and their adjacent non-neoplastic tissues (Hafez et al. 2012). Upregulation of miRNAs (mir-155, mir-10, mir-21, and mir-373) with an upregulation of *MMP2*, *MMP9*, and *VEGF* genes was noticed in tumors. These miRNAs and genes were significantly associated with an increase in tumor size. Compared to non-malignant tumor-adjacent tissue, tumor tissue downregulated mir-17p, mir-126, mir-335, mir-30b, and also *TIMP3*, *TMP1*, and *PDCD4* genes. Mir-21 was significantly overexpressed in ER–/PR– cases. These results imply that specific miRNAs are associated with tumor metastasis and other clinical characteristics of breast cancer, facilitating identification of individuals who are at risk.

The gene expression and miRNA profiling studies in breast cancer patients of the African population are scarce. It remains to be investigated whether the same signature gene set or microRNA set identified in other populations hold promise for the use in diagnosis or prognosis in an African population. It is also not evaluated whether the diagnostic or predictive power of tests that are already marketed, such as Oncotype DX and MammaPrint, can be of use in African breast cancer patients as these tests are designed based on the tumor molecular characteristics specific to US or European populations.

9.8 Conclusion

While breast cancer incidence is on a downward trend in recent years in developed countries, it is rather steeply on the rise in developing and underdeveloped countries such as those in Africa. The increased incidence may be due to a variety of risk factors, including Westernized lifestyles and dietary habits. Breast cancer is the most frequent cancer in African women and affects this population at least ten years earlier than it does women in developed countries, and is diagnosed at advanced stages. The collection of epidemiological data on breast cancer is hampered by the lack of cancer registries at the national and regional level in most African countries. Countries have to plan resources for breast cancer early detection and treatment. Increasing awareness and breast screening, which are simple and relatively inexpensive measures, could help in early detection and decrease the mortality of breast cancer in African populations.

There is a need to carry out more detailed research on breast cancer at the molecular level. There is a paucity of studies evaluating the genetic basis of breast cancer in African populations. Therefore, there is a need to perform population screening to determine the impact of germline mutations in African breast cancer patients. Similarly, the genetic basis of early-onset clinically aggressive disease is incompletely understood in African women, and genetic studies will likely contribute to our understanding of this disease presentation.

Ethnic variability in genetic differences seems to be more common than previously anticipated in breast cancer patients. With the advancement of high-throughput technologies for genomic studies, a significant amount of genomic profiling data are obtained for Caucasian, Asian, and other populations, with much breast cancer data available from various public repository sites, such as Gene Expression Omnibus (GEO). However, lack of systematic studies of breast cancer in African populations, especially by using current high-throughput technologies such as gene expression profile, array CGH for copy number changes, and miRNA for gene regulation, significantly hinders the progress of understanding the onset and evolution mechanisms, diagnosis, and treatment of breast cancer in African populations. Thus, more resources should be spent to support the genetic research to unravel the biological, pathological, and genetic features of breast cancer according to the uniqueness and specificity of African populations. The results of the breast cancer expression profiling studies will help in the development of novel and better diagnostics, prognostics, and therapies over conventional approaches for breast cancer in African populations.

References

Abdulrahman JGO and Rahman GA (2012). Epidemiology of breast cancer in Europe and Africa. *J Cancer Epidemiol.* DOI: 10.1155/2012/915610.

Amir H, Kaaya EE, Kwesigabo G, and Kiitinya JN (2000). Breast cancer before and during the AIDS epidemic in women and men: a study of Tanzanian Cancer Registry Data 1968 to 1996. *J Natl Med Assoc* 92: 301–305.

Anderson BO, Cazap E, El Saghir NS, et al. (2011). Optimisation of breast cancer management in low-resource and middle-resource countries: executive summary of the Breast Health Global Initiative consensus, 2010. *Lancet Oncol* 12: 387–398.

Anderson WF, Schairer C, Chen BE, Hance KW, and Levine PH (2005–2006). Epidemiology of inflammatory breast cancer (IBC). *Breast Dis* 22: 9–23.

Azaiza F and Cohen M (2006). Health beliefs and rates of breast cancer screening among Arab women. *J Womens Health (Larchmt)* 15: 520–530.

Bener A, Ayoubi HR, Ali AL, Al-Kubaisi A, and Al-Sultani H (2010). Does consanguinity lead to decreased incidence of breast cancer? *Cancer Epidemiol* 34: 413–418.

Bertwistle D, Swift S, Marston NJ, et al. (1997). Nuclear location and cell cycle regulation of the BRCA2 protein. *Cancer Res* 57: 5485–5488.

Bird PA, Hill AG, and Houssami N (2008). Poor hormone receptor expression in East African breast cancer: evidence of a biologically different disease? *Ann Surg Oncol* 15: 1983–1988.

Biunno I, Acetao G, Awadelkarim KD, et al. (2014). BRCA1 point mutations in premenopausal breast cancer patients from Central Sudan. *Familial Cancer* 13: 437–444.

Boder JM, ElmabroukAbdalla FB, Elfageih MA, et al. (2011). Breast cancer patients in Libya: comparison with European and Central African patients. *Oncol Lett* 2: 323–330.

Boulos S, Gadallah M, Neguib S, et al. (2005). Breast screening in the emerging world: high prevalence of breast cancer in Cairo. *Breast* 14: 340–346.

Boussen H, Bouzaiene H, Ben Hassouna J, et al. (2008). Inflammatory breast cancer in Tunisia: reassessment of incidence and clinicopathological features. *Semin Oncol* 35: 17–24.

Boyle P (2012). Triple-negative breast cancer: epidemiological considerations and recommendations. *Ann Oncol* 23(Suppl. 6): vi7–12.

Chalabi N, Bernard-Gallon DJ, Bignon YJ, et al. (2008). Comparative clinical and transcriptomal profiles of breast cancer between French and South Mediterranean patients show minor but significative biological differences. *Cancer Genomics Proteomics* 5: 253–261.

Chen Y, Farmer AA, Chen CF, et al. (1996). BRCA1 is a 220-kDa nuclear phosphoprotein that is expressed and phosphorylated in a cell cycle-dependent manner. *Cancer Res* 56: 3168–3172.

Chokunonga E, Borok MZ, Chirenje ZM, Nyakabau AM, and Parkin DM (2013). Trends in the incidence of cancer in the black population of Harare, Zimbabwe 1991–2010. *Int J Cancer* 133: 721–729.

Chouchane L, Boussen H, and Sastry KS (2013). Breast cancer in Arab populations: molecular characteristics and disease management implications. *Lancet Oncol* 14: e417–e424.

Corbex M, Bouzbid S, and Boffetta P (2014). Features of breast cancer in developing countries, examples from North-Africa. *Eur J Cancer* 50: 1808–1818.

Dawood S, Merajver SD, Viens P, et al. (2011). International expert panel on inflammatory breast cancer: consensus statement for standardized diagnosis and treatment. *Ann Oncol* 22(3): 515–523.

Easton DF, Pooley K, Dunning AM, et al. (2007). Genome-wide association study identifies novel breast cancer susceptibility loci. *Nature* 447: 1087–1093.

Edge J, Buccimazza I, Cubasch H, and Panieri E (2014). The challenges of managing breast cancer in the developing world: a perspective from sub-Saharan Africa. *S Afr Med J* 104: 377–379.

Elalaoui SC, Jaouad IC, Laarabi FZ, and Elgueddari BK (2013). Low level of consanguinity in Moroccan families at high risk of breast cancer. *Asian Pac J Cancer Prev* 14: 723–726.

Elgaili EM, Abuidris DO, Rahman M, Michalek AM, and Mohammed SI (2010). Breast cancer burden in central Sudan. *Int J Womens Health* **2**: 77–82.

El-Hawary AK, Abbas AS, Elsayed AA, and Zalata KR (2012). Molecular subtypes of breast carcinoma in Egyptian women: clinicopathological features. *Pathol Res Pract* **208(7)**: 382–386.

El Saghir NS, Khalil MK, Eid T, et al. (2007). Trends in epidemiology and management of breast cancer in developing Arab countries: a literature and registry analysis. *Int J Surg* **5**: 225–233.

El Saghir SN and Abulkhair O (2010). Epidemiology, prevention and management guidelines for breast cancer in Arab countries. *Pan Arab J Oncol* **3**: 12–18.

Engle LJ, Simpson CL, and Landers JE (2006). Using high-throughput SNP technologies to study cancer. *Oncogene* **25**: 1594–1601.

Ermiah E, Abdalla F, Buhmeida A, et al. (2012). Diagnostic delay in Libyan female breast cancer. *BMC Res Notes* **5**: 452.

Fackenthal JD, Zhang J, Zhang B, et al. (2012). High prevalence of BRCA1 and BRCA2 mutations in unselected Nigerian breast cancer patients. *Int J Cancer* **131**: 1114–1123.

Ferlay J, Shin HR, Bray F, et al. (2010). Estimates of worldwide burden of cancer in 2008: GLOBOCAN 2008. *Int J Cancer* **127**: 2893–2917.

Ferlay J, Soerjomataram I, Ervik M, et al. (2013). GLOBOCAN 2012 v1.0, Cancer Incidence and Mortality Worldwide: IARC CancerBase No. 11. Lyon: International Agency for Research on Cancer. Available at https://publications.iarc.fr/Databases/Iarc-Cancerbases/GLOBOCAN-2012-Estimated-Cancer-Incidence-Mortality-And-Prevalence-Worldwide-In-2012-V1.0-2012 (accessed May 5 2015).

Forae GD, Nwachokor FN, and Igbe AP (2014). Histopathological profile of breast cancer in an African population. *Ann Med Health Sci Res* **4**: 369–373.

Fourati A, Boussen H, El May MV, et al. (2012). Descriptive analysis of molecular subtypes in Tunisian breast cancer. *Asia Pac J ClinOncol* **10**: e69–e74.

Galukande M and Kiguli-Malwadde E (2010). Rethinking breast cancer screening strategies in resource-limited settings. *Afr Health Sci* **10**: 89–92.

Gao Q, Adebamowo CA, Fackenthal J, et al. (2000). Protein truncating BRCA1 and BRCA2 mutations in African women with pre-menopausal breast cancer. *Hum Genet* **107**: 192–194.

Garner CP, Ding YC, John EM, et al. (2008). Genetic variation in *IGFBP2* and *IGFBP5* is associated with breast cancer in populations of African descent. *Hum Genet* **123**: 247–255.

Gilani GM and Kamal S (2004). Risk factors for breast cancer in Pakistani women aged less than 45 years. *Ann Hum Biol* **31**: 398–407.

Gotzsche PC and Jorgensen KJ (2013). Screening for breast cancer with mammography. *Cochrane Database Sys Rev* **6**: CD001877.

Hafez MM, Hassan ZK, Zekri AR, et al. (2012). MicroRNAs and metastasis-related gene expression in Egyptian breast cancer patients. *Asian Pac J Prev* **13**: 591–598.

Hamrita B, Ben Nasr H, Gabbouj S, et al. (2011). Apolipoprotein A1 -75 G/A and +83 C/T polymorphisms: susceptibility and prognostic implications in breast cancer. *Mol Biol Rep* **38**: 1637–1643.

Hayanga AJ and Newman LA (2007). Investigating the phenotypes and genotypes of breast cancer in women with African ancestry: the need for more genetic epidemiology. *SurgClin N Am* **87**: 551–568.

Honrado E, Benitez J, and Palacioas J (2004). The pathology of hereditary breast cancer. *Hereditary Cancer Clin Pract* **2**: 131–138.

Hunter DJ, Kraft P, Jacobs KB, et al. (2007). A genome-wide association study identifies alleles in *FGFR2* associated with risk of sporadic postmenopausal breast cancer. *Nat Genet* **39**: 870–874.

Huo D, Ikpatt F, Khramtsov A, et al. (2009). Population differences in breast cancer: survey in indigenous African women reveals overrepresentation of triple-negative breast cancer. *J Clin Oncol* **27**: 4515–4521.

Ibrahim SS, Hafez EE, and Hashishe MM (2010). Presymptomatic breast cancer in Egypt: role of BRCA1 and BRCA2 tumor suppressor genes mutations detection. *J Exp Clin Cancer Res* **29**: e82.

Ismaili N, Elyaakoubi H, Bensouda Y, and Errihani H (2014). Demographic, clinical, pathological, molecular, treatment characteristics and outcomes of nonmetastatic inflammatory breast cancer in Morocco: 2007 and 2008. *Exp Hematol Oncol* **3**: 1.

Joukov V, Groen AC, Prokhorova T, et al. (2006). The BRCA1/BARD1 heterodimer modulates ran-dependent mitotic spindle assembly. *Cell* **127**: 539–552.

Kallel I, Rebai M, Khabir A, Farid NR, and Rebaï A (2009). Genetic polymorphisms in the EGFR (R521K) and estrogen receptor (T594T) genes, EGFR and ErbB-2 protein expression, and breast cancer risk in Tunisia. *J Biomed Biotechnol.* DOI: 10.1155/2009/753683.

Kallel I, Kharrat N, Al-Fadhly S, et al. (2010). HER2 polymorphisms and breast cancer in Tunisian women. *Genet Test Mol Biomarkers* **14**: 29–35.

Kestler DP, Foster JS, Bruker CT, et al. (2011). ODAM expression inhibits human breast cancer tumorigenesis. *Breast Cancer (Auckl)* **5**: 73–85.

Khoury MJ, Cohen BH, Chase GA, and Diamond EL (1987). An epidemiologic approach to the evaluation of the effect of inbreeding on prereproductive mortality. *Am J Epidemiol* **125**: 251–262.

Labidi I, Mrad K, Mezlini A, et al. (2008). Inflammatory breast cancer in Tunisia in the era of multimodality therapy. *Ann Oncol* **19**: 473–480.

Laraqui A, Uhrhammer N, Lahlou-Amine I, et al. (2013). Mutation screening of the BRCA1 gene in early onset and familial breast/ovarian cancer in Moroccan population. *Int J Med Sci* **10**: 60–67.

Levine PH, Pogo BG, Klouj A, et al. (2004). Increasing evidence for a human breast carcinoma virus with geographic differences. *Cancer* **101**: 721–726.

Lo AC, Kleer CG, Banerjee M, et al. (2008). Molecular epidemiologic features of inflammatory breast cancer: a comparison between Egyptian and US patients. *Breast Cancer Res Treat* **112**(1): 141–147.

Maalej M, Hentati D, Messai T, et al. (1999). Breast cancer in Tunisia: clinical and epidemiological study. *Bull Cancer* **86**: 302–306.

Mahfoudh W, Bouaouina N, Ahmed SB, et al. (2012a). Hereditary breast cancer in Middle Eastern and North African (MENA) populations: identification of novel, recurrent and founder *BRCA1* mutations in the Tunisian population. *Mol Biol Rep* **39**(2): 1037–1046.

Mahfoudh W, Bouaouina N, Gabbouj S, and Chouchane L (2012b). FASL-844 T/C polymorphism: a biomarker of good prognosis of breast cancer in the Tunisian population. *Hum Immunol* **73**: 932–938.

Missaoui N, Jaidene L, Abdelkrim SB, et al. (2011). Breast cancer in Tunisia: clinical and pathological findings. *Asian Pac J Cancer Prev* **12**: 169–172.

Missaoui N, Landolsi H, Jaidaine L, et al. (2012). Breast cancer in Central Tunisia: an earlier age at diagnosis and incidence increase over a 15-year period. *Breast J* **18**: 289–291.

Molckovsky A, Fitzgerald B, Freedman O, Heisey R, and Clemons M (2009). Approach to inflammatory breast cancer. *Can Fam Physician* **55**: 25–31.

Morris GJ and Mitchell EP (2008). Higher incidence of aggressive breast cancers in African-American women: a review. *J Natl Med Assoc* **100**: 698–702.

Morse EP, Maegga B, Joseph G, and Miesfeldt S (2014). Breast cancer knowledge, beliefs, and screening practices among women seeking care at district hospitals in Dar es Salaam, Tanzania. *Breast Cancer (Auckl)* **8**: 73–79.

Najjar H and Easson A (2010). Age at diagnosis of breast cancer in Arab nations. *Int J Surg* **8**: 448–452.

Nggada HA, Yawe KD, Abdulazeez J, and Khalil MA (2008). Breast cancer burden in Maiduguri, north eastern Nigeria. *Breast J* **14**: 284–286.

Okobia M, Bunker C, Zmuda J, et al. (2005). Cytochrome P4501A1 genetic polymorphisms and breast cancer risk in Nigerian women. *Breast Cancer Res Treat* **94**: 285–293.

Okobia MN, Bunker CH, Okonofua FE, and Osime U (2006). Knowledge, attitude and practice of Nigerian women towards breast cancer: a cross-sectional study. *World J Surg Oncol* **4**: 11.

Okobia MN, Bunker CH, Garte SJ, et al. (2008). Leptin receptor Gln223Arg polymorphism and breast cancer risk in Nigerian women: a case control study. *BMC Cancer* **8**: 338.

Othieno-Abinya NA, Nyabola LO, Abwao HO, and Ndege P (2002). Postsurgical management of patients with breast cancer at Kenyatta National Hospital. *East Afr Med J* **79**: 156–162.

Paik S, Shak S, Tang G, et al. (2004). A multigene assay to predict recurrence of tamoxifen-treated, node-negative breast cancer. *N Engl J Med* **351**: 2817–2826.

Parkin DM, Sitas F, Chirenje M, et al. (2008). Part I: cancer in indigenous Africans – burden, distribution, and trends. *Lancet Oncol* **9**: 683–692.

Patel KJ, Yu VP, Lee H, et al. (1998). Involvement of Brca2 in DNA repair. *Mol Cell* **1**: 347–357.

Peng S, Bingjian L, Wenjing R, et al. (2011). Genetic polymorphisms and breast cancer risk: evidence from meta-analyses, pooled analyses, and genome-wide association studies. *Breast Cancer Res Treat* **127**: 309–324.

Perou CM, Jeffrey SS, van de Rijn M, et al. (1999). Distinctive gene expression patterns in human mammary epithelial cells and breast cancers. *PNAS* **96**: 9212–9217.

Perou CM, Sørlie T, Eisen MB, et al. (2000). Molecular portraits of human breast tumours. *Nature* **406**: 747–752.

Rahoui J, Laraqui A, Sbitti Y, et al. (2014). Investigating the association of vascular endothelial growth factor polymorphisms with breast cancer: a Moroccan case–control study. *Med Oncol* **31**: 193.

Rais G, Raissouni S, Aitelhaj M, et al. (2012). Triple negative breast cancer in Moroccan women: clinicopathological and therapeutic study at the National Institute of Oncology. *BMC Womens Health* **12**: 35.

Reeves MD, Yawitch TM, van der Merwe NC, et al. (2004). *BRCA1* mutations in South African breast and/or ovarian cancer families: evidence of a novel founder mutation in Afrikaner families. *Int J Cancer* **110**: 677–682.

Romdhane L, Ben Halim N, Rejeb I, et al. (2014). Specific aspects of consanguinity: some examples from the Tunisian population. *Hum Hered* **77**: 167–174.

Roth C, Rack B, Muller V, et al. (2010). Circulating microRNAs as blood-based markers for patients with primary and metastatic breast cancer. *Breast Cancer Res* **12**: R90.

Salhia B, Tapia C, Ishak EA, et al. (2011). Molecular subtype analysis determines the association of advanced breast cancer in Egypt with favorable biology. *BMC Womens Health* **11**: 44.

Schlichting J, Solima AS, Schairer C, et al. (2012a). Association of inflammatory and noninflammatory breast cancer with socioeconomic characteristics in the Surveillance, Epidemiology, and End Results Database, 2000–2007. *Cancer Epidemiol Biomarkers Prev* **21**: 155–165.

Schlichting JA, Soliman AS, Schairer C, et al. (2012b). Inflammatory and non-inflammatory breast cancer survival by socioeconomic position in the Surveillance, Epidemiology, and End Results database, 1990–2008. *Breast Cancer Res Treat* **134**: 1257–1268.

Shan J, Mahfoudh W, Dsouza SP, et al. (2012). Genome-wide association studies (GWAS) breast cancer susceptibility loci in Arabs: susceptibility and prognostic implications in Tunisians. *Breast Cancer Res Treat* **135**: 715–724.

Siddig A, Mohamed AO, Kamal H, et al. (2008). HER-2/neu Ile655Val polymorphism and the risk of breast cancer. *Ann N Y Acad Sci* **1138**: 84–94.

Singletary SE and Cristofanilli M (2008). Defining the clinical diagnosis of inflammatory breast cancer. *Semin Oncol* **35**: 7–10.

Slaoui M, Razine R, Ibrahimi A, et al. (2014). Breast cancer in Morocco: a literature review. *Asian Pac J Cancer Prev* **15**: 1067–1074.

Snoussi K, Mahfoudh W, Bouaouina N, et al. (2006). Genetic variation in IL-8 associated with increased risk and poor prognosis of breast carcinoma. *Hum Immunol* **67**: 13–21.

Soliman AS, Banerjee M, Lo AC, et al. (2009). High proportion of inflammatory breast cancer in the population-based cancer registry of Gharbiah, Egypt. *Breast J* 15: 432–434.

Soliman AS, Kleer CG, Mrad K, et al. (2011). Inflammatory breast cancer in north Africa: comparison of clinical and molecular epidemiologic characteristics of patients from Egypt, Tunisia, and Morocco. *Breast Dis* 33: 159–169.

Steven AN (2011). Genetic variants associated with breast-cancer risk. *Lancet Oncol* 12: 415–416.

Stoppa-Lyonnet D, Laurent-Puig P, Essioux L, et al. (1997). *BRCA1* sequence variations in 160 individuals referred to a breast/ovarian family cancer clinic. Institut Curie Breast Cancer Group. *Am J Hum Genet* 60(5): 1021–1030.

Tazzite A, Jouhadi H, Nadifi S, et al. (2012). *BRCA1* and *BRCA2* germline mutations in Moroccan breast/ovarian cancer families: novel mutations and unclassified variants. *Gynecologic Oncol* 125(3): 687–692.

Trifa F, Karray-Chouayekh S, Mabrouk I, et al. (2010). Haplotype analysis of p53 polymorphisms: Arg72Pro, Ins16bp and G13964C in Tunisian patients with familial or sporadic breast cancer. *Cancer Epidemiol* 34: 184–188.

Troudi W, Uhrhammer N, Romdhane KB, et al. (2008). Complete mutation screening and haplotype characterization of *BRCA1* gene in Tunisian patients with familial breast cancer. *Cancer Biomark* 4: 11–18.

Uhrhammer N, Abdelouahab A, Lafarge L, et al. (2008). *BRCA1* mutations in Algerian breast cancer patients: high frequency in young, sporadic cases. *Int J Med Sci* 5(4): 197–202.

Van Bogaert LJ (2013). Breast cancer molecular subtypes as identified by immunohistochemistry in South African black women. *Breast J* 19: 210–211.

Vanderpuye V, Grover S, Hammad N, et al. (2017). An update on the management of breast cancer in Africa. *Infect Agent Cancer* 12: 13.

van 't Veer LJ, Dai H, van de Vijver MJ, et al. (2002). Gene expression profiling predicts clinical outcome of breast cancer. *Nature* 415: 530–536.

World Populations Review (2015). Africa population 2018. Available at http://worldpopulationreview.com/continents/africa-population.

Wu M, Wu ZF, Kumar-Sinha C, Chinnaiyan A, and Merajver SD (2004). RhoC induces differential expression of genes involved in invasion and metastasis in MCF10A breast cells. *Breast Cancer Res Treat* 84: 3–12.

Zaanouni E, Ben Abdallah M, Bouchlaka A, et al. (2009). Preliminary results and analysis of feasibility of mammography breast cancer screening in women younger than 50 years in the l'Ariana state of Tunisia. *Tunis Med* 87: 443–449.

Zhang B, Beeghly-Fadiel A, Long J, and Zheng W (2011). Genetic variants associated with breast-cancer risk: comprehensive research synopsis, meta-analysis, and epidemiological evidence. *Lancet Oncol* 12: 477–488.

10 Sociobiological Transition and Cancer

Prospects for Africa

Sulma I. Mohammed, Khalid O. Alfarouk, Ahmed M. Elhassan, Kamal Hamad, and Muntaser E. Ibrahim

10.1 Introduction

This chapter addresses key questions about the basis of cancer susceptibility in African people, as well as the uniqueness of the population genetics and genotype–phenotype relationship that resulted during social transition. Social transition encompasses the change of lifestyle from a rural subsistence one into an urban market-oriented, technology dependent, and predominantly sedentary mode of life, as well as changes that occur in the population genetics as a result of changes from ethnically defined structured populations into panmictic populations of an urban nature, as currently seen in megacities. Although humans passed through several periods of transition in history, the current one is the most profound.

Understanding the genetic basis of the present human traits from an evolutionary perspective is based on the idea that in the relatively short history of human emergence as a species, a large amount of the current traits came into existence within the past 10,000–20,000 years following the advent of agriculture and pastoralism. Most relatively recent changes in human culture and adaptation to their environment that followed are consequences of these events. Since the East Africa region is believed to play a critical role in these changes, it is expected that the population of this area may display many unique genetic and phenotypic traits. This may help in explaining the evolutionary basis of human adaptation to their environment, including that against infectious diseases, as well as understanding the etiology and pathogenesis of acute and chronic non-infectious diseases. Furthermore, human adaptation is an ongoing process exemplified even more recently in the changes in our lifestyles, which are believed to be the trigger for the current wave of chronic diseases known as lifestyle diseases (Newsom et al. 2012; Willett et al. 2006).

Given the pace at which lifestyle has been changing in Africa over the past decades and the steady growth of the urban population, the uniqueness of the disease focality original genetic makeup of the population may disappear in the very near future. This includes the unique characteristics of populations and their lifestyles that left marks and legacies on individual genomes.

The transition has a number of risk factors believed to contribute to etiologic determinants of cancer. Some of these transitions are not yet common in Africa, at least in rural areas, and may explain the differential distribution and burden of cancer in the continent and/or individual countries (Elamin et al. 2015; Jemal et al. 2010). The potential impact of this transition on a modern lifestyle with the accompanying risks of the emergence and distribution of cancer, that is complicated by Africa's unique range of environments that have shaped and continue to shape the biology of its inhabitants, requires urgent attention. Cancer is registering an alarming rise in incidence in Africa (Jemal et al. 2010; Kanavos 2006). A dual escalation of infectious and non-infectious diseases is draining the already meager resources of African countries and their populations. For example, cancer has risen from the seventh to the third cause of mortality in Sudan in the past decades (Ahmed et al. 2014).

The increase in cancer incidence in East Africa could be attributed to several factors, including the higher frequencies of common and population-specific genetic variants that are potentially predisposing to cancer (Campbell and Tishkoff 2008; Choudhury et al. 2013; Gurdasani et al. 2015; Parkin 2004; Tishkoff and Williams 2002). Similarly, epigenetic modifications act on the interfaces between the genome and environmental risk factors (including oncogenic viruses and pathogens) (Ferrari et al. 2009; Flanagan 2007; Paschos and Allday 2010; Poreba et al. 2011), predisposing individuals to cancer.

The study of the genetic epidemiology of cancer necessitates the examination of the uniqueness of gene–environment interactions in Africa, and the transition of African communities from rural subsistence to an urban, market-oriented lifestyle. Another factor is the state of socioeconomic underdevelopment that applies to most African countries (Gebreab et al. 2015; Houston et al. 2005; Mersha and Abebe 2015; Newman et al. 2006; Schlesinger et al. 2013) and the plethora of extreme environments, including a wide range of climatic conditions and cultures (El Fadli et al. 2013; Laux et al. 2008, 2009). Environmental risk factors, however, fall short of explaining the focality of cancer type and the aggressive course of disease in some cancers in Africa, such as breast cancer. For example, the presence of well-defined ethnic and geographic clusters of breast cancer in certain parts of Sudan hints at genetic determinants or a specific pattern of local interaction of genes and environment.

Some key issues and questions should be addressed, including why cancer is emerging during social transition; why different types of cancer cluster differently; and why cancer manifests in such a wide range of phenotypes.

10.2 African Population Genetics and Cancer Biology

The molecular basis of cancer segregation and manifestations are not yet fully understood (Hanahan and Weinberg, 2000, 2011; Suleiman et al. 2015). For many years, cancer genetics has focused on mutational events that have their primary effects on the cell. Recently that focus has widened, with evidence for the importance of epigenetic occurrences and cellular interactions in cancer development (Suleiman et al. 2015). The role of common genetic variation in determining individuals' susceptibility within the population is increasingly being recognized (Ponder 2001). Several types of cancer may appear in families and communities at high risk of developing the disease; therefore, the study of a single cancer does not guarantee explanation of the molecular etiology of the disease because risk variants and etiologic factors may be common among populations.

Analysis of whole-exome sequences of colorectal cancer and control samples from an extended colorectal cancer pedigree in Sudan provided one of the first datasets of exome sequencing of cancer in an African population and allowed investigation of an issue of pertinence to the local epidemiological milieu in Africa (Suleiman et al. 2015), where populations have a large effective size, typically with an excess of variants (Elhassan et al. 2014). A combination of familial and population studies might help ameliorate such caveats and shed more light on cancer etiology until whole-exome analysis becomes more affordable, and hence more individual/population data become available in the public domain. This is quite important as the field of genomics continues to debate the basis of susceptibility in complex inheritance and whether it is caused by common mutations of weak or modest effects or rare alleles with major and strong effects. We report some examples in which these paradigms were violated or shown to hold in certain instances. Most notable are the common polymorphism p53 Arg/Pro in which the Arg allele has been shown to pose a major risk factor for breast cancer. The association of p53 codon 72 Arg/Pro polymorphism with different types of cancer was examined in 264 cancers and normal controls (n = 235) in Sudan (Eltahir et al. 2012). Significant differences in frequency and genotype association between different types of cancers were reported. For example, breast carcinoma patients prominently showed an excess of homozygous Arg genotype as compared to controls with an odds ratio (OR) of 19.44 (95 percent CI: 6.6–78.3, $p < 0.0001$). Less prominently, cervical cancer showed genotype effect of 2.4 OR (95 percent CI: 1.12–5.33, $p = 0.015$), while esophageal cancer had an OR of 0.57 (95 percent CI: 0.23–1.42, $p = 0.1$). In Burkett's lymphoma, however, in contrast, the homozygous Arg accounted for only 6.9 percent (OR 0.18, 95 percent CI: 0.02–0.89, $p = 0.018$). The study showed that p53 Arg/Pro polymorphism has a different pattern of frequency in different types of cancer among Sudanese patients, indicating perhaps different etiology and biology of these tumors (Eltahir et al. 2012). The result was supported by a study conducted by Elamin et al. (2013), which reported differences in peroxiredoxin5 (Prdx5) expression (associated with susceptibility to breast cancer) in breast cancer from Sudanese and Chinese patients. Prdx5 is also known to communicate with the mitochondria, thus bringing into focus the role of mitochondrial DNA polymorphism in breast cancer (Masri et al. 2015).

Another class of markers known to exhibit population differences is the drug-metabolizing enzymes glutathione S-transferases (GSTs). Glutathione S-transferases are a family member of genes encoding the enzymes involved in the metabolism of many endogenous and exogenous chemicals (Hayes et al. 2005; Mannervik et al. 1985). These genes are shown to be polymorphic with *GSTM1*, and *GSTT1* is deleted in the proportion of individuals wherein the homozygous state (null genotype) results in a phenotypic absence of the corresponding enzyme. These enzymes are of considerable importance in the detoxification of many environmental compounds and reactive oxygen species, and their association with many cancers (Garcia-Closas et al. 1999; Ihsan et al. 2014; McIlwain et al. 2006; Pongtheerat et al. 2009; Strange and Fryer 1999), including leukemia, has been studied in several populations (Allan et al. 2001; Davies et al. 2001; He et al. 2014; Stanulla et al. 2000; Tang et al. 2014; Ye and Song 2005; Yuille et al. 2002). In Sudan, the distribution of the *GSTM1* and *GSTT1* genotype was evaluated in 77 leukemic patients and 107 controls. *GSTM1* null genotype was found in 31 percent of patients and 10.3 percent of controls, yielding a statistically significant positive association in patients ($p = 0.001$ [OR 3.45, 95 percent CI: 1.65–7.19]). *GSTT1* null genotype was detected in 41 percent of patients and 7.5 percent of the control group, producing an even higher association ($p < 0.0001$, OR = 8.57 [95 percent CI: 3.68–19.93]). When the cases were separated according to the type of the disease, different frequencies were obtained. The *GSTM1* null genotype was most marked in the AML group ($p < 0.0001$) and the *GSTT1* null genotype was most evident in the AML and CML group ($p < 0.0001$) (Ibrahim and Osman 2011). We extended this analysis to colorectal cancer (CRC). The *GSTT1* null allele was more frequent among CRC patients at 33 (86.8 percent) versus 26 (65 percent) in controls (OR: 3.55, 95 percent CI: 1.02–14.08). The result is statistically significant ($p = 0.025$). The *GSTM1* null genotype was not significantly associated with risk of CRC (OR = 1.39, 95 percent CI: 0.51–3.76, $p = 0.475$). In cervical cancer, the genotyping of 23 cases and 24 controls showed that the frequency of *GSTM1* null genotype was significantly higher in healthy controls (66.7 percent) compared to cervical cancer patients (34.8 percent) ($p = 0.029$, OR 0.27, 95 percent CI 0.07–1.03), while *GSTT1* null genotype was significantly ($p = 0.005$, OR 0.17, 95 percent CI 0.04–0.72) detected in normal controls (79.2 percent) compared to cancer patients (39.1 percent). *GSTM1* null allele was more frequent among breast cancer patients: 27 percent, versus 7.5 percent in controls (OR: 4.6, 95 percent CI: 1.9–11.5, $p = 0.0007$). The *GSTT1* null genotype was also significantly associated with risk of breast cancer ($p = 0.01$). These findings suggest that the *GSTM1* null genotype may strongly contribute to breast cancer development in Sudanese (Garcia-Closas et al. 1999).

Polymorphisms in metabolizing enzymes extend beyond cancer susceptibility, as it also contributes to cancer resistance to chemotherapy drugs, which is a multifactorial phenomenon (Alfarouk et al. 2015). The genetic polymorphism in the population is one key factor for this chemotherapy drugs resistance. For example, tamoxifen is an estrogen receptor antagonist that has been used in the treatment of estrogen-positive breast tumors (Lloyd et al. 2014) and as prophylaxis in prevention of recurrence (Jordan 1993). Tamoxifen is administered as a pro-drug and requires biological activation by CY2D6 and 3A4 cytochrome to 4-hydroxytamoxifen (afimoxifene) and N-desmethyl-4-hydroxytamoxifen (tamoxifen), respectively, which antagonize estrogen hormone activity (Barrière et al. 2010; Desta et al. 2004; García-Becerra et al. 2012; Hoskins et al. 2009; Wang et al. 2004). CYP2D6 is subject to genetic polymorphism (Bernard et al. 2006; Hoskins et al. 2009). In East Africa, e.g., Ethiopia, the *CYP2D6* gene is thought to be duplicated due to dietary selection (Ingelman-Sundberg 2005) and consumption of certain dietary alkaloids that inhibit the *CYP2D6* gene. It seems that the gene duplication is an adaptive evolution that might have led to evolutionary trade-off seen in fluctuation of tamoxifen responses.

CYP3A4 is also subject to genetic polymorphism in African populations (Novillo et al. 2015). This polymorphism will influence the tamoxifen response, but may also have a direct impact on carcinogenesis and tumor growth (Keshava et al. 2004).

Thiopurinemethyltransferase or thiopurine S-methyltransferase (TPMT) is an enzyme that in humans is encoded by the *TPMT* gene (Lee et al. 1995). TPMT methylates thiopurine compounds (Sahasranaman et al. 2008; Weinshilboum and Sladek 1980). Thiopurine is a purine antimetabolite that is intended to work as anti-cancer and autoimmune disorders (Zaza et al. 2010). Examples of these drugs include mercaptopurine and 6-mercaptopurine (6-MP). 6-MP is an anti-metabolite (purine base analog adenine and hypoxanthine) active against human leukemia (Freireich et al. 1963; Hyman et al. 1957; Podsiadlo et al. 2008). 6-MP causes myelosuppression, inhibiting the production of white and red blood cells, and it may be toxic to bone marrow. TPMT catalyzes the methylation of 6-MP into the inactive metabolite 6-methyl mercaptopurine. This methylation prevents 6-MP from further conversion into active, cytotoxic thioguanine nucleotide (TGN) metabolites (Fujita and Sasaki 2007; Zaza et al. 2010). The *TPMT* gene is also polymorphic (Evans et al. 2001) in humans, and is less so in black individuals (Jones et al. 1993; Krynetski and Evans 2003; McLeod et al. 1994). Therefore, this drug may cause myelotoxicity among darker-skinned people (Hon et al. 1999).

Another drug, thioguanine or 6-thioguanine (6-TG), is a purine analog of the nucleobase guanine. 6-TG utilizes the enzyme hypoxanthine-guanine phosphoribosyltransferase (HGPRTase) to be converted to 6-thioguanine monophosphate (TGMP). High concentrations of TGMP may accumulate intracellularly and hamper the synthesis of guanine nucleotides via the enzyme inosine monophosphate dehydrogenase (IMP dehydrogenase). In addition to inhibiting de novo purine synthesis, thioguanine may also inhibit purine nucleotide interconversions. TPMT metabolizes thioguanine to its methyl thioguanine base – this methylation prevents thioguanine from a further conversion into active, cytotoxic thioguanine nucleotide (TGN) metabolites (Relling et al. 2011; Zaza et al. 2010). Certain genetic variations within the *TPMT* gene can lead to decreased or absent TPMT enzyme activity. Individuals who are homozygous or heterozygous for these types of genetic variations may have increased levels of TGN metabolites and increased risk of severe bone marrow suppression (myelosuppression) when receiving thioguanine (Relling et al. 2011). Therefore, this drug should also be monitored among subjects of darker skin before administration to prevent myelotoxicity (Hon et al. 1999).

Interaction with the environment extends beyond the instant and short-term modulation and evolutionary processes exemplified in the genes that reside at the interface with the environment, to traits that are products of long-term evolutionary adaptation, like skin color. The skin displays a range of melanin tones, believed to be a product of natural selection, primarily processes associated with vitamin D synthesis. Vitamin D refers to a group of fat-soluble secosteroids responsible for enhancing intestinal absorption of calcium, iron, magnesium, phosphate, and zinc. The active vitamin D metabolite calcitriol mediates its biological effects by binding to the vitamin D receptor (VDR), which is principally located in the nuclei of target cells.

The binding of calcitriol to the VDR allows the VDR to act as a transcription factor that modulates the gene expression of transport proteins (such as TRPV6 and calbindin), which are involved in calcium absorption in the intestine (Bouillon et al. 2003; Van Cromphaut et al. 2003; van de Graaf et al. 2004). Vitamin D receptor activation in the intestine, bone, kidney, and parathyroid gland cells leads to the maintenance of calcium and phosphorus levels in the blood and the maintenance of bone content (Cranney et al. 2007; Holick, 2004, 2005, 2008; Holick and Chen 2008). Vitamin D3 is produced photochemically from 7-dehydrocholesterol in the skin of most vertebrates, including humans. The precursor of vitamin D3, 7-dehydrocholesterol, is produced in relatively large quantities (Blann 2015; Chandra et al. 2007; Correia et al. 2014; Crissey et al. 2003; Dawson-Hughes 2004; Harris et al. 2000; Looker et al. 2002). Melanin protects the skin from the hazardous effects of UVB, known to induce skin cancer. However, adaptation to sunlight exposure, reflected in darker skin, results in vitamin D deficiency that is correlated with several diseases in African populations, e.g., tuberculosis, cardiovascular diseases, diabetes (Davis 2011; Forrest and Stuhldreher 2011; Harris 2006; Saccone et al. 2015; Wejse et al. 2007), and cancer, where it has been shown to have a protective role as well (Wu et al. 2015). Vitamin D is one of several examples that signifies the challenge of understanding cancer etiology within a context of genetic networks instead of a single gene contribution to genetic epidemiology and cancer susceptibility within a multiple two-hit hypothesis (Knudson 1971) or lever and ball model (Ibrahim and Osman 2011). In the network analysis of the exome data mentioned earlier, multiple hub proteins of centrality were identified. Most of these proteins are blatant tumor suppressors, among which is the RNA-binding protein ELAVL1/HuR. The centrality of ELAVL1 was shown to associate with non-synonymous mutations rather than other classes of single nucleotide variants (SNVs). This poses an interesting proposition of organized and targeted mutagenesis of cancer signaling pathways. The insertions/deletions contribution was more significant in dictating the molecular changes toward carcinogenesis and features of identity by state between tumor samples (Suleiman et al. 2015). A likely explanation of such organization is DNA/RNA editing, suggested by nucleotide transition to transversion ratio that departs from the expected distribution ($p = 5e^{-6}$). A model is proposed by which viral etiology is a key factor based on known molecular counterparts like the RNA binding protein ELAVL1 and NF-kβ, both of which appeared to be central in interaction networks. Allele frequency differences in natural populations in genes of centrality, such as networks like the p53 example mentioned above, and the genes that influence and control epigenetic status itself (discussed below), such as *DNMT* and *TET*, are worth considering as potential underlying causes of population and interindividual differences in cancer profiles.

Unpublished data of a partial sequence of both the DNMT3b and the cofactor-binding site in tissue samples from histopathologically confirmed cancer cases ($n = 20$) detected two polymorphisms. These polymorphisms were previously reported (reviewed in Veeck and Esteller [2010]). One of the reported polymorphisms, C/T (rs2424913), was found in the promoter region and shown to increase promoter activity. There are indications from our studies and other published work that genetic mutations are occurring in the *DNMT3b* gene and possibly other relevant genes. Viruses that utilize restriction modification systems of certain classes such as the BART (BamH1A rightward transcription) are believed to cause these mutations. In a previous study by Giha (MSc thesis, unpublished), it was shown that the common modification system utilized by the mismatch repair is differentially distributed in breast cancer tissues compared to controls. Also, differences in the Alu repeat sequence between cancer and normal control samples were observed (unpublished). These results put together suggest a role for these restriction modification sites in cancer and possibly in disrupting the mismatch repair machinery in particular.

10.3 Epigenetic Silencing and Expression Profiling

The fact that methylation lies prominently at the interface of genes and the environment makes it particularly interesting for understanding both short- and long-term evolutionary effects. By analyzing epigenetic tags and DNA differential methylation, mainly methylation of CpG in promoters, basic questions about the presence of population-specific epigenetics tags and expression patterns could be answered. The focality of diseases mentioned earlier may well be partly influenced by such a pattern, which is also quite intriguing, given the recent reports of epigenetic transgenerational inheritance (Jablonka and Raz 2009). Collective stressful episodes of famine and war are believed to have an effect on the human epigenome and psychosomatic state of well-being, which is worth investigating, given the recent rampant history of famine and civil war in this part of Africa. Furthermore, epigenetic changes and effects on the expression of genes, unlike evolutionary processes at the level of the gene, are relevant to the problem of transition, which is a short-term process on a time span of years or decades rather than centuries or millennia, as is the case of natural selection at the gene level. There is a marked scarcity of epigenetics studies in Africa and even fewer on population epigenetic profiles worldwide, despite the recent publication of the human reference epigenome (Roadmap Epigenomics Consortium et al., 2015).

One early study on a limited set of genes to investigate their methylation status came in response to findings of a limited contribution of mutational events in tumor suppressors to breast cancer etiology (Masri et al. 2002). Epigenetic silencing was therefore the next obvious candidate to look at. The findings were, from six tumor suppressors (BRCA1, BRCA2, p14, p16, eMLH, MGM), that methylation was nearly 100 percent in the first three genes. The fact that some genes are methylated in one type of cancer rather than others, as is the case in that set of data, and the potential of identifying common epigenetics pathways shared between different cancers, may provide insights into the biology of various tumors and shed further light on the role of methylation and epigenetic silencing in cancer. Using Illumina Infinium Human Methylation 450 Bead chip assay, the methylome of eight breast cancer Sudanese patients and unpaired adjacent breast tissue were investigated (Abdallah et al. 2015). Differential methylation analysis corroborated the previous results and revealed epigenetic dysregulation of major developmental pathways, including the Hippo signaling pathway. This provided several clues to a possible role for Epstein–Barr virus (EBV) infection in breast cancer. A study showed 100 percent positivity of EBV in breast sample by *in situ* hybridization and 54 percent positivity by PCR (Yahia et al. 2014). The results combined present unequivocal evidence for a major etiologic role of EBV in breast cancer pathogenesis in Sudanese women. We, therefore, set out to answer a plethora of questions, including the specific role and molecular basis of epigenetic silencing in inactivation of tumor suppressors, how the virus may affect the genetic mismatch machinery, and why some individuals tend to develop cancer while others do not. Oncogenic viruses like human papilloma virus (HPV) and EBV are common infections; should we speculate population-specific susceptibility factors that predispose to cancer? Epidemiological studies indicate that infections by certain types of HPV are causally linked to the development of cervical cancer and particular EBV strains have been found to be more oncogenic (Palser et al. 2015; Tzellos and Farrell 2012). However, it is also known that HPV infections alone do not cause progression to cervical cancer, as additional genetic changes such as loss of distinct chromosomal regions, inactivation of tumor-suppressor genes, and activation of oncogenes must also occur for malignant transformation to take place. A study in Sudanese cervical cancer patients investigating the p53 Arg/Pro polymorphism and Rb LOH showed significant association of the chromosomal allelic loss of Rb. The relative risk (RR) of LOH and Arg/Arg alone was 1.7 and 1.1, respectively, while the combined RR for Rb LOH and p53 Arg/Arg was 2.5, while no such association was observed with other parameters, such as clinical stage and degree of differentiation; hence, it cannot be a determinant of tumor behavior in cervical carcinoma. Although the p53 Arg allele is itself an important risk factor for cervical cancer, the combined risk with LOH of Rb, which appears to be greater, might indicate a possible epistatic effect of the two genes/polymorphisms.

For EBV, establishing an etiologic link in a wider cohort of patients may carry therapeutic impacts influencing both management and prophylaxis because of the possible role of EBV-specific cytotoxic T cells in targeting EBV-associated tumor cells (Herrmann and Niedobitek 2003; Ricciardelli et al. 2014). Similar to HPV, vaccination could be introduced for EBV. Recently, different vaccine formulations against EBV, including therapeutic vaccines, have been advertised (Hui et al. 2013; Ruiss et al. 2011). Epstein–Barr virus has also been found in breast tissue and milk. Lymphoblastoid cell lines bearing EBV can infect mammary epithelial cells *in vitro*. Also, transfection of the p31 fragment of EBV DNA immortalizes epithelial cells, including mammary epithelial cells (Glaser et al. 2004).

10.4 Trade-Off Parasites and Cancer

Particularly interesting in the transition from rural and subsistence economies, mentioned above, is the decreased exposure to infectious parasitic diseases. The surge of a horde of diseases from autoimmune disease to cancer has been attributed to the change of immunological patterns in human societies associated with new cultures and ways of living. Cancer is exemplary as a disease of complex inheritance characterized by multiple phenotypes and signaling processes, and by virtue is an outcome of multiple interactions of proteins and nucleic acids in relation to the environment. Henceforth, one has to consider the notorious prevalence of both infectious and non-infectious diseases that may have a bearing on the initiation and prognosis of cancer and its management; these include tuberculosis (Omoti et al. 2009), malaria, HIV, other chronic infections, and sickle cell disease (Ahmed and Ibrahim 2007). Common adaptive traits that could modulate the course of the disease, as well as the role of oncogenic viruses, are discussed above and below.

Malaria, in particular, is believed to rank high among pathogens that engraved deep evolutionary marks in our genome, a consequence of past high death toll leading to allele frequency change. This pernicious malignant nature of the disease motivated control attempts in the past even when the culprit was of an uncertain nature. Last century, plans were up and running to eradicate malaria on a global scale. This particular line of thought is a vivid example of how it is possible to completely disregard the evolutionary context of trade-off and cost, an essential component of a Darwinian paradigm. We consider here recent and past evidence that weighs the cost of malaria eradication versus its potential benefits, particularly against cancer. Observational and experimental clues have accumulated through longitudinal cohort studies of two ethnic populations, Hausa and Masalit, both being recent migrants from the Sahel to the malaria-endemic area of the Rahad River in eastern Sudan. Those populations were found to possess considerable frequencies of malaria-protective polymorphisms, including sickle cell anemia, thalassemia, G6PD, and possibly others yet to be identified, consistent with a prolonged history of repeated episodes of natural selection by the parasite. Intriguingly, but not surprisingly, the overall morbidity and mortality from malaria seem to be low, despite the disease being endemic to the area; in addition, there is a spectacular absence of chronic diseases. Interestingly, no cancer cases in the population of these adjacent villages were recorded in almost two decades of close follow-up. Both genotype and expression data of malaria candidate cytokine genes suggest differences in the evolutionary trajectories and shed light on the intricacy of the immunogenetic responses in these populations as compared to a control sample from urban populations. A total of 52 malaria candidate genes, among which are a number of obvious cancer candidates, e.g., *FASL*, *CTL4*, *RAD50*, were found. These are polymorphic and differentially distributed between and within the populations (including a few of the cancer candidates). Moreover, expression levels of genes like CD40, a TNF ligand, were found to be 30-fold higher than urban controls (Eid, unpublished). *TNFα* is known to be elevated in clinical and subclinical infections by both malaria and visceral leishmaniasis (VL) endemic in the area. *TNFα* is a known tumor-associated gene, and has been suggested recently as a main molecule sending tumor cells to senescence (Braumüller et al. 2013).

An example of polymorphisms selected for by malaria parasites in this context is CD36, a multi-ligand receptor associated with a broad array of physiological processes. It is believed to be under selective pressure from *Plasmodium falciparum* and is deficient or polymorphic in several African populations. The role of CD36 in sickle cell crises and in cerebral malaria is debatable. However, as a receptor for thrombospondin 1, CD36 plays a role in the regulation of angiogenesis, which may potentially be a therapeutic strategy for controlling the dissemination of malignant neoplasm. Moreover, it is commonly expressed on blasts in acute monocytic leukemia, megakaryoblastic leukemia, and erythroleukemia (Ge and Elghetany 2005). However, CD36-negative AML cells could be found in several African populations that usually do not express CD36. CD36-negative cells appeared less susceptible to trombospondin-1 induced apoptosis, which makes leukemic cells less vulnerable to death through this therapeutic strategy (Li et al. 2003).

G6PD is a cytosolic enzyme that utilizes glucose through the pentose phosphate pathway to pool nucleic acid building blocks and maintain redox status of the cell (Jain et al. 2003). So, it would not be surprising if it had been found in peroxisomes (Corpas et al. 1998). G6PD deficiency had been observed in sickle cell anemia (Benkerrou et al. 2013), which is associated with malaria resistance (Fico et al. 2004; Greene 1993; Ruwende et al. 1995; Stanton 2012; Verrelli et al. 2006).

In cancer, G6PD has a vital function in preventing cellular death through interruption of cellular death pathways (Fico et al. 2004; Stanton 2012; Yang et al. 2013; Zhang et al. 2010), a hallmark of cancer (Hanahan and Weinberg, 2000, 2011). On the contrary, it has been shown that *G6PD* deletion has prevented tumor growth (Cocco 1987). The *G6PD* gene is widely deleted in African populations (Cappellini and Fiorelli 2008; Lin et al. 2015).

A very interesting example is *Plasmodium falciparum* protein, VAR2CSA. VAR2CSA binds a distinct type of chondroitin sulfate (CS), exclusively expressed in the placenta and present in a high proportion in malignant cells. This protein can be specifically targeted by recombinant VAR2CSA (rVAR2).

These examples and others are of common polymorphisms that manifest intermediate to major effect under selection by the malaria parasite, and could act individually or epistatically to influence resistance/susceptibility networks. This may extend to include many other candidates, providing, in an evolutionary context, a glimpse of the possible mechanism that could explain the complex epidemiology of susceptibility to different diseases, including cancer.

Sahel populations have existed and adapted to the malaria parasite for centuries. The disease should often be considered more of a chronic infection from a public health point of view. Emphasis should shift to possible aspects of the development of biological fitness that resulted from millennia of succumbing to malaria burden. We should not sacrifice this acquired adaptation by methods that eliminate such natural adaptation.

10.5 Conclusion

Africa's unique genetic population structure, current state of socioeconomic transition, and the peculiarity of environmental challenges renders the understanding of cancer epidemiology and etiology a rather daunting and complex undertaking. A contextual approach fostering an appreciation of the significance and interplay of these elements is a necessity for a meaningful understanding of disease susceptibility and progression, and in mounting effective management and control measures.

References

Abdallah MO, Algizouli UK, Suliman MA, et al. (2015). Methylome analysis reveals dysregulated developmental and viral pathways in breast cancer. *bioRxiv (online)*. DOI: 10.1101/034322.

Ahmed HG, Doumi MA, Eltybe MM, Mahmoud TA, and Ebnoof SO (2014). Burden of cancer in North Sudan: a community-based survey. *Manag Health*. DOI: 10.5233/mih. v18i3.329.

Ahmed SG and Ibrahim UA (2007). Significance of haemoglobin-S in the pathogenesis of hyperleucocytic syndromes in Nigerian patients with chronic myeloid leukaemia. *Eur J Haematol* **79** (2): 174–176.

Alfarouk KO, Stock C-M, Taylor S, et al. (2015). Resistance to cancer chemotherapy: failure in drug response from ADME to P-gp. *Cancer Cell Int* **15**(71): 71.

Allan JM, Wild CP, Rollinson S, et al. (2001). Polymorphism in glutathione S-transferase P1 is associated with susceptibility to chemotherapy-induced leukemia. *PNAS* **98**(20): 11592–11597.

Barrière J, Formento J-L, Milano G, and Ferrero J-M (2010). CYP2D6 polymorphisms and tamoxifen: therapeutic perspectives in the management of hormone dependent breast cancer patients. *Bulletin du Cancer* **97**(3): 311–320.

Benkerrou M, Alberti C, Couque N, et al. (2013). Impact of glucose-6-phosphate dehydrogenase deficiency on sickle cell anaemia expression in infancy and early childhood: a prospective study. *Br J Haematol* **163**(5): 646–654.

Bernard S, Neville KA, Nguyen AT, and Flockhart DA (2006). Interethnic differences in genetic polymorphisms of CYP2D6 in the U.S. population: clinical implications. *Oncologist* **11**(2): 126–135.

Blann A (2015). An update on vitamin D deficiency and at risk groups. *J Family Health* **25**(3): 16–19.

Bouillon R, Van Cromphaut S, and Carmeliet G (2003). Intestinal calcium absorption: molecular vitamin D mediated mechanisms. *J Cell Biochem* **88**(2): 332–339.

Braumüller H, Wieder T, Brenner E, et al. (2013). T-helper-1-cell cytokines drive cancer into senescence. *Nature* **494**: 361–365.

Campbell MC and Tishkoff SA (2008). African genetic diversity: implications for human demographic history, modern human origins, and complex disease mapping. *Annu Rev Genomics Hum Genet* **9**: 403–433.

Cappellini MD and Fiorelli G (2008). Glucose-6-phosphate dehydrogenase deficiency. *Lancet* **371** (9606): 64–74.

Chandra P, Wolfenden LL, Ziegler TR, et al. (2007). Treatment of vitamin D deficiency with UV light in patients with malabsorption syndromes: a case series. *Photodermatol, Photo Immunol Photomed* **23**(5): 179–185.

Choudhury A, Hazelhurst S, Meintjes A, et al. (2013). Population-specific common SNPs reflect demographic histories and highlight regions of genomic plasticity with functional relevance. *BMC Genomics* **15**(1): 437.

Cocco P (1987). Does G6PD deficiency protect against cancer? A critical review. *J Epidemiol Community Health* **41**(2): 89–93.

Corpas FJ, Barroso JB, Sandalio LM, et al. (1998). A dehydrogenase-mediated recycling system of NADPH in plant peroxisomes. *Biochem J* **330** (Pt 2): 777–784.

Correia A, Azevedo M, Do S, Gondim F, and Bandeira F (2014). Ethnic aspects of vitamin D deficiency. *Arquivos brasileiros de endocrinologia e metabologia* **58**(5): 540–544.

Cranney A, Horsley T, O'Donnell S, et al. (2007). Effectiveness and safety of vitamin D in relation to bone health. *Evid Rep Technol Assess* **158**: 1–235.

Crissey SD, Ange KD, Jacobsen KL, et al. (2003). Serum concentrations of lipids, vitamin D metabolites, retinol, retinyl esters, tocopherols and selected carotenoids in twelve captive wild felid species at four zoos. *J Nutr* **133**(1): 160–166.

Davies SM, Robison LL, Buckley JD, et al. (2001). Glutathione S-transferase polymorphisms and outcome of chemotherapy in childhood acute myeloid leukemia. *J Clin Oncol* **19**(5): 1279–1287.

Davis SV (2011). Vitamin D deficiency and type 2 diabetes in African Americans: the common denominators. *Diabetes Spectrum* **24**(3): 148–153.

Dawson-Hughes B (2004). Racial/ethnic considerations in making recommendations for vitamin D for adult and elderly men and women. *Am J Clin Nutr* **80**(6 Suppl.): 1763S–1766S.

Desta Z, Ward BA, Soukhova NV, and Flockhart DA (2004). Comprehensive evaluation of tamoxifen sequential biotransformation by the human cytochrome P450 system in vitro: prominent roles for CYP3A and CYP2D6. *J Pharmacol Exp Ther* **310**(3): 1062–1075.

Elamin A, Zhu H, Hassan AME, Xu N, and Ibrahim ME (2013). Peroxiredoxin V: a candidate breast tumor marker of population specificity. *Mol Clin Oncol* **1**(3): 541–549.

Elamin A, Ibrahim ME, Abuidris D, Mohamed KEH, and Mohammed SI (2015). Part I: cancer in Sudan – burden, distribution, and trends breast, gynecological, and prostate cancers. *Cancer Med* **4**(3): 447–456.

El Fadli KI, Cerveny RS, Burt CC, et al. (2013). World Meteorological Organization assessment of the purported world record 58°C temperature extreme at El Azizia, Libya (13 September 1922). *Bull Am Meteorol Soc* **94**(2): 199–204.

Elhassan N, Gebremeskel EI, Elnour MA, et al. (2014). The episode of genetic drift defining the migration of humans out of Africa is derived from a large East African population size. *PLoS One* **9**(5): e97674.

Eltahir HA, Adam AAM, Yahia ZA, et al. (2012). p53 codon 72 arginine/proline polymorphism and cancer in Sudan. *Mol Biol Rep* **39**(12): 10833–10836.

Evans WE, Hon YY, Bomgaars L, et al. (2001). Preponderance of thiopurine S-methyltransferase deficiency and heterozygosity among patients intolerant to mercaptopurine or azathioprine. *J Clin Oncol* **19**(8): 2293–2301.

Ferrari R, Berk AJ, and Kurdistani SK (2009). Viral manipulation of the host epigenome for oncogenic transformation. *Nat Rev Genet* **10**(5): 290–294.

Fico A, Paglialunga F, Cigliano L, et al. (2004). Glucose-6-phosphate dehydrogenase plays a crucial role in protection from redox-stress-induced apoptosis. *Cell Death Differ* **11**(8): 823–831.

Flanagan JM (2007). Host epigenetic modifications by oncogenic viruses. *Br J Cancer* **96**(2): 183–188.

Forrest KYZ and Stuhldreher WL (2011). Prevalence and correlates of vitamin D deficiency in US adults. *Nutr Res* **31**(1): 48–54.

Freireich EJ, Gehan E, Frei E, et al. (1963). The effect of 6-mercaptopurine on the duration of steroid-induced remissions in acute leukemia: a model for evaluation of other potentially useful therapy. *Blood* **21**(6): 699–716.

Fujita K-I and Sasaki Y (2007). Pharmacogenomics in drug-metabolizing enzymes catalyzing anticancer drugs for personalized cancer chemotherapy. *Curr Drug Metab* **8**(6): 554–562.

García-Becerra R, Santos N, Díaz L, and Camacho J (2012). Mechanisms of resistance to endocrine therapy in breast cancer: focus on signaling pathways, miRNAs and genetically based resistance. *Int J Mol Sci* **14**(1): 108–145.

Garcia-Closas M, Kelsey KT, Hankinson SE, et al. (1999). Glutathione S-transferase mu and theta polymorphisms and breast cancer susceptibility. *JNCI* **91**(22): 1960–1964.
Ge Y and Elghetany MT (2005). CD36: a multiligand molecule. *Lab Hematol* **11**(1): 31–37.
Gebreab SY, Diez Roux AV, Brenner AB, et al. (2015). The impact of lifecourse socioeconomic position on cardiovascular disease events in African Americans: the Jackson Heart Study. *J Am Heart Assoc* **4**(6): e001553–e001553.
Glaser SL, Hsu JL, and Gulley ML (2004). Epstein–Barr virus and breast cancer: state of the evidence for viral carcinogenesis. *Cancer Epidemiol Biomarkers Prev* **13**(5): 688–697.
Greene L (1993). G6PD deficiency as protection against *falciparum* malaria: an epidemiologic critique of population and experimental studies. *Yearbook Phys Anthropol* **36**: 153–178.
Gurdasani D, Carstensen T, Tekola-Ayele F, et al. (2015). The African Genome Variation Project shapes medical genetics in Africa. *Nature* **517**(7534): 327–332.
Hanahan D and Weinberg RA (2000). The hallmarks of cancer. *Cell* **100**(1): 57–70.
Hanahan D and Weinberg RA (2011). Hallmarks of cancer: the next generation. *Cell Elsevier* **144**(5): 646–674.
Harris SS (2006). Vitamin D and African Americans. *J Nutr* **136**(4): 1126–1129.
Harris SS, Soteriades E, Coolidge JA, Mudgal S, and Dawson-Hughes B (2000). Vitamin D insufficiency and hyperparathyroidism in a low income, multiracial, elderly population. *J Clin Endocrinol Metab* **85**(11): 4125–4130.
Hayes JD, Flanagan JU, and Jowsey IR (2005). Glutathione transferases. *Annu Rev Pharmacol Toxicol* **45**: 51–88.
He H, Zhang X, Sun J, et al. (2014). Glutathione S-transferase gene polymorphisms and susceptibility to chronic myeloid leukemia. *Tumour Biol* **35**(6): 6119–6125.
Herrmann K and Niedobitek G (2003). Epstein–Barr virus-associated carcinomas: facts and fiction. *J Pathol* **199**(2): 140–145.
Holick MF (2004). Sunlight and vitamin D for bone health and prevention of autoimmune diseases, cancers, and cardiovascular disease. *Am J Clin Nutr* **80**(6 Suppl.): 1678S–1688S.
Holick MF (2005). The vitamin D epidemic and its health consequences. *J Nutr* **135**(11): 2739S–2748S.
Holick MF (2008). Vitamin D and sunlight: strategies for cancer prevention and other health benefits. *Clin J Am Soc Nephrol* **3**(5): 1548–1554.
Holick MF and Chen TC (2008). Vitamin D deficiency: a worldwide problem with health consequences. *Am J Clin Nutr* **87**(4): 1080S–1086S.
Hon YY, Fessing MY, Pui CH, et al. (1999). Polymorphism of the thiopurine S-methyltransferase gene in African-Americans. *Hum Mol Genet* **8**(2): 371–376.
Hoskins JM, Carey LA, and McLeod HL (2009). CYP2D6 and tamoxifen: DNA matters in breast cancer. *Nat Rev Cancer* **9**(8): 576–586.
Houston TK, Scarinci IC, Person SD, and Greene PG (2005). Patient smoking cessation advice by health care providers: the role of ethnicity, socioeconomic status, and health. *Am J Public Health* **95**(6): 1056–1061.
Hui EP, Taylor GS, Jia H, et al. (2013). Phase I trial of recombinant modified vaccinia ankara encoding Epstein–Barr viral tumor antigens in nasopharyngeal carcinoma patients. *Cancer Res* **73**(6): 1676–1688.
Hyman CB, Brubaker C, and Sturgeon P (1957). 6-mercaptopurine in childhood leukemia: comparison of large dose interrupted with small dose continuous therapy. *Cancer Res* **17**(9): 851–856.
Ibrahim ME and Osman E-AI (2011). Myeloid leukemia: a molecular focus on etiology and risk within Africa. In Koschmieder S and Krug U, eds., *Myeloid Leukemia: Basic Mechanisms of Leukemogenesis*. Rijeka: InTech.
Ihsan R, Chauhan PS, Mishra AK, et al. (2014). Copy number polymorphism of glutathione-S-transferase genes (*GSTM1* & *GSTT1*) in susceptibility to lung cancer in a high-risk population from north-east India. *Ind J Med Res* **139**(5): 720–729.

Ingelman-Sundberg M (2005). Genetic polymorphisms of cytochrome P450 2D6 (CYP2D6): clinical consequences, evolutionary aspects and functional diversity. *Pharmacogenomics J* **5**(1): 6–13.

Jablonka E and Raz G (2009). Transgenerational epigenetic inheritance: prevalence, mechanisms, and implications for the study of heredity and evolution. *Quart Rev Biol* **84**(2): 131–176.

Jain M, Brenner DA, Cui L, et al. (2003). Glucose-6-phosphate dehydrogenase modulates cytosolic redox status and contractile phenotype in adult cardiomyocytes. *Circulation Res* **93**(2): e9–e16.

Jemal A, Center MM, DeSantis C, and Ward EM (2010). Global patterns of cancer incidence and mortality rates and trends. *Cancer Epidemiol Biomarkers Prev* **19**(8): 1893–1907.

Jones CD, Smart C, Titus A, et al. (1993). Thiopurine methyltransferase activity in a sample population of black subjects in Florida. *Pharmacol Ther, Part C* **53**(3): 348–353.

Jordan VC (1993). Fourteenth Gaddum Memorial Lecture: a current view of tamoxifen for the treatment and prevention of breast cancer. *Br J Pharmacol* **110**(2): 507–517.

Kanavos P (2006). The rising burden of cancer in the developing world. *Ann Oncol* **17**(Suppl. 8): viii15–viii23.

Keshava C, McCanlies EC, and Weston A (2004). CYP3A4 polymorphisms: potential risk factors for breast and prostate cancer – a HuGE review. *Am J Epidemiol* **160**(9): 825–841.

Knudson AG (1971). Mutation and cancer: statistical study of retinoblastoma. *PNAS* **68**(4): 820–823.

Krynetski E and Evans WE (2003). Drug methylation in cancer therapy: lessons from the TPMT polymorphism. *Oncogene* **22**(47): 7403–7413.

Laux P, Kunstmann H, and Bárdossy A (2008). Predicting the regional onset of the rainy season in West Africa. *Int J Climatol* **28**(3): 329–342.

Laux P, Wagner S, Wagner A, et al. (2009). Modelling daily precipitation features in the Volta Basin of West Africa. *Int J Climatol* **29**(7): 937–954.

Lee D, Szumlanski C, Houtman J, et al. (1995). Thiopurine methyltransferase pharmacogenetics: cloning of human liver cDNA and a processed pseudogene on human chromosome 18q21.1. *Drug Metab Dispos* **23**(3): 398–405.

Li K, Yang M, Yuen PMP, et al. (2003). Thrombospondin-1 induces apoptosis in primary leukemia and cell lines mediated by CD36 and Caspase-3. *Int J Med Mol Adv Sci* **12**(6): 995–1001.

Lin M, Yang LY, Xie DD, et al. (2015). G6PD deficiency and hemoglobinopathies: molecular epidemiological characteristics and healthy effects on malaria endemic Bioko Island, Equatorial Guinea. *PLoS One* **10**(4): e0123991.

Lloyd MC, Alfarouk KO, Verduzco D, et al. (2014). Vascular measurements correlate with estrogen receptor status. *BMC Cancer* **14**(1): 279.

Looker AC, Dawson-Hughes B, Calvo MS, Gunter EW, and Sahyoun NR (2002). Serum 25-hydroxyvitamin D status of adolescents and adults in two seasonal subpopulations from NHANES III. *Bone* **30**(5): 771–777.

Mannervik B, Alin P, Guthenberg C, et al. (1985). Identification of three classes of cytosolic glutathione transferase common to several mammalian species: correlation between structural data and enzymatic properties. *PNAS* **82**(21): 7202–7206.

Masri MA, Abdel Seed NM, Fahal AH, et al. (2002). Minor role for BRCA2 (exon11) and p53 (exon 5–9) among Sudanese breast cancer patients. *Breast Cancer Res Treat* **71**(2): 145–147.

Masri MA, Elhassan NM, Hiba SM, Wasunna CSA, and Ibrahim ME (2015). In-vitro "depletion" of mitochondrial cytochrome C oxidase subunit II affects the patterns of gene expression across multiple cancer pathways. *J Mol Biomark Diagn*. DOI: 10.4172/2155-9929.S2-011.

McIlwain CC, Townsend DM, and Tew KD (2006). Glutathione S-transferase polymorphisms: cancer incidence and therapy. *Oncogene* **25**(11): 1639–1648.

McLeod HL, Lin JS, Scott EP, Pui CH, and Evans WE (1994). Thiopurine methyltransferase activity in American white subjects and black subjects. *Pharmacol Ther, Part C* **55**(1): 15–20.

Mersha TB and Abebe T (2015). Self-reported race/ethnicity in the age of genomic research: its potential impact on understanding health disparities. *Hum Genomics* **9**(1): 1.

Newman LA, Griffith KA, Jatoi I, et al. (2006). Meta-analysis of survival in African American and white American patients with breast cancer: ethnicity compared with socioeconomic status. *J Clin Oncol* 24(9): 1342–1349.

Newsom JT, Huguet N, McCarthy MJ, et al. (2012). Health behavior change following chronic illness in middle and later life. *J Gerontol B* 67(3): 279–288.

Novillo A, Romero-Lorca A, Gaibar M, et al. (2015). Genetic diversity of *CYP3A4* and *CYP3A5* polymorphisms in North African populations from Morocco and Tunisia. *Int J Biol Markers* 30 (1): e148–e151.

Omoti CE, Olu-Eddo AN, and Nwannadi AI (2009). Co-existence of TB and adult haematological cancers in Benin City, Nigeria. *Tropical Doctor* 39(4): 205–207.

Palser AL, Grayson NE, White RE, et al. (2015). Genome diversity of Epstein–Barr virus from multiple tumour types and normal infection. *J Virol* 89: 03614.

Parkin DM (2004). International variation. *Oncogene* 23(38): 6329–6340.

Paschos K and Allday MJ (2010). Epigenetic reprogramming of host genes in viral and microbial pathogenesis. *Trends Microbiol* 18(10): 439–447.

Podsiadlo P, Sinani VA, Bahng JH, et al. (2008). Gold nanoparticles enhance the anti-leukemia action of a 6-mercaptopurine chemotherapeutic agent. *Langmuir* 24(2): 568–574.

Ponder BA (2001). Cancer genetics. *Nature* 411(6835): 336–341.

Pongtheerat T, Treetrisool M, and Purisa W (2009). Glutathione s-transferase polymorphisms in breast cancers of Thai patients. *APJCP* 10(1): 127–132.

Poreba E, Broniarczyk JK, and Gozdzicka-Jozefiak A (2011). Epigenetic mechanisms in virus-induced tumorigenesis. *Clinical Epigenetics* 2(2): 233–247.

Relling MV, Gardner EE, Sandborn WJ, et al. (2011). Clinical Pharmacogenetics Implementation Consortium guidelines for thiopurine methyltransferase genotype and thiopurine dosing. *Pharmacol Ther, Part C* 89(3): 387–391.

Ricciardelli I, Blundell MP, Brewin J, et al. (2014). Towards gene therapy for EBV-associated posttransplant lymphoma with genetically modified EBV-specific cytotoxic T cells. *Blood* 124 (16): 2514–2522.

Roadmap Epigenomics Consortium, Kundaje A, Meuleman W, et al. (2015). Integrative analysis of 111 reference human epigenomes. *Nature* 518(7539): 317–330.

Ruiss R, Jochum S, Wanner G, et al. (2011). A virus-like particle-based Epstein–Barr virus vaccine. *J Virol* 85(24): 13105–13113.

Ruwende C, Khoo SC, Snow RW, et al. (1995). Natural selection of hemi- and heterozygotes for G6PD deficiency in Africa by resistance to severe malaria. *Nature* 376(6537): 246–249.

Saccone D, Asani F, and Bornman L (2015). Regulation of the vitamin D receptor gene by environment, genetics and epigenetics. *Gene* 561(2): 171–180.

Sahasranaman S, Howard D, and Roy S (2008). Clinical pharmacology and pharmacogenetics of thiopurines. *Eur J Clin Pharmacol* 64(8): 753–767.

Schlesinger D, Grinberg LT, Alba JG, et al. (2013). African ancestry protects against Alzheimer's disease-related neuropathology. *Mol Psychiatry* 18(1): 79–85.

Stanton RC (2012). Glucose-6-phosphate dehydrogenase, NADPH, and cell survival. *IUBMB Life* 64 (5): 362–369.

Stanulla M, Schrappe M, Brechlin AM, Zimmermann M, and Welte K (2000). Polymorphisms within glutathione S-transferase genes (*GSTM1, GSTT1, GSTP1*) and risk of relapse in childhood B-cell precursor acute lymphoblastic leukemia: a case–control study. *Blood* 95(4): 1222–1228.

Strange RC and Fryer AA (1999). The glutathione S-transferases: influence of polymorphism on cancer susceptibility. *IARC Sci Publ* 148: 231–249.

Suleiman SH, Koko ME, Nasir WH, et al. (2015). Exome sequencing of a colorectal cancer family reveals shared mutation pattern and predisposition circuitry along tumor pathways. *Front Genet* 6: 288.

Tang Z-H, Zhang C, Cheng P, et al. (2014). Glutathione-S-transferase polymorphisms (*GSTM1*, *GSTT1* and *GSTP1*) and acute leukemia risk in Asians: a meta-analysis. *APJCP* **15**(5): 2075–2081.
Tishkoff SA and Williams SM (2002). Genetic analysis of African populations: human evolution and complex disease. *Nat Rev Genet* **3**(8): 611–621.
Tzellos S and Farrell PJ (2012). Epstein–Barr virus sequence variation: biology and disease. *Pathogens* **1**: 156–175.
Van Cromphaut SJ, Rummens K, Stockmans I, et al. (2003). Intestinal calcium transporter genes are upregulated by estrogens and the reproductive cycle through vitamin D receptor-independent mechanisms. *J Bone Miner Res* **18**(10): 1725–1736.
van de Graaf SFJ, Boullart I, Hoenderop JGJ, and Bindels RJM (2004). Regulation of the epithelial Ca2+ channels TRPV5 and TRPV6 by 1alpha,25-dihydroxy vitamin D3 and dietary Ca2+. *J Steroid Biochem Mol Biol* **89–90**(1–5): 303–308.
Veeck J and Esteller M (2010). Breast cancer epigenetics: from DNA methylation to microRNAs. *J Mammary Gland Biol Neoplasia* **15**(1): 5–17.
Verrelli BC, Tishkoff SA, Stone AC, and Touchman JW (2006). Contrasting histories of G6PD molecular evolution and malarial resistance in humans and chimpanzees. *Mol Biol Evol* **23**(8): 1592–1601.
Wang D-Y, Fulthorpe R, Liss SN, and Edwards EA (2004). Identification of estrogen-responsive genes by complementary deoxyribonucleic acid microarray and characterization of a novel early estrogen-induced gene: *EEIG1*. *Mol Endocrinol* **18**(2): 402–411.
Weinshilboum RM and Sladek SL (1980). Mercaptopurine pharmacogenetics: monogenic inheritance of erythrocyte thiopurine methyltransferase activity. *Am J Hum Genet* **32**(5): 651–662.
Wejse C, Olesen R, Rabna P, et al. (2007). Serum 25-hydroxyvitamin D in a West African population of tuberculosis patients and unmatched healthy controls. *Am J Clin Nutr* **86**(5): 1376–1383.
Willett WC, Koplan JP, Nugent R, et al. (2006). Prevention of chronic disease by means of diet and lifestyle changes. In Jamison DT, Breman JG, Measham AR, et al., eds., *Disease Control Priorities in Developing Countries*, 2nd edn. World Bank.
Wu X, Zhou T, Cao N, Ni J, and Wang X (2015). Role of vitamin D metabolism and activity on carcinogenesis. *Oncol Res* **22**(3): 129–137.
Yahia ZA, Adam AA, Elgizouli M, et al. (2014). Epstein Barr virus: a prime candidate of breast cancer aetiology in Sudanese patients. *Infect Agent Cancer* **9**(1): 9.
Yang H-C, Chen T-L, Wu Y-H, et al. (2013). Glucose 6-phosphate dehydrogenase deficiency enhances germ cell apoptosis and causes defective embryogenesis in *Caenorhabditis elegans*. *Cell Death Dis* **4**: e616.
Ye Z and Song H (2005). Glutathione s-transferase polymorphisms (*GSTM1*, *GSTP1* and *GSTT1*) and the risk of acute leukaemia: a systematic review and meta-analysis. *Eur J Cancer* **41**(7): 980–989.
Yuille M, Condie A, Hudson C, et al. (2002). Relationship between glutathione S-transferase M1, T1, and P1 polymorphisms and chronic lymphocytic leukemia. *Blood* **99**(11): 4216–4218.
Zaza G, Cheok M, Krynetskaia N, et al. (2010). Thiopurine pathway. *Pharmacogenet Genomics* **20**(9): 573–574.
Zhang Z, Liew CW, Handy DE, et al. (2010). High glucose inhibits glucose-6-phosphate dehydrogenase, leading to increased oxidative stress and beta-cell apoptosis. *FASEB J* **24**(5): 1497–1505.

11 The Genetic Epidemiology of Orphan Diseases in North Africa

Lilia Romdhane, Olfa Messaoud, Rym Kefi, Afaf Tiar, Ahlem Amouri, Mourad Mokni, Neji Tebib, Mohamed Zghal, Abdelhamid Barakat, Ahmed Houmeida, Mariem Bozguiya, Mohamed Othman, Ghada El-Kamah, and Sonia Abdelhak

11.1 Introduction

North Africa (NA) is defined as the northernmost geographical region of the African continent, linked by the Sahara from the south and separated from Europe by the Mediterranean Sea. It encompasses five countries from east to west: Egypt, Libya, Tunisia, Algeria, and Morocco; these are also known as Maghreb countries. From the political point of view, at the regional level, Mauritania is considered part of the Maghreb. The current population size is estimated to be around 75 million. A set of features affects markedly the prevalence and distribution of the genetic diseases in the region. The demographic features include high population growth rates, high birth rates, and high infant mortality rates. The family structure was as large as that of Arab families, with more than six children per woman until the middle of the twentieth century. Nowadays, North African (NAf) families show reduced rates of fertility that reach their lowest point at 1.06 children per decade in Tunisia (United Nations, Department of Economic and Social Affairs, Population Division, 2010).

Genetic information generated so far from some NAf countries has provided a valuable indicator of the prevailing disorders in the region. However, such information is quite fragmented and often based on case reports and case study series, rather than prospective epidemiological and molecular studies. In this chapter, we describe major historical events that shaped the NAf population structure and report on the spectrum of genetic disease altered genes as well as mutations responsible for these disorders.

11.2 North Africa Settlement and Major Prehistorical and Historical Events

The most ancient human presence in NA is dated at 700,000 years before present (YBP) and corresponds to bone remains of *Atlanthropus mauretanicus* discovered in Ternifine in Algeria (Arambourg 1963). Several other remains have been exhumed from archaeological sites in Morocco, highlighting the transition of *Homo erectus* to *Homo sapiens sapiens*. Around 20,000 YBP to 10,000 YBP, western NA was occupied by Iberomaurusian populations (Mechta-Afalou Men), who replaced the Aterian aborigines in reference to middle Paleolithic industry (Aouraghe 2006; Ferembach 1986).

In the eighth millennium, a new type of *Homo sapiens sapiens* had appeared in the oriental part of NA (Tunisia and eastern Algeria), called "proto-Mediterranean," who is considered as author of the Capsian culture and the ancestor of Amazigh, also known as Berbers (Camps, 1974, 1989).

The historical period of the region began with the settlement of Carthage City in Tunisia by Phoenician traders coming from Tyre in 814 BC (Lebanon region today). The newcomers established many colonies on the Mediterranean coasts. For many centuries, the Punic civilization has displaced Berbers to the mountains or city peripheries, or integrated them. Carthage took control of the western Mediterranean and most of NA except Egypt, which was colonized by Alexander the Great in 332 BC.

After its defeat during the third Punic war (149–146 BC), Carthage lost its dominion on the region and became a part of the Roman province. During the fifth century, the Roman Empire was dislocated and the Germanic nation, the so-called Vandals, dominated over the coastal region of NA for nearly 100 years. In AD 533, the Byzantine Empire re-established control until the Islamic conquest in the seventh century (Jamil 1975).

The new invaders arrived from the Arab Peninsula and brought with them their language, their culture, and the Islamic religion (Jamil 1975). The first Arab-Muslim city, Kairouan, was founded in AD 670 in "Ifriqya," which is nowadays Tunisia, eastern Algeria, and western Libya. A few centuries later, Ifriqya witnessed a second large invasion wave of two Arab tribes: Banu Hilal and Banu Sulaym. In AD 1135, some cities of Ifriqya were colonized by Normands and Sicilians, who were expelled under the Almohad dynasty (AD 1147–1269) (Jamil 1975). The Hilalian invasion had a drastic impact on the country's administration and accelerated the dissolution of the Almohad dynasty, which was replaced by the Hafsid dynasty (AD 1230–1574) (Jamil 1975). During that period, the NAf population suffered from painful and extensive human losses due to the Black Death from Tunis to Tanger. A part of the region was under Spanish occupation (AD 1534) until its incorporation in the Ottoman Empire (Jamil 1975).

In the seventeenth century, 5.4 million inhabitants lived in the NAf region after the massive migrations of hundreds of thousands of Moorish and Jews from Spain (Guyon 1855; Jamil 1975; Tabutin et al. 2002). The eighteenth and nineteenth centuries were marked by successive epidemic episodes, the decline of the Ottoman Empire, and the establishment of European colonization forces in NA.

11.3 Ethnic Groups in NA and Genetic Discovery Revealed by Anthropological Data

There are two major ethnic groups in the NAf region: the Amazigh and the Arabs. Despite the deep Arabization and Islamization of the region, the Amazigh culture and language still survive in some restricted regions, mostly because of little intermixture with the new settlers. They remained highly endogamous in order to preserve their ethnicity. The current Amazigh population in NA is around 20 million, of whom 20 percent live mainly in Algeria and 40 percent in Morocco. A minority is settled in Tunisia and Libya (around 100,000) (Brett and Fentress 1997). Nevertheless, these figures are probably underestimates, and after the so-called Arab spring and recent political changes in the region, more and more families that have conserved oral and written transmission of Amazigh are coming to light. In addition to the major ethnic groups, Jews from Tunisia, Libya, Morocco, and Algeria, called North African Jews and/or Sephardic Jews and/or non-Ashkenazi Jews, constitute a minority, together with Christians.

The richness of the prehistoric and historic cultural inheritance of NAf civilizations has contributed to the richness and complexity of the genetic pool of its populations. According to molecular studies (mitochondrial DNA, Y chromosome, autosomal STRs, etc.), the NAf population is a mosaic, composed of three components: North African, Eurasian, and sub-Saharan (Bekada et al. 2013; Bosch et al. 2000; Ennafaa et al. 2011; Esteban et al. 2004; Flores et al. 2000; Gonzalez et al. 2006; Kefi et al. 2005; Plaza et al. 2003).

Regarding mitochondrial data, the distribution of these three components is very variable, not only between NA populations but also between localities that are geographically very close, belonging to the same country. The high level of heterogeneity is also observed between ethnic groups and even between populations belonging to the same ethnic group. Moreover, phylogenetic analysis highlighted the presence of some NA groups that are genetically very distant from other NA groups, and behave as outliers (Kefi et al. 2015).

11.4 Consanguinity and Endogamy

Like other Arab countries, NAf populations show high rates of consanguinity and endogamy (Table 11.1). The most common form of intermarriage is between first-cousins, particularly paternal first-cousins, and includes double first-cousin marriage (Ben Halim et al. 2013; Talbi et al. 2007). Although monotheist religions such as Judaism, Christianity, and Islam condemn incestuous matings, some minorities of Moroccan Jews still practice uncle–niece mating (Fried and Davies 1974). Contrary to popular ideas, Islam does not encourage nor prohibit consanguineous marriages. The combination of religious, cultural, social, geographical, and economical factors contribute to maintaining such marriages between relatives. The marital choice is preferentially performed inside the family or the clan, thus installing endogamy and consanguinity (Ben M'rad and Chalbi 2006). The choice of a first-cousin or a distant relative or a neighbor could be considered as an obligation toward the family, since any deviation could provoke social reprobation and even familial sanction (Ben M'rad and Chalbi 2006). Traditional, social, and educational levels, especially of women, are the main factors that favor consanguineous marriages, especially in rural surroundings (Kerkeni et al. 2006; Romdhane et al. 2014; Zaoui and Biémont 2002).

Table 11.1 Consanguineous marriage in North Africa

Country	Location	Consanguinity (percentage)	Consanguinity type	References
Algeria	Urban	27.5	1C, 2C	Benallègue and Kedji (1984)
	Rural	34.0	1C, 2C	Benallègue and Kedji (1984)
	All Algeria	22.6	1C, 2C	Benallègue and Kedji (1984)
	Tlemcen	34.04	1C, >1C	Zaoui and Biémont (2002)
Egypt	Alexandria	32.8	1C	Stevenson et al. (1966)
	Nubia	60.5	1C	Hussien (1971)
	Nubia (Fadetchi)	61.4	D1C, $1^{1/2}$C, 2C	Badr (1972)
	Nubia (Kenuzi)	62	D1C, 1C, $1^{1/2}$C, 2C	Badr (1972)
	Nubia (Arabs)	69.8	D1C, 1C, $1^{1/2}$C, 2C	Badr (1972)
	All Nubia	63.6	D1C, 1C, $1^{1/2}$C, 2C	Badr (1972)
	All Egypt	23.2	1C	Habib and Böök (1983)
	All Egypt (urban)	22.1	D1C, 1C, $1^{1/2}$C, 2C	Hafez et al. (1983)
	All Egypt (rural)	39.1	D1C, 1C, $1^{1/2}$C, 2C	Hafez et al. (1983)
Libya	Benghazi	37.6	1C, 2C	Abudejaja et al. (1987)
Morocco	All Morocco	19.9	1C, 2C	Lamdouar Bouazzaoui (1994)
	All Morocco	22.8	1C, >1C	Talbi et al. 2007
Mauritania	All Mauritania	47.20	1C, >1C	Hammami et al. 2005
Tunisia	North	26.9	1C, $1^{1/2}$C, 2C	Riou et al. (1989)
	Bizerte	39.33	1C, >1C	Ben Arab et al. 2004
	Monastir	20.1	1C, 2C	Kerkeni et al. 2006
	Monastir	24.8	1C, 2C	Kerkeni et al. 2007
	Different localities of Tunisia	29.9	1C, D1C, >1C, 2C, >2C	Ben Halim et al. (2013)

D1C = double first-cousin; 1C = first-cousin; >1C = beyond first-cousin; $1^{1/2}$C = first-cousin once removed; 2C = second cousin; >2C = beyond second cousin.

11.5 Spectrum of Orphan Genetic Disease in North Africa

Data on orphan diseases among NAfs are becoming available following the introduction of diagnostic facilities in these countries, and as a result of the development of biomedical health research activities and international collaborations. Nevertheless, due to limited resources and the orientation of public health and research priorities, few genetic diseases have been concerned with complete epidemiological studies. Therefore, in order to assess the burden of these disorders in the region, we have performed a systematic manual text mining of available data in the literature, including gray literature, as performed previously for genetic diseases in Tunisia (Romdhane and Abdelhak 2012). Our study revealed that more than 530 genetic diseases are reported. Unfortunately, no updated database, be it national or regional, is currently available. A few years ago, a general database on Arab populations was constructed (Tadmouri et al. 2006), as well as one on the Tunisian population (The Tunisian National Genetic Database). A Moroccan population-specific database has been developed (Charoute et al. 2013; http://mgdd.pasteur.ma). Nevertheless, these databases are not comprehensive and not always updated, and do not reflect the spectrum of inherited disorders in these populations.

In the absence of updated databases and of precise epidemiological data, one solution is the establishment of national patient registries. An initiative has been realized recently in Tunisia for Fanconi anemia (Hadiji Mseddi et al. 2012), bleeding disorders (Elmahmoudi et al. 2012), and for Gaucher disease (http://maladie-gaucher-tunisie.org).

We have classified the genetic diseases reported in NA according to their modes of transmission, and have shown that autosomal recessive ones reach 60 percent while autosomal dominant and X-linked diseases are only observed at frequencies of 27.2 percent and 5.8 percent, respectively. This emphasizes the impact of the high rate of endogamy and consanguinity, combined with the relatively large family sizes in NA, which increase the risk of emergence of an autosomal recessive disease in a family. Other modes of transmission are also observed at lower frequencies (Romdhane and Abdelhak 2012).

Particular modes of transmission, constellations of undescribed symptoms, extreme rare syndromes and comorbidity are specific features of genetic diseases in North Africa (Romdhane and Abdelhak, 2011, 2012; Romdhane et al. 2016). A similar pattern of disease group distribution exists as in the Arab world. Indeed, three major disease classes are the most described among NAf populations (Figure 11.1) (Romdhane and Abdelhak 2012; Tadmouri et al. 2006). The following disease classification was facilitated by the use of the World Health Organization International Classification of Disease in its tenth version (WHO ICD-10). The most frequent disease set encompasses congenital malformations, deformations, and chromosomal abnormalities whose detection is relatively obvious at birth and represents 29.4 percent of the genetic diseases. The second major group is formed by the endocrine, nutritional, and metabolic diseases (22.4 percent). Finally, diseases of the nervous system represent the third most reported disease group (14.3 percent) (Romdhane and Abdelhak 2012).

For several decades until now, there have been important migration waves from NA to the Middle East and/or Europe for trade, work, or study purposes. Therefore, consanguineous marriages continue to be contracted within the NA diaspora in host countries.

Hence, the reported rates of inherited diseases in these communities are comparable to those in their countries of origin, including a large proportion of autosomal recessive diseases and developmental disorders (Alif et al. 1999; Chkioua et al. 2007; Richard et al. 2008).

Approximately 10 percent of these genetic diseases are reported in more than three countries, suggesting a common genetic background. Other diseases are specific to one NA country or one ethnic and/or religious group (Romdhane and Abdelhak 2012).

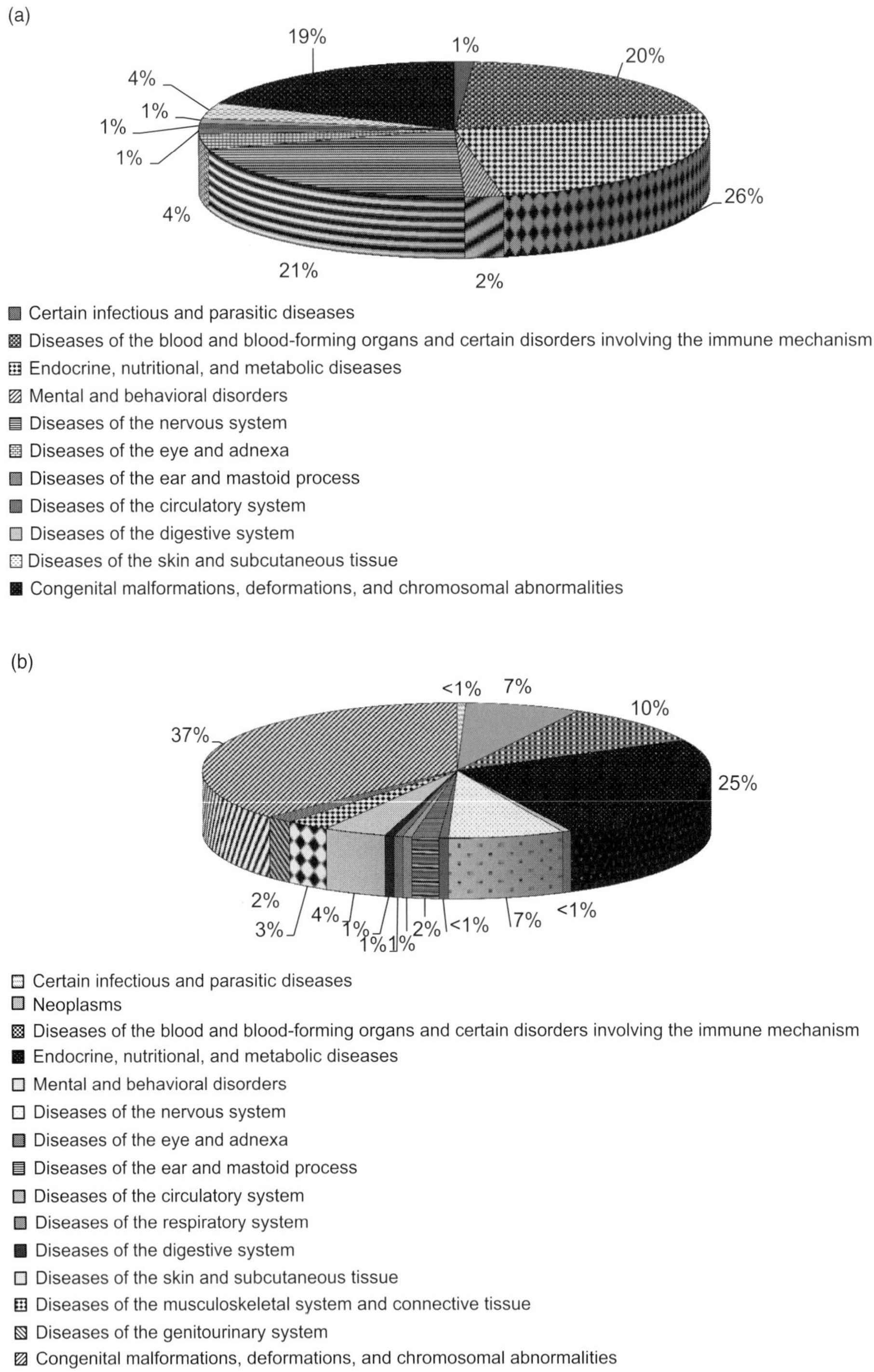

Figure 11.1 WHO ICD classification of genetic diseases in North Africa: (a) Algeria, (b) Egypt, (c) Libya, (d) Morocco, and (e) Tunisia.

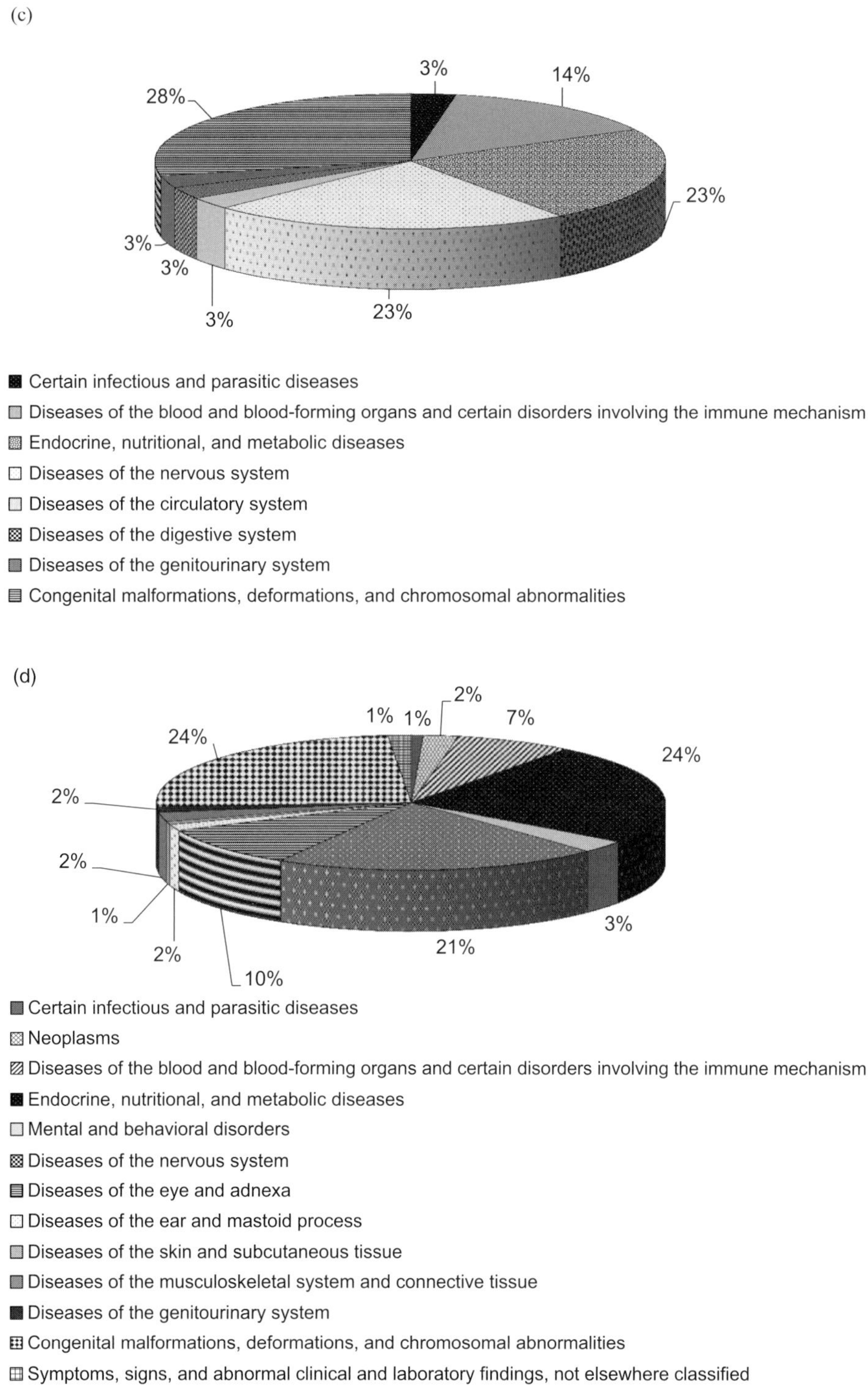

Figure 11.1 (*cont.*)

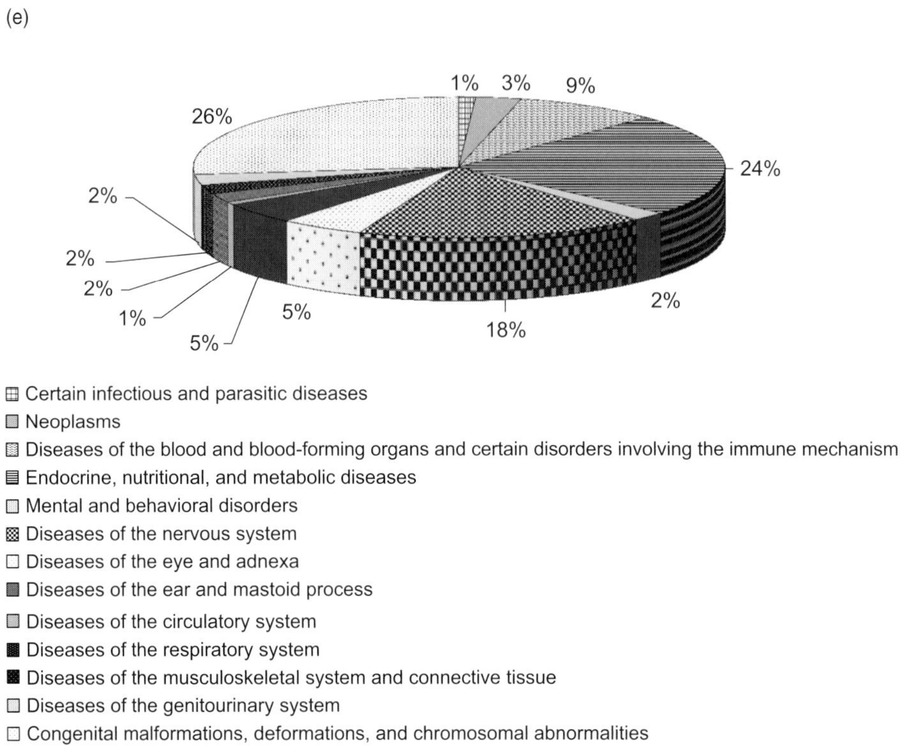

Figure 11.1 (*cont.*)

Underlying molecular etiology is further reported in this chapter. In the following section, examples of diseases in some morbid groups are reported.

11.5.1 Congenital Malformations, Deformations, and Chromosomal Abnormalities

Within this category we have chosen to report on the most frequent diseases or those relatively specific to NA. These include congenital bone disorders and severe genodermatosis, encompassing the groups of hereditary epidermolysis bullosa, xeroderma pigmentosum, and keratinization disorders.

Congenital Bone Disorders

The Algerian type of spondylometaphyseal dysplasia (ICD-10: Q77.8) is a rare and severe malformation and an example of NA-specific genetic disease. It was reported for the first time in five members of an Algerian family (Kozlowski et al. 1988). Other diseases characterized by a unique clinical and radiological set of features were also described, mainly in Tunisian families. Persistent torticollis, facial asymmetry, grooved tongue, and dolicho-odontoid process in connection with atlas malformation complex was reported in three patients from the same family by Al Kaissi et al. (2007a). The same author also reported rare variants of SEMDJL and Desbuquois syndromes (ICD-10: Q78.8), in addition to new clinical features associated with already known syndromes (Al Kaissi et al. 2005, 2007b, 2008, 2009). Moreover, co-existence of a genodermatoses, called hypohidrotic ectodermal dysplasia (ICD-10: Q82.4), with limb deformities was highlighted twice in members from two unrelated and highly inbred families (Al Kaissi et al. 2002). Pycnodysostosis (ICD-10: Q78.8) is a rare autosomal recessive disorder of osteoclast dysfunction having an incidence of 1.7 per one million births and fewer than 200 patients reported (Mujawar et al. 2009; Ornetti et al. 2008). In NA, eight patients were reported with the following distribution: one from Algeria (Osimani et al. 2010), one from Tunisia (Donnarumma et al. 2007), one from Egypt (Donnarumma et al. 2007), and five from Morocco (Donnarumma et al. 2007; Gelb et al. 1996).

Hereditary Epidermolysis Bullosa (ICD-10: Q81.0, Q81.2, Q81.9)

Hereditary epidermolysis bullosa (EB) comprises a group of genetically determined skin fragility disorders, which are characterized by blistering of the skin and mucosa in response to little or unapparent trauma. The disease is clinically and genetically heterogeneous and is associated with a variable range of complications, from localized skin fragility to neonatal death (Ouragini et al. 2008). According to the mode of inheritance and the histological characteristics, three major forms of EB could be distinguished: EB simplex (EBS), junctional EB (JEB), and dystrophic EB (DEB). In Tunisia, the EB frequency was evaluated to 115/1,000,000 (Cherif et al. 2005b), which is more elevated than the frequency in the USA (8.22/1,000,000) (Fine et al. 1999). Among Tunisian EB patients, 46 percent present the dystrophic form, among whom 90 percent present the recessive form, while in the rest of the world only 5 percent have DEB (Cherif et al. 2005b). Few reports are available on EB in other NA countries. In Algeria, two EB girls with laryngeal involvement, one with JEB and another with DEB, have been so far reported (Nègre et al. 1990). Junctional EB with pyloric atresia has also been reported in a few cases from Tunisia (Cherif et al. 2005a), Morocco (Ruzzi et al. 1997), and Libya (Gahukamble et al. 2003). In our experience, several cases of severe hereditary EB originating from Libya have been observed. Unfortunately, it was not possible to classify those cases to a specific clinical entity due to inaccessibility to ultrastructural analysis. For multiplex informative families, it was possible to classify the patients by genotyping and homozygosity by descent analysis (Ouragini et al. 2008).

Xeroderma Pigmentosum (ICD-10: Q82.1)

Xeroderma pigmentosum (XP) is a rare genodermatosis characterized by a high skin sun-sensitivity predisposing patients to skin cancers at an early age. Eight genes (*XPA* to *XPG* and *POLH*) are responsible for the disease, which seems to be frequent in NA as large patient series have been reported in the region (Ben Rekaya et al. 2009; Bouadjar et al. 1996; Messaoud et al. 2010a; Senhaji et al. 2013; Soufir et al. 2010). The XPC gene is known to be the most mutated in Mediterranean patients. Using molecular epidemiological tools, Doubaj et al. estimated for the first time in an Arab country the carrier frequency for the founder mutation c.1643_1644delTG to be 1/250, which would imply that the prevalence of the XPC form would be approximately of 1/80,504, considering the effect of consanguinity in Morocco (Doubaj et al. 2012). In Tunisia, the prevalence of XP was evaluated to 1/10,000, thus 30 times as high as in Europe (Zghal et al. 2006). In addition to the main cutaneous manifestations and the malignancy outcomes accounting for 9–14 percent of childhood cancers in Tunisia, other features were also described that associated with the disease as neurological abnormalities that characterize the XPA form (El Fékih et al. 2009; Maalej et al. 2007; Messaoud et al. 2010b; Zghal et al. 2005), as well as ocular manifestations (Touzri et al. 2008). A few cases with another form of XPV have also been described from Tunisia and Algeria (Ben Rekaya et al. 2011; Broughton et al. 2002; Johnson et al. 1999).

Mal de Meleda (ICD-10: Q82.8)

Mal de Meleda (MDM), also referred to as keratosis palmoplantaris transgrediens of Siemens, is a rare autosomal recessive skin disorder with a prevalence of 1 in 100,000 in the general population of Yugoslavia, where it was reported for the first time (island of Meleda) (Hovorka and Ehlers 1897). Linkage analysis of large consanguineous families from Algeria localized the responsible gene to *8qter* (Bouadjar et al. 2000; Fischer et al. 1998). Mal de Meleda seems to be frequent in regions where consanguinity is common, such as NA. The largest MDM series were reported in NA: in Tunisia, 10 families with 55 patients were reported (Charfeddine et al. 2003; Marrakchi et al. 2003) and 16 other cases (Marrakchi et al. 2006). In Algeria, 16 families with 21 affected members have been described (Marrakchi et al. 2006) and more recently one family from Libya (Bchetnia et al. 2015).

11.5.2 Endocrine, Nutritional, and Metabolic Disease

Inherited metabolic diseases (IMDs) are a significant cause of morbidity and mortality in North Africa, and their prevalence remains to be defined. The first estimation of one carrier for a mutation in an IMD gene per 32 Egyptians was made by Hashem (1978). Inherited metabolic diseases play an important role in mental retardation predisposition and were detected among 280 out of 450 individuals with mental disability (Shawky et al. 2001). In Libya, Eldeep et al. [2009] studied 1503 (856 neonates and 647 children) patients (Eldeep et al. 2009). IMD accounted for 3 percent of total admissions. A similar figure was found in Tunisia (2.8 percent) among 13,171 requests for analysis of patients with symptoms suggestive of IMD (Hadj-Taieb et al. 2012).

Cystic Fibrosis (ICD-10: E84)

Cystic fibrosis (CF) seems to be relatively frequent in NA. Based on patient records, 390 children from 383 Tunisian families were diagnosed with CF (Messaoud et al. 2005). The disease is less frequent in Algeria as only 81 affected individuals have been reported (Loumi et al. 2008). In Morocco, CF prevalence was estimated as equal to that in Europe after screening of 150 healthy individuals (Ratbi et al. 2008). The disease occurs at a similar frequency of 1/2700 among Jews from Libya, but seems to be rare in Jews from Morocco (1/15,000) (Kerem et al. 1995).

Familial Mediterranean Fever (MIM249100; ICD-10: E85.0)

Familial Mediterranean fever (FMF) is an autosomal recessive autoinflammatory periodic disorder characterized by febrile and painful attacks due to inflammation involving the serosal membranes. The disease is caused by mutations in the *MEFV* gene. It was reported that the FMF carrier rate can reach one carrier per three individuals in the commonly affected ethnic groups, namely non-Ashkenazi Jews of NAf origin, Armenians, Turks, and Arabs, raising the possibility of selective heterozygote advantage (Al-Alami et al. 2003; Gershoni-Baruch et al. 2001; Kogan et al. 2001; Stoffman et al. 2000; Touitou 2001; Yilmaz et al. 2001). In Algeria, the *M694I* mutation is the most common and its carrier rate is about 1:5 (Ait-Idir et al. 2011).

Gaucher Disease (ICD-10: E75.2)

Gaucher disease is an autosomal recessive lysosomal storage disorder caused by a deficiency of the acid β-glucosidase enzyme. Its prevalence is underestimated, though it does not seem exceptional. In Tunisia, a retrospective study over an 18-year period revealed 27 cases (Chaabouni et al. 2004). Another multicenter study, performed in 2005, reported 55 cases and a prevalence of 0.48 affected individuals per 100,000 inhabitants (Ben Turkia et al. 2005), which is higher than that in Turkey (0.23 cases/100,000 inhabitants) (Ozkara and Topçu 2004). Gaucher disease type I was by far the most frequent form (Chaabouni et al. 2004). In Egypt, a study of 22 Egyptian children with Gaucher disease revealed that two-thirds of patients were consanguineous (El-Beshlawy et al. 2006). Additional Egyptian patients have also been reported (El-Morsy et al. 2011). In Libya, 49 patients diagnosed with Gaucher disease have been described (Ben-Omran 2010). Adult forms of Gaucher disease have also been reported and remain under-diagnosed, thus rendering the disease incidence estimation difficult to establish (Ben Rhouma et al. 2012).

Glycogen Storage Disease I, III, IV (ICD-10: E74.0)

Glycogen storage diseases (GSDs) are a group of IMDs characterized by the deficiency of one enzyme among the enzymatic complex involved in glycogen metabolism, and subsequent build-up of excessive levels of glycogen in the body tissues. Cases with different subtypes of GSD have been reported in NA. In Tunisia, the pediatric prevalence of GSD I in the north can be estimated at 7.93 cases per one million inhabitants, with an incidence of 1/100,000 births (Ben Chehida et al. 2008). These values are underestimates, as death rapidly occurs in the neonatal period before the establishment of the diagnosis. GSD III is usually frequent among NAf Jews in Israel, where the prevalence reaches 1/54,000 and the carrier prevalence has been estimated at 1/35 (Parvari et al. 1997). In Tunisia, the precise disease prevalence is still unknown. A study, performed in 1988, concluded that the GSD III form is predominant (39 cases, 83 percent) (Ayadi 1988). This conclusion is supported by recent studies on large patient series (Cherif et al. 2012; Mili et al. 2012a, 2012b; Ouachani 2001). Additional cases with GSD III have been reported (Endo et al. 2005, 2009; Kotb et al. 2004).

Mucopolysaccharidosis (ICD-10: E76.2)

The mucopolysaccharidoses (MPS) are a devastating heterogeneous group of lysosomal storage disorders. In Tunisia, a retrospective epidemiological survey covering the period 1970–2005 was conducted. Among the 132 suspected cases, 96 confirmed MPS cases were identified. The crude prevalence subtypes in the collected data were estimated at 0.63, 0.7, and 0.45 per 100,000 live births, respectively. The cumulative incidence of MPS type VI (0.3 per 105 live births) was higher than that reported in European countries (Ben Turkia et al. 2009). In another recent study, the Tunisian MPS type IVA incidence was approximately 0.45 in 100,000 live births, which is considerably higher than the rate in Japan (1 in 500,000), Australia (1 in 201,000), or Portugal (1 in 450,000) (Khedhiri et al. 2012). In Egypt, 77 families were reported with MPS. Among them, 23 cases had Hurler syndrome, 16 had Morquio disease, 12 had Hunter disease, 10 had Hurler/Scheie, 9 had San Philippo, 4 had Maroteau-Lamy, and 3 had Scheie (Hashem 1978). Later, out of 1240 patients referred to the Human Genetic Clinic between 1997 and 1998, 248 (20 percent) had inborn errors of metabolism, 36 (14 percent) of whom were diagnosed as MPS. Deficiency of pha-L-iduronidase (IDUA) enzyme in leukocytes and increased urinary mucopolysaccharide excretion were detected in 17 patients (El-Bassyouni et al. 1999).

Phenylketonuria (PKU) (ICD-10: E70.0)

Phenylketonuria is an autosomal recessive disease caused by liver phenylalanine hydroxylase (PAH) enzyme deficiency, which if untreated causes mental retardation. Phenylketonuria patients constituted 2.3 percent in Egypt and 1.32 percent in Tunisia of the developmentally disabled persons (Khemir et al. 2009; Temtamy et al. 1991). Recent studies showed that its prevalence is approximately the same in Tunisia (1/7631) and in Egypt (1/7000), which is clearly higher than that of the Caucasian population, assessed to be approximately 1:10,000 (Khemir et al. 2011; Scriver and Kaufman 2001). In Libya, its incidence was evaluated at 1/7000 (Ben-Omran 2010), and it seems to be fewer than 1/100,000 in Morocco (Lamzouri et al. 2012).

Tay–Sachs Disease (ICD-10: E75.0)

Tay–Sachs disease is a common disorder among Jews, particularly Ashkenazi and Moroccan Jews, but is rarely described among Arabs (Jacoub 1938; Navon et al.

1981). Indeed, in Egypt a study performed on 59 children with seizures and psychomotor retardation showed that only one patient had Tay–Sachs disease (Shatla et al. 1999). According to the age at onset and the type of hexosaminidase deficiency, patients are classified into: Type I, "Tay–Sachs disease" – 58 percent of cases; Type II, "Sandhoff disease" – 10 percent of cases; and Type III, "Juvenile GM2 gangliosidosis" – 32 percent of cases. Consanguinity was reported in 73.3 percent of cases, and familial cases with affected sibs were detected among 23 percent of the studied patients (El-Harouni et al. 2002). Studies on Tay–Sachs disease in other NA populations are lacking.

Other IMDs

Other IMDs have also been reported to be frequent among NAf populations. In Tunisia, 212 cases with aminoacidopathies other than PKU (57.3 percent) and 158 cases with organic acidurias (42.7 percent) have been reported. The most frequent aminoacidopathies were maple syrup disease (32.5 percent), tyrosinemia type I (28.8 percent), and nonketotic hyperglycinemia (16 percent), while methylmalonic aciduria (33.5 percent), propionic aciduria (18.4 percent), and 2-hydroxy glutaric aciduria (10.8 percent) were the most frequent organic acidurias. Their incidences were calculated using the Hardy–Weinberg formula and were estimated at 1/13,716 for maple syrup disease, 1/14,804 for tyrosinemia type I, 1/16,144 for methylmalonic aciduria, and 1/23,176 for propionic aciduria (Hadj-Taieb et al. 2012). A prevalence of 1.5 percent of aminoacidopathies was reported by Shawky et al. (2007) in Egypt, among which PKU was found to be the most common (1.11 percent), while the remaining diagnosed cases (urea cycle defects, hyperglycemia, maple syrup urine disease, alkaptonuria, trimethylaminuria) represented 0.07 percent each (Shawky et al. 2007).

11.5.3 Diseases of the Nervous System

The foundation of the Institut National de Neurologie of La Rabta in Tunisia in the middle of the 1980s by the late Professor Mongi Ben Hamida represented an excellent opportunity for neuromuscular disease study in the country and even in the wider NAf region. This was illustrated by the numerous scientific publications in bibliographic databases on this kind of disorder originating mainly from Tunisia.

Duchenne muscular dystrophy (DMD; ICD-10: G71.0) and Becker muscular dystrophy (BMD) are X-linked recessive disorders caused by mutations in the *DMD* gene located at chromosome Xp21. In DMD patients, dystrophin is virtually absent, whereas BMD patients have 10–40 percent of the normal dystrophin amount. Both forms are encountered in NA, with large patient series (Ben Hamida et al. 1992; Mhiri et al. 1996). Molecular investigation consisting of dystrophin deletion pattern determination has been performed in Egypt and Morocco (El Sherif et al. 2007; Itto et al. 2013).

Limb-Girdle Muscular Dystrophy (ICD-10: G71.0)

Formerly called "Duchenne-like muscular dystrophy" then "Maghrebian myopathy" by Ben Hamida and Fardeau in 1980, limb-girdle muscular dystrophy (LGMD2C) neuromuscular disease is frequent in the NA region, especially in Tunisia and has been reported among Egyptian patients (Kefi et al. 2003). Molecular investigation of Tunisian patients led to the localization of the *SGCG* gene in chromosome 13 (Ben Othmane et al. 1992; Teebi and Farag 1997). In Tunisia, this myopathy, along with other forms of limb-girdle muscular dystrophy (LGMD), are the most common form of hereditary muscular diseases, with a frequency of 61.3 percent. Among 95 patients who have had a molecular diagnosis of the 564 who were referred to the Institut National de Neurologie between 2000 and 2009, 55.8 percent had LGMD2C; 14.7 percent had LGMD2I; 5.3 percent had LGMD2B; 4.2 percent had LGMD2D; and 20 percent had another form of myopathy (Nabli et al. 2010). Studies on LGMD are rare in the other NAf countries. LGMD2B seems to be frequent among Libyan Jews, with a prevalence of 1/3000 (Argov et al. 2000). In Tunisia, other types of muscular dystrophy are also encountered with variable frequencies, such as metabolic myopathy (16.5 percent); congenital muscular dystrophy (9.4 percent); facioscapulohumeral muscular dystrophy (7.6 percent); congenital myopathy (3.4 percent); and inclusion hereditary myopathy (1.8 percent) (Nabli et al. 2010). In Morocco, the mode of inheritance of LGMD is mainly autosomal recessive. It was shown that nearly 50 percent of the Moroccan LGMD is due to the del521T mutation of the *SGCG* gene responsible for LGMD2C, with a Maghrebian founder effect as in Tunisian and Algerian families (Elkerch et al. 2001). Few families have been reported with LGMD2D (Piccolo et al. 1995).

Parkinson Disease (ICD-10: G20)

Various studies have demonstrated that the frequency of Parkinson disease (PD) in the region is high, as reported by the large patient series mainly from Tunisia, Morocco, and Algeria (Belarbi et al. 2010; Healy et al. 2008; Ishihara et al. 2007; Lesage et al. 2005). Since the first clinical report by Ben Hamida and Gouider-Khouja, it was clearly noted that in NA a very frequent form of PD exists and is characterized by the association of rest with postural tremor (Gouider-Khouja 1991). This phenotype is much more frequent in Tunisia and NA than in Europe. It was called "Maghrebian form of PD" (Gouider-Khouja and Rebai 2012). A recent comparative study by Nishioka et al. demonstrated that the genetic background of this phenotype is yet to be identified (Nishioka et al. 2010).

11.5.4 Disease of the Blood and the Immune Mechanism

These diseases encompass essentially beta- and alpha-thalassemias (ICD-10: D56.0, D56.1), sickle cell anemia (ICD-10: D57), and glucose-6-phosphate dehydrogenase deficiency (G6PD) (ICD-10: 55.0). Their high frequencies in the NA region are considered as a real public health problem and therefore they have been widely studied. Beta-thalassemia carrier rates range from 1.12 to 9 percent; those for alpha-thalassemia carriers were estimated to be less than 1–9.25 percent (Hamamy and Al-Allawi 2013). In Tunisia, beta-thalassemia and sickle cell are mostly prevalent in the northwest and the southwest (Haj Khelil et al. 2004). Sickle cell anemia and beta-thalassemia are for the moment the only monogenic diseases investigated in Mauritania (Deyde et al. 2002; Veten et al. 2012). These erythrocyte disorders are known examples of how natural selection confers adaptive advantage to heterozygous individuals against malaria infection (Denic and Nicholls 2007). Other types of hematological diseases like Fanconi anemia and megaloblastic anemia, as well as bare lymphocyte syndrome, are quite common and genetic investigations have been conducted.

Bare Lymphocyte Syndrome Type II (ICD-10: D81.7)

Bare lymphocyte syndrome type II is a rare autosomal recessive defect of the expression of the major histocompatibility complex class II genes, caused by mutations in the *RFXANK* gene. Fewer than 200 patients have been reported (Al-Herz et al. 2013; Djidjik et al. 2012; Klein et al. 1993; Naamane et al. 2010; Ouederni et al. 2011; Reda et al. 2009; Reith and Mach 2001; Rezaei et al. 2006; Wiszniewski et al. 2000). The prevalence of this deficiency is highest in Mediterranean areas, especially in NA, where the majority of the patients were reported: 15 Moroccan (Djidjik et al. 2012; Naamane et al. 2010; Ouederni et al. 2011), 36 Algerian (Ouederni et al. 2011), and 39 Tunisian patients out of 551 with primary immunodeficiencies diagnosed from 1993 to 2011 (Ben-Mustapha et al. 2013; Ouederni et al. 2011). Bare lymphocyte syndrome type II is also frequent in the Maghrebian community living in France (Ouederni et al. 2011). This high prevalence is due to founder effect in the NAf region (Ouederni et al. 2011).

Fanconi Anemia (ICD-10: D61.0)

Fanconi anemia (FA) is a rare autosomal recessive disease that is clinically and genetically heterogeneous with at least eight complementation groups that have been defined (A–H). In Tunisia, a relatively high incidence of FA is observed: 1.4/million/year (Bouchlaka et al. 2003). In order to identify to which complementation group Tunisian patients belong, Bouchlaka et al. (2003) analyzed gene mutations in 39 unrelated families from different regions in Tunisia. Thirty-four families were assigned to the FA-A group (94 percent), whereas one family was probably not linked to the *FANCA* gene or to any known FA gene. Recently, data from a Tunisian national registry on FA revealed that 142 patients belonging to 118 families were collected between January 1983 and December 2008 (Hadiji Mseddi et al. 2012). The mean age at diagnosis was 11 years. Consanguinity was present in 86 percent of cases (Hadiji Mseddi et al. 2012). Of 28 patients, 95 percent belonged to the FANCA group. Androgen treatment was given for 109 cases and genoidentical bone marrow transplantation (BMT) was performed for 27 patients (Hadiji Mseddi et al. 2012).

The patient geographic distribution showed a heterogeneous pattern with a gradient prevalence from south to north (Hadiji Mseddi et al. 2012). Epidemiological studies are not available in the other NAf countries. Nevertheless, in Egypt, among 48 patients clinically suspected to have FA, 31 cases were found to have FA, with a consanguinity rate of 97 percent (Temtamy et al. 2007). Jewish FA patients have been reported, mainly from Morocco and Tunisia (Tamary et al. 2000). Fanconi anemia cases from Algeria and Libya have been recently investigated and showed evidence for linkage to *FANCA* (unpublished data).

Megaloblastic Anemia (ICD-10: D53)

The majority of cases of megaloblastic anemias result from deficiencies of either cobalamin (vitamin B12) or folate (vitamin B9). The relative etiological role of each vitamin deficiency is well established in Western countries (Chanarin 1990) but few data have been reported so far from developing countries. From January 1999 to December 2001, 478 consecutive patients from two hematology departments in Tunisia with megaloblastic changes in bone marrow smears were prospectively studied. This study is one of the largest series reported so far on megaloblastic anemia in non-Western countries. It revealed several interesting features of the disease in the southern Mediterranean Arab region. It also revealed that megaloblastic anemia appears to be a rather common hematologic disorder in NA. With almost 500 affected patients prospectively identified over a three-year period in a country with ten million inhabitants, a crude incidence of 16.6 cases per million per year could be estimated (Maktouf et al. 2006). Megaloblastic anemia type 1 is a rare disease (prevalence about 1 in 200,000) (Grasbeck 2006) and a rare form of macrocytic vitamin B12-deficient anemia has been reported in nine Tunisian patients from six unrelated consanguineous families. The molecular investigations revealed genetic heterogeneity (Bouchlaka et al. 2007).

11.5.5 Disease of the Ear and Mastoid Process (H60-H95)

Hereditary hearing impairment is so frequent that there are some villages concentrating a large number of families in Mauritania (unpublished data) as well as in Morocco and in Tunisia where the prevalence is estimated to be 0.0007 in the governorates of Nabeul (Ben Arab et al. 1990) and 2–8 percent in Bizerte (Ben Arab et al. 2004). Segregation analysis performed in 29 pedigrees containing 415 subjects with 129 affected cases provided evidence for simple recessive inheritance with no sporadic cases (Ben Arab et al. 1990). In Egypt, 35 patients with sensorineural hearing loss and a high rate of parental consanguinity were studied (Ismail et al. 1996). They were classified according to their diagnoses into: four cases with dominant inheritance; 21 cases with congenital deafness and autosomal recessive inheritance; one case with childhood deafness which also showed autosomal recessive inheritance; and one case with X-linked inheritance. Six cases were isolated; their mode of inheritance was indefinite (Ismail et al. 1996). Furthermore,

Snoeckx et al. (2005) reported 159 additional Egyptians from 111 families with non-syndromic mild to profound hearing impairment (Snoeckx et al. 2005). The *DFNB1* locus is the most important locus for non-syndromic autosomal recessive deafness and was originally identified by linkage analysis in two consanguineous Tunisian families with prelingual and profound deafness (Guilford et al. 1994). The 35delG mutation was the most frequent *GJB2* gene lesion in NA (Ammar-Khodja et al. 2007; Denoyelle et al. 1997; Lucotte 2007; Masmoudi et al. 2000; Ratbi et al. 2007; Riahi et al. 2013; Snoeckx et al. 2005). Additional loci implying hearing impairment were further identified in families mainly from Tunisia and Morocco (Romdhane and Abdelhak 2012). Furthermore, progressive hereditary hearing loss, in particular, affecting adults in middle-age, is sometimes considered as a natural sign of "aging." Despite the relatively large number of molecular studies on hereditary hearing impairment, no epidemiological studies have been conducted so far in NA.

11.5.6 Other Diseases

Creutzfeldt–Jakob Disease (ICD-10: A81.0)

Creutzfeldt-Jakob Disease (CJD) is the most common transmissible spongiform encephalopathy found in humans. Clinically, it is a degenerative neurological disorder, quite similar to other neurological disorders such as Alzheimer's disease. The CJD occurs frequently among Libyan Jews, where its prevalence accounts for 1/10,000, which is 100 times higher than in other populations (Colombo 2000) but much less than among Jews born in Egypt for whom the CJD rate is 3.5 per million (Zilber et al. 1991).

11.6 Molecular Epidemiology of Orphan Genetic Disease

Approximately half of the inherited diseases described in the region are not genetically studied. Genetic investigation facilities are not equally distributed in the region due to lack of resources. Thanks to the international collaboration that helped to build local capacities, and the introduction of molecular tools in the last decades in most of the North African countries, neurosensorial, neuromuscular, and dermatological diseases of genetic origin have become active research domains. This was illustrated by the multiple new loci and subsequent causative mutation identifications, mainly from Tunisian and Moroccan families (Ebermann et al. 2007; Li et al. 2010; Masmoudi et al. 2003; Tlili et al. 2008). Founder effects were also identified (Romdhane et al. 2012).

Among the diseases with known molecular etiology, one mutation in one gene has been reported in addition to allelic and genetic heterogeneity. Reasons for such genetic diversity could be due to natural selection whenever genetic variations might confer an adaptive advantage to individuals in a specific environment. The second factor is random changes of the gene pool due to gene drift (Zlotogora 2007).

Founder effect is an example of how gene drift operates to increase frequency of deleterious alleles in isolated populations. These gene lesions are called founder mutations. They are inherited from a common ancestor and are transmitted through generations in isolated communities. Isolation could be of geographic, linguistic, ethnic, or religious orders; sometimes a combination of all of these. Usually, a period of endogamy follows, thus increasing the probability that two gametes harboring the same mutation meet each other. In its broader definition, a founder mutation is a genetic lesion identified in two unrelated individuals (Romdhane et al. 2012).

Identification of founder mutations is possible by haplotype analysis. In this situation, the surrounding genomic regions are inherited identically by descent. Founder mutations are of particular importance in the field of genetic disease research and diagnosis as they provide a rapid molecular diagnosis tool, especially in countries with limited resources (Romdhane et al. 2012). Therefore, establishing the mutational spectrum of a population with a focus on founder lesions is a suitable diagnosis strategy for diseases with unknown molecular etiology, in particular some metabolic diseases (Barkaoui et al. 2007) or genodermatosis (Ouragini et al. 2008), and thus circumvent invasive biopsies, to confirm a biochemical or a histological test (Romdhane et al. 2012).

Distribution of founder mutations in the region has shown some differences that could be assigned to two main categories. The first class encompasses country-specific founder mutations in which the genetic variations are not reported elsewhere. In this case, this kind of lesion is of autochthonous origin. They could be restricted to a region of the country, or to an ethnic group, or both. They could also be dispersed in the country, thus giving insights into regional exodus. The second class includes founder mutations shared with other populations. In a recent study it was shown that specific founder mutations are the causes of 73 genetic diseases in Tunisia (Romdhane et al. 2012). Among these, 43 are caused by founder mutations reported also in neighboring populations, mainly from NA (Table 11.2), but also from the Middle East and Europe.

Allelic heterogeneity has also contributed to the emergence of the diseases due to founder mutations in the NAf region. This genetic diversity suggests gene flows due to population movements in the region. Two possible explanations could shed light on this observation. In the first situation, one major frequent mutation is encountered due essentially to founder effect and genetic drift. In addition to this mutation, another less frequent mutation is also observed. This is, for example, the case of the predominant founder mutation c.521delT in the *SGCG* gene that leads to LGMD2C in the Tunisian population. Another private mutation was reported in two families: one from Tunisia and the other from Libya (Hentati 2008). In the second observation, more than one frequent mutation is responsible for the disease. The most illustrating examples are mutations causing beta-thalassemia (Haj Khelil et al. 2010), G6PD deficiency (Haj Khelil et al. 2010), CF (Messaoud et al. 2005), and FMF (Chaabouni et al. 2007). Natural selection could also have acted as a factor for recurrence of these mutations in the region.

Table 11.2 Diseases with founder mutations in North Africa

Disease	OMIM	Gene	Nucleotide change	Amino acid change	Populations
21-hydroxylase deficiency	201910	*CYP21A2*	c.955C>T	p.Q318X	Tunisian, Moroccan
			Large deletions		Tunisian, Moroccan
			IVS2-13A/C-G		Tunisian, Moroccan
			c.515T>A	p.I172N	Tunisian, Moroccan
3-M syndrome	273750	*CUL7*	4449_4450delTG	p.V1484GfsX69	Tunisian, Algerian
Adenomatous polyposis of the colon	175100	*MUTYH*	c.1227_1228dup	p.E396GfsX43	Tunisian, Moroccan, Algerian
Ataxia-telangiectasia	208900	*ATM*	c.103C>T	p.R35X	Jews from North Africa: Morocco and Tunisia
Ataxia with isolated vitamin E deficiency	277460	*TTPA*	c.744delA	p.E249NfsX15	Tunisian, Moroccan, North African
Autosomal recessive inclusion body myopathy 2	600737	*GNE*	c.2186T>C	p.M412T	Tunisian, Egyptian
Autosomal recessive non-syndromal hearing loss	220290	*GJB2*	c.35delG	p.G12VfsX12	Tunisian, Moroccan, Egyptian
Autosomal recessive neurosensory deafness 7	600974	*TMC1*	c.100C>T	p.R34X	Tunisian, Algerian
Autosomal recessive non-syndromic optic atrophy	612989	*TMEM126A*	c.163C>T	p.R55X	Tunisian, Algerian, Moroccan
Autosomal recessive spastic paraplegia with thinning of corpus callosum	604360	*SPG11*	c.6100C>T	p.R2034X	Tunisian, Algerian, Moroccan
Bare lymphocyte syndrome type II	209920	*RFXANK*	752delG-25		North African: Tunisian, Algerian, Moroccan
Beta-thalassemia	141900	*HBB*	c.118C>T	p. Q39X	Tunisian, Moroccan, Algerian
			c.118C>T (haplotype I)	p. Q39X	Tunisian, Algerian, Moroccan

Table 11.2 (cont.)

Disease	OMIM	Gene	Nucleotide change	Amino acid change	Populations
			c.118C>T (haplotype II)	p.Q39X	Tunisian, Algerian, Moroccan
			c.IVS1–110G>A	Aberrant splicing	Tunisian, Algerian, Moroccan
			c.IVS1–110G>A	Aberrant splicing	Tunisian, Moroccan
			c.IVS1–110G>A (HT1 haplotype)	Aberrant splicing	Tunisian, Algerian
			c.IVS1–110G>A (HT2 haplotype)	Aberrant splicing	Tunisian, Algerian
			c.92+2T>G (IX haplotype)	Aberrant splicing	Tunisian, Moroccan
Carbonic anhydrase II deficiency	259730	*CA2*	c.IVS2+1G>A	Aberrant splicing	Tunisian
Congenital myasthenia syndrome	608931	*CHRNE*	c.1293insG	p.E432GfsX24	North African: Tunisian, Algerian, Moroccan, Libyan
Creutzfeldt–Jakob disease	123400	*PRNP*	c.598G>A	p.E200K	Libyan Jews, Tunisian
Cystic fibrosis	219700	*CFTR*	c.1521_1523del	p.F508del	Tunisian, Algerian,
			c.1624G>T	p.G542X	Tunisian, Mediterranean, Algerian
			c.3486G>A	p.W1282X	Tunisian, Mediterranean, North African
			c.3090C>G	N1303K	Tunisian, Western and Mediterranean, Northern African
			711 +1G>T	Aberrant splicing	Tunisian, North African, Algerian
Distal renal tubular acidosis	267300	*ATP6V1B1*	1149–1155insC	I386fsX441	Tunisian, Algerian, Moroccan
			c.IVS2–1G>C	Aberrant splicing	Tunisian, Algerian

Familial Mediterranean fever	249100	*MEFV*	c.2040G>C / c.2040G>A	p.M680I	Tunisian, Moroccan, Algerian
			c.2080A>G	p.M694V	Tunisian, North African Jews, Egyptian, Moroccan, Algerian
			c.2177T>C	p.V726A	Tunisian, Egyptian
			c.2082G>A	p.M694I	Tunisian, Egyptian, Maghrebian, Moroccan, Algerian
			c.442G>C	p.E148Q	Tunisian, Egyptian, North African Jews
Fragile X syndrome	300624	*FMR1*	Large CGG repeats on a common haplotype DXS548-FRAXAC1-FRAXAC2 (7+ -4- 6+)		Tunisian Jews, Sephardic Jews
Gaucher disease	230800	*GBA*	c.1448G>C	p.L444P	Tunisian, Egyptian
Glucose-6-phosphate dehydrogenase deficiency	305900	*G6PD*	202 GTG>ATG (Aurès mutation)	68V>M	Tunisian, Algerian
			143T>C (Aurès mutation)	48I>T	Tunisian, Algerian
Glycogen storage disease Ia	232200	*G6PC*	c.247C>T	p.R83C	Tunisian, Moroccan
Glycogen storage disease type III	232400	*AGL*	c.3908G>A	p.W1327X	Tunisian, Egyptian, Libyan
			4455delT		North African Jews
Hereditary breast and ovarian cancer	114480	*BRCA1*	c.798_799delTT	p.V266fs	Non-Jewish from North Africa: Algerian, Tunisian
		BRCA2	c.1310_1313delAAGA	p.K437IfsX22	Tunisian, Algerian
Hereditary elliptocytosis	130600	*SPTA1*	c.444_445insTTG	p.L148dup	Tunisian, North African
Hereditary hemochromatosis	235200	*HFE*	c.187C>G	p.H63D	Tunisian, Algerian, Moroccan
			c.845G>A	p.C282Y	Tunisian, Algerian, Moroccan

Table 11.2 (cont.)

Disease	OMIM	Gene	Nucleotide change	Amino acid change	Populations
Hepatocerebral mitochondrial DNA depletion syndrome	251880	*DGUOK*	c.444–62C>A	Aberrant splicing	Tunisian, Moroccan
Hurler syndrome	607014	*IDUA*	c.1598C>G	p.P533R	Tunisian, Moroccan, Maghrebian population living in France
Leber congenital amaurosis	204000	*GUCY2D*	c.306delC	p.E103SfsX26	Tunisian, Moroccan, North African
			1767T>C	p.F589S	North African, Algerian
Limb-girdle muscular dystrophy, type 2C	253700	*SGCG*	c.521delT	Frame shift	Tunisian, Moroccan, North African
Mal de Meleda	248300	*SLURP-1*	c.82delT	p.C28Afs32X	Tunisian, Algerian
Male infertility	243060	*AURKC*	c.144delC	p.L49WfsX22	Tunisian, Algerian, Moroccan
Megaloblastic anemia 1	261100	*AMN*	c.208–2A>G	Aberrant splicing	Tunisian, Tunisian Jews, Mediterranean basin
Niemann–Pick disease type B	607616	*SMPD1*		p.ΔR608	Tunisian, Algerian, Moroccan, North African Jews
Parkinson disease	607060	*LRRK2*	c.6055G>A	p.G2019S	Tunisian, Moroccan, Algerian
Phenylketonuria	261600	*PAH*		p.E280K	North African
				p.S349P	Tunisian and Moroccan Jews
			c.IVS10–11G>A	Aberrant splicing	Mediterranean
Primary hyperoxaluria type 1	259900	*AGXT*	734T>C	p.I244T	Tunisian, Libyan Jews, North African

Sickle cell anemia	603903	*HBB*	β6, A>T (RFLP Benin haplotype)	E6V	Tunisian, Algerian, Moroccan, Egyptian
			β6, A>T (SNP Benin haplotype)	E6V	Tunisian, Algerian
Susceptibility to cutaneous malignant melanoma	155601	*CDKN2A*	c.176T>G	p.V59G	Tunisian Jews, Moroccan Jews
Triple-A syndrome	231550	*AAAS*	c.IVS14+1G>A		Tunisian, Algerian
Unverricht–Lundborg disease	254800	*CSTB*	Expansion of a dodecamer repeat (CCCCGCCCCGCG)*n* common haplotype: 1(D21S1885)-1(D12S2040)-A(D12S1259)-1(D12S59)		Tunisian, Moroccan, Algerian
Xeroderma pigmentosumgroup A	278700	*XPA*	c.682C>T	p.R228X	Tunisian, Algerian, North African, Egyptian, Libyan
Xeroderma pigmentosumgroup C	278720	*XPC*	c.1643_1644delTG	p.V548AfsX25	Tunisian, Moroccan, Algerian, Egyptian, Libyan

11.7 Founder Mutations and the Origin of NA Populations

Distribution of a founder mutation in the region is the consequence of population migrations and new settlements. Its geographic repartition and frequency could give insights on its age and history. Indeed, older founder mutations tend to be widely distributed and the geographic location harboring its highest frequency is considered as its origin. Basically, the age of the mutation is the time since it has appeared. Some genetic disorders have shown that mutations can be traced to a founder whose name is unknown but his/her existence can be inferred through chromosomal background and coalescence time calculation (Zeegers et al. 2004). This calculation is based on many observations of the studied population, mainly the average reproductive age (Zeegers et al. 2004). The age of the founder event was estimated at approximately 700 years for 1293insG mutation in the *CHRNE* gene (Richard et al. 2008) and calculated to rise in the second half of the thirteenth century in the case of the Creutzfeldt–Jakob disease (Colombo 2000; Lee et al. 1999). Triple-A syndrome (Tullio-Pelet et al. 2000), Unverricht–Lundborg disease (Moulard et al. 2002), and Leber congenital amaurosis (Hanein et al. 2008) are thought to be more ancient – their ages have been estimated to be between 2400 and 2600 years ago or more.

In the same context, multiple debates occurred regarding the p.G2019S mutation located in the *LRRK2* gene in an attempt to determine whether it has a Berber origin or not (Benamer and de Silva 2010). Some authors supposed that this mutation dates back to Phoenician times rather than Arabic times, to the founding of Carthage in 814 BC (Farrer et al. 2009). The founder mutation 752delG-25 responsible for bare lymphocyte syndrome type II seems to have arisen 2250 years ago (Ouederni et al. 2011). Of interest, the same mutation, 35delG, is responsible for the most frequent form of deafness and seems to have emerged 111 generations ago in the Tunisian and the Moroccan populations and since 216 generations ago in the Lebanese population (Belguith et al. 2005). This age difference could be due to two different founder effects or to the delay in the introduction of the same mutational event from Lebanon to the NAf region. Another filiation study has allowed accurately tracing the Tunisian families affected by carbonic anhydrase deficiency II back to a common Arabic tribe of "Banu Hilal" that settled in the Maghreb in the tenth century (Fathallah et al. 1997). Moreover, based on the comparison of frequencies of two *HBB* gene mutations, historical events linked to the introduction of each mutation have been well identified. The IVS-I-110G>A mutation, which is predominant in the western part of Tunisia and in northern Algeria (Haj Khelil et al. 2010) has been attributed to the Ottoman introduction that does not reach Moroccan territory, thus explaining its rarity in Morocco. However, the codon 39 mutation, which is largely predominant in Tunisia, in Morocco, and in northwestern Algeria, is thought to have an occidental origin that was introduced in the region during the Roman period (Haj Khelil et al. 2010). Another example is illustrated by the mutation frequencies for PKU around the Mediterranean basin. In fact, the most frequent mutation observed in Tunisia, the p.E280K, accounts for 0.17 (Khemir et al. 2012), which is a higher frequency than that described in the northern Mediterranean: France (0.06), Greece (0.05), and Spain (0.04) (Zschocke 2003), thus suggesting a Tunisian origin of this mutation. The second most frequent mutation, G352VfsdelG, has been reported in Morocco as the most frequent (Dahri et al. 2010). The same mutation is also common in Spain and probably originated from the Spanish and the Andalusian migration flows during the seventh century. In fact, after the Reconquista, the Arab Andalusians migrated extensively from Spain to Morocco and Tunisia. Finally, regarding the IVS10-11G>A mutation, Scriver proposed that its distribution was suggestive of a Turkish origin that had subsequently spread to the Mediterranean basin (Scriver and Kaufman 2001).

11.8 Conclusion

Rare genetic diseases constitute a serious health problem in NA. We still refer to these diseases as orphan, as they are relatively neglected compared to other diseases. With the current situation of the so-called democratic transition and the climate of social and economic instability, they are even more neglected and affected families have to face several challenges. Their actual landscape is the result of the combination of multiple factors: founder effect, genetic drift, selective pressure, consanguinity, and endogamy that favor the occurrence of autosomal recessive conditions at higher frequencies. The effect on health is particularly dramatic, in particular with the epidemiological transition and the rise of non-communicable diseases, although the impact of consanguinity on multifactorial diseases remains controversial. A potential benefit of consanguinity has even started to be hypothesized (Denic and Nicholls 2007). During the last decades, health professionals via learning societies such as the Tunisian Association for Inherited Metabolic Diseases have played an important role in educating and communicating on genetic diseases, and thus raising awareness among all stakeholders. In addition, patient support groups like the Association of Support to Children with Xeroderma Pigmentosum, which is active at the regional level, has improved the protection of affected children and accelerated the diagnosis of new cases. Several tentative attempts to control the burden of genetic disorders in the region are being made. Unfortunately, no concrete measures such as systematic neonatal screening are being made in NA (Krotoski et al. 2009), and initiatives that have been started have also been practically abandoned. Oriented carrier screening, targeting the most frequent mutations or the founder mutations specific to each locality and cascade screening, could be a more efficient and cost-effective alternative to systematic neonatal screening. Efforts should be made at the national and regional levels to increase the interaction between researchers from the biomedical domain and other fields, in particular the social sciences and humanities, in addition to the involvement of patient support groups and decision-makers. This must be more than wishful thinking in order to take concrete steps toward reducing the burden of these conditions in the region.

References

Abudejaja AH, Khan MA, Singh R, et al. (1987). Experience of a family clinic at Benghazi, Libya, and sociomedical aspects of its catchment population. *Fam Pract* 4: 19–26.

Ait-Idir D, Khilan A, Djerdjouri B, and El-Shanti H (2011). Spectrum of mutations and carrier frequency of familial Mediterranean fever gene in the Algerian population. *Rheumatology (Oxford)* 50: 2306–2310.

Al-Alami JR, Tayeh MK, Najib DA, et al. (2003). Familial Mediterranean fever mutation frequencies and carrier rates among a mixed Arabic population. *Saudi Med J* 24: 1055–1059.

Al-Herz W, Alsmadi O, Melhem M, et al. (2013). Major histocompatibility complex class II deficiency in Kuwait: clinical manifestations, immunological findings and molecular profile. *J Clin Immunol* 33(3): 513–519.

Alif N, Hess K, Straczek J, et al. (1999). Mucopolysaccharidosis type I: characterization of a common mutation that causes Hurler syndrome in Moroccan subjects. *Ann Hum Genet* 63: 9–16.

Al Kaissi A, Ghachem MB, Necib MN, et al. (2002). Hypohidrotic ectodermal dysplasia with tibial aplasia. *Clin Dysmorphol* 11: 175–178.

Al Kaissi A, Nessib N, Ghachem MB, et al. (2005). A novel syndrome resembling Desbuquois dysplasia. *Am J Med Genet* 132: 68–75.

Al Kaissi A, Ben Chehida F, Gharbi H, et al. (2007a). Persistent torticollis, facial asymmetry, grooved tongue, and dolicho-odontoid process in connection with atlas malformation complex in three family subjects. *Eur Spine J* 16: 265–270.

Al Kaissi A, Grill F, Safi H, et al. (2007b). Craniocervical junction malformation in a child with oroman-dibular-limb hypogenesis: Meobius syndrome. *Orphanet J Rare Dis* 8: 2.

Al Kaissi A, Chehida FB, Ghachem MB, Klaushofer K, and Grill F (2008). A hypoplastic atlas and long odontoid process in a girl manifesting pheno-typic features resembling spondyloepimetaphyseal dysplasia joint laxity syndrome. *Skeletal Radiol* 37: 469–473.

Al Kaissi A, Chehida FB, Ghachem MB, Grill F, and Klaushofer K (2009). Atlanto-axial segmentation defects and os odontoideum in two male siblings with opsismodysplasia. *Skeletal Radiol* 38: 293–296.

Ammar-Khodja F, Makrelouf M, Malek R, Ibrahim H, and Zenati A (2007). Frequency of the 35delG allele causing nonsyndromic recessive deafness in the Algerian patients. *Genet Couns* 18(4): 383–391.

Aouraghe H (2006). Histoire du peuplement paléolithique de l'Afrique du Nord et dynamique des interactions entre l'homme et son environnement. *CR Pale* 5: 237–242.

Arambourg C (1963). Le gisement de Ternifine II-l'Atlanthrope mauretanicus. *Archives de l'IPH* 32: 37–190.

Argov Z, Sadeh M, Mazor K, et al. (2000). Muscular dystrophy due to dysferlin deficiency in Libyan Jews: clinical and genetic features. *Brain* 123 (Pt 6): 1229–1237.

Ayadi A (1988). Les glycogènoses hépatiques. Thèse de Médecine, Faculté de Médecine de Tunis.

Badr FM (1972). Genetic studies of Egyptian Nubian population: frequency and types of consanguineous marriages. *Human Heredity* 22: 387–398.

Barkaoui E, Cherif W, Tebib N, et al. (2007). Mutation spectrum of glycogen storage disease type Ia in Tunisia: implication for molecular diagnosis. *J Inherit Metab Dis* 30(6): 989.

Bchetnia M, Bozgia M, Laroussi N, et al. (2015). The first Mal de Meleda case in Libya: identification of a SLURP1 mutation. *Int J Dermatol* 54: 1426–1428.

Bekada A, Fregel R, Cabrera VM, et al. (2013). Introducing the Algerian mitochondrial DNA and Y-chromosome profiles into the North African landscape. *PLoS One* 8(2): e56775.

Belarbi S, Hecham N, Lesage S, et al. (2010). LRRK2 G2019S mutation in Parkinson's disease: a neuropsychological and neuropsychiatric study in a large Algerian cohort. *Parkinsonism Relat Disord* 16(10): 676–679.

Belguith H, Hajji S, Salem N, et al. (2005). Analysis of GJB2 mutation: evidence for a Mediterranean ancestor for the 35delG mutation. *Clin Genet* **68**: 188–189.

Ben Arab S, Bonaïti-Pellié C, and Belkahia A (1990). An epidemiological and genetic study of congenital profound deafness in Tunisia (governorate of Nabeul). *J Med Genet* **27**(1): 29–33.

Ben Arab S, Masmoudi S, Beltaief N, Hachicha S, and Ayadi H (2004). Consanguinity and endogamy in Northern Tunisia and its impact on non-syndromic deafness. *Genet Epidemiol* **27**: 74–79.

Ben Chehida A, Tebib N, Cherif W, et al. (2008). Glycogen storage disease type I in Tunisia: an epidemiological analysis. *J Inherit Metab Dis* **31**(Suppl. 2): S199–S204.

Ben Halim N, Ben Alaya Bouafif N, Romdhane L, et al. (2013). Consanguinity, endogamy, and genetic disorders in Tunisia. *J Community Genet* **4**: 273–284.

Ben Hamida M, Miladi N, Turki I, and Zaiem H (1992). Duchenne muscular dystrophy in Tunisia: a clinical and morphological study of 77 cases. *J Neurol Sci* **107**(1): 60–64.

Ben M'rad L and Chalbi N (2006). Milieu de résidence origine des conjoints et consanguinité en Tunsie. *Antropo* **12**: 63–71.

Ben Othmane K, Ben Hamida M, Pericak-Vance MA, et al. (1992). Linkage of Tunisian autosomal recessive Duchenne-like muscular dystrophy to the pericentromeric region of chromosome 13q. *Nat Genet* **2**: 315–317.

Ben Rekaya M, Messaoud O, Talmoudi F, et al. (2009). High frequency of the V548AfsX572 XPC mutation in Tunisia: implication for molecular diagnosis. *J Hum Genet* **54**: 426–429.

Ben Rekaya M, Messaoud O, Mebazaa A, et al. (2011). A novel *POLH* gene mutation in a xeroderma pigmentosum-V Tunisian patient: phenotype–genotype correlation. *J Genet* **90**(3): 483–487.

Ben Rhouma F, Kallel F, Kefi R, et al. (2012). Adult Gaucher disease in southern Tunisia: report of three cases. *Diagn Pathol* **10**: 4.

Ben Turkia A, Chemli J, and Chaabouni M (2005). Les maladies lysosomales en Tunisie: étude multicentrique sur les mucopolysaccharidoses, la maladie de Gaucher et la maladie de Niemann Pick type B. Communication. 6ème école des maladies héréditaires du métabolisme.

Ben Turkia H, Tebib N, Azzouz H, et al. (2009). Incidence of mucopolysaccharidoses in Tunisia. *Tunis Med* **87**(11): 782–785.

Benallègue A and Kedji F (1984). Consanguinité et santé publique: etude algérienne. *Archives Francaises de Pediatrie* **41**: 435–440.

Benamer HT and de Silva R (2010). LRRK2 G2019S in the North African population: a review. *Eur Neurol* **63**: 321–325.

Ben-Mustapha I, Ben-Farhat K, Guirat-Dhouib N, et al. (2013). Clinical, immunological and genetic findings of a large Tunisian series of major histocompatibility complex class II deficiency patients. *J Clin Immunol* **33**(4): 865–870.

Ben-Omran T (2010). Genetic disorders in Libya. In Teebi AS, ed., *Genetic Disorders Among Arab Populations*. Springer.

Bosch E, Calafell F, Perez-Lezaun A, et al. (2000). Genetic structure of north-west Africa revealed by STR analysis. *Eur J Hum Gene* **8**: 360–366.

Bouadjar B, Aït-Belkacem F, Daya-Grosjean L, et al. (1996). [Xeroderma pigmentosum: a study in 40 Algerian patients]. *Ann Dermatol Venereol, 1* **23**(5): 303–306.

Bouadjar B, Benmazouzia S, Prud'homme JF, Cure S, and Fischer J (2000). Clinical and genetic studies of 3 large, consanguineous, Algerian families with Mal de Meleda. *Arch Dermatol* **136**: 1247–1252.

Bouchlaka C, Abdelhak S, Amouri A, et al. (2003). Fanconi anemia in Tunisia: high prevalence of group A and identification of new FANCA mutations. *J Hum Genet* **48**(7): 352–361.

Bouchlaka C, Maktouf C, Mahjoub B, et al. (2007). Genetic heterogeneity of megaloblastic anaemia type 1 in Tunisian patients. *J Hum Genet* **52**(3): 262–270.

Brett M and Fentress E (1997). *The Berbers*. Blackwell.

Broughton BC, Cordonnier A, Kleijer WJ, et al. (2002). Molecular analysis of mutations in DNA polymerase eta in xeroderma pigmentosum variant patients. *PNAS* **99**: 815–820.

Camps G (1974). *Les civilisations préhistoriques de l'Afrique du Nord et du Sahara*. Drouin Edition.

Camps G (1989). *Les Berbères: Mémoire et identité*. Errance Edition.

Chaabouni HB, Ksantini M, M'rad R, et al. (2007). MEFV mutations in Tunisian patients suffering from familial Mediterranean fever. *Semin Arthritis Rheum* **36**: 397–401.

Chaabouni M, Aoulou H, Tebib N, et al. (2004). [Gaucher's disease in Tunisia (multicenter study)]. *Rev Med Interne* **25**(2): 104–110.

Chanarin I (1990). *The Megaloblastic Anaemias*. 3rd edn. Blackwell Scientific.

Charfeddine C, Mokni M, Ben Mousli R, et al. (2003). A novel missense mutation in the gene encoding SLURP-1 in patients with Mal de Meleda from northern Tunisia. *Brit J Derm* **149**: 1108–1115.

Charoute H, Nahili H, Abidi O, et al. (2013). The Moroccan Genetic Disease Database (MGDD): a database for DNA variations related to inherited disorders and disease susceptibility. *Eur J Hum Genet*. DOI: 10.1038/ejhg.2013.151.

Cherif A, Cherif F, Siala-Guagi F, et al. (2005a). Épidermolyse bulleuse jonctionnelle congénitale, aplasie cutanée et atrésie pylorique. *Ann Dermatol Venereol* **132**(3): 298–299.

Cherif F, Mnajja N, Feriani S, et al. (2005b). Hereditary epidermolysis bullosa in Tunisia: an epidemio-clinical and ultrastructural study. *Arch Inst Pasteur Tunis* **821**(4): 53–58.

Cherif W, Ben Rhouma F, Messai H, et al. (2012). High frequency of W1327X mutation in glycogen storage disease type III patients from central Tunisia. *Ann Biol Clin (Paris)* **70**(6): 648–650.

Chkioua L, Khedhiri S, Turkia HB, et al. (2007). Mucopolysaccharidosis type I: identification of alpha-L-iduronidase mutations in Tunisian families. *Arch Pediatr* **14**: 1183–1189.

Colombo R (2000). Age and origin of the PRNP E200K mutation causing familial Creutzfeldt-Jacob disease in Libyan Jews. *Am J Hum Genet* **67**: 528–531.

Dahri S, Desviat LR, Pérez B, et al. (2010). Mutation analysis of phenylketonuria patients from Morocco: high prevalence of mutation G352fsdelG and detection of a novel mutation p.K85X. *Clin Biochem* **43**: 76–81.

Denic S and Nicholls MG (2007). Genetic benefits of consanguinity through selection of genotypes protective against malaria. *Hum Biol* **79**: 145–158.

Denoyelle F, Weil D, Maw MA, et al. (1997). Prelingual deafness: high prevalence of a 30delG mutation in the connexin 26 gene. *Hum Mol Genet* **6**(12): 2173–2177.

Deyde VM, Lo BB, Khalifa IO, et al. (2002). Epidemiological profile of hemoglobinopathies in the Mauritanian population. *Ann Hematol* **81**: 320–321.

Djidjik R, Messaoudani N, Tahiat A, et al. (2012). Clinical, immunological and genetic features in eleven Algerian patients with major histocompatibility complex class II expression deficiency. *Allergy Asthma Clin Immunol* **8**(1): 14.

Donnarumma M, Regis S, Tappino B, et al. (2007). Molecular analysis and characterization of nine novel CTSK mutations in twelve patients affected by pycnodysostosis. *Hum Mutat* **28**(5): 524.

Doubaj Y, Laarabi FZ, Elalaoui SC, Barkat A, and Sefiani A (2012). Carrier frequency of the recurrent mutation c.1643_1644delTG in the XPC gene and birth prevalence of the xeroderma pigmentosum in Morocco. *J Dermatol* **39**(4): 382–384.

Ebermann I, Walger M, Scholl HP, et al. (2007). Truncating mutation of the *DFNB59* gene causes cochlear hearing impairment and central vestibular dysfunction. *Hum Mutat* **28**(6): 571–577.

El Fékih N, Fredj M, Aounallah-Skhiri H, et al. (2009). [Neurological abnormalities in xeroderma pigmentosum]. *Rev Neurol (Paris)* **165**(11): 967–970.

El-Bassyouni HT, Afifi HH, El-Awady MK, and Meguid NA (1999). Mucopolysaccharidosis type I: clinical and biochemical study. *East Mediterr Health J* **5**: 259–366.

El-Beshlawy A, Ragab L, Youssry I, et al. (2006). Enzyme replacement therapy and bony changes in Egyptian paediatric Gaucher disease patients. *J Inherit Metab Dis* **29**: 92–98.

Eldeep A, Benyones R, Dougman K, and Esahli H (2009). A five years review of admission patterns and outcomes in a pediatric intensive care unit in Elkhadra hospital. *Jamahiriya Med J* **9**(3): 205–209.

El-Harouni AA, Zaki MS, Fateen EM, and Meguid NA (2002). GM2 gangliosidosis: clinical, genetic and biochemical studies among Egyptian patients. *Egypt J Neurol Psych Neurosurg* **39**: 87–97.

Elkerch F, Sbiti A, Azibi K, et al. (2001). La gamma-sarcoglycanopathie par la mutation del521T auMaroc: À propos de 20 cas. *Rev Magh Pédiat* **11**(4): 189–192.

Elmahmoudi H, Chalbi A, Ben-Lakhal F, et al. (2012). Regional registry of bleeding disorders in Tunisia. *Haemophilia* **18**(6): e400–e403.

El-Morsy Z, Khashaba MT, Soliman O-S, Yahia S, and El-Hady DA (2011). Glucosidase acid beta gene mutations in Egyptian children with Gaucher disease and relation to disease phenotypes. *World J Pediatr* **7**(4): 326–330.

El Sherif RM, Fahmy NA, Nonaka I, and Etribi MA (2007). Patterns of dystrophin gene deletion in Egyptian Duchenne/Becker muscular dystrophy patients. *Acta Myol* **26**: 145–150.

Endo Y, Fateen E, Aoyama Y, et al. (2005). Molecular characterization of Egyptian patients with glycogen storage disease type IIIa. *J Hum Genet* **50**(10): 538–542.

Endo Y, Fateen E, El Shabrawy M, et al. (2009). Egyptian glycogen storage disease type III: identification of six novel AGL mutations, including a large 1.5 kb deletion and a missense mutation p.L620P with subtype IIId. *Clin Chem Lab Med* **47**(10): 1233–1238.

Ennafaa H, Fregel R, Khodjet-El-Khil H, et al. (2011). Mitochondrial DNA and Y-chromosome microstructure in Tunisia. *J Hum Genet* **56**: 734–741.

Esteban E, Gonzalez-Perez E, Harich N, et al. (2004). Genetic relationships among Berbers and South Spaniards based on CD4 microsatellite/Alu haplotypes. *Ann Hum Biol* **31**: 202–212.

Farrer MJ, Gibson R, and Hentati F (2009). The ancestry of LRRK2Gly2019Ser parkinsonism. *Lancet Neurol* **7**: 770.

Fathallah DM, Bejaoui M, Lepaslier D, et al. (1997). Carbonic anhydrase II (CA II) deficiency in Maghrebian patients: evidence for founder effect and genomic recombination at the CA II locus. *Hum Genet* **99**: 634–647.

Ferembach D (1986). Les hommes du Paléolithique supérieur autour du bassin méditerranéen. *L'Anthropologie Paris* **90**: 579–587.

Fine JD, Johnson LB, Suchindran C, Moshell A, and Gedde-Dahl T, Jr. (1999). The epidemiology of inherited epidermolysis bulllosa: findings in the US, Canadian and European study populations. In Fine JD, Bauer EA, McGuire J, and Moshell A, eds., *Clinical, Epidemiological and Laboratory Advances, and the Findings of the National Epidermolysis Bullosa Registry*. John's Hopkins University Press, pp. 101–113.

Fischer J, Bouadjar B, Heilig R, et al. (1998). Genetic linkage of Meleda disease to chromosome 8qter. *Eur J Hum Genet* **6**: 542–547.

Flores C, Maca-Meyer N, Gonzalez AM, and Cabrera VM (2000). Northwest African distribution of the CD4/Alu microsatellite haplotypes. *Ann Hum Genet* **64**: 321–327.

Fried K and Davies AM (1974). Some effects on the offspring of uncle–niece marriage in the Moroccan Jewish community in Jerusalem. *Am J Hum Genet* **26**: 65–72.

Gahukamble DB, Adnan AR, and Al Gadi M (2003). Distal foregut atresias in consecutive siblings and twins in the same family. *Pediatr Surg Int* **19**(4): 288–292.

Gelb BD, Shi GP, Chapman HA, and Desnick RJ (1996). Pycnodysostosis, a lysosomal disease caused by cathepsin K deficiency. *Science* **273**: 1236–1238.

Gershoni-Baruch R, Shinawi M, Leah K et al. (2001). Familial Mediterranean fever: prevalence, penetrance and genetic drift. *Eur J Hum Genet* **9**: 634–637.

Gonzalez AM, Cabrera VM, Larruga JM, et al. (2006). Mitochondrial DNA variation in Mauritania and Mali and their genetic relationship to other Western Africa populations. *Ann Hum Genet* **70**: 631–657.

Gouider-Khouja N (1991). [Parkinson disease in Tunisia: Epidemiological, clinical, follow-up and therapeutic aspects.] La maladie de Parkinson. Aspects épidémiologiques, cliniques, évolutifs et thérapeutiques. Faculty of Medicine of Tunis.

Gouider-Khouja N and Rebai I (2012). Genetic neurological diseases in Tunisia. In Kumar D, ed., *Genomics and Health in the Developing World.* Oxford University Press.

Grasbeck R (2006). Imerslund–Gräsbeck syndrome (selective vitamin B12 malabsorption with proteinuria). *Orphanet J Rare Dis* **1**: 17. DOI: 10.1186/1750-1172-1-17.

Guilford P, Ben Arab S, Blanchard S, et al. (1994). A non-syndrome form of neurosensory, recessive deafness maps to the pericentromeric region of chromosome 13q. *Nat Genet* **6**(1): 24–28.

Guyon JLG (1855). *Histoire chronologique des épidémies du nord de l'Afrique.* Alger.

Habib Z and Böök JA (1983). Consanguinity and the incidence of thalassaemia in Egypt. *Hereditas* **99**: 215–217.

Hadiji Mseddi S, Kammoun L, Bellaaj H, et al. (2012). Creation and report of the Tunisian Fanconi Anemia Registry (TFAR). *Arch Pediatr* **19**: 467–475.

Hadj-Taieb S, Nasrallah F, Hammami MB, et al. (2012). Aminoacidopathies and organic acidurias in Tunisia: a retrospective survey over 23 years. *Tunis Med* **90**(3): 258–261.

Hafez M, El-Tahan H, Awadalla M, et al. (1983). Consanguineous matings in the Egyptian population. *J Med Genet* **20**: 58–60.

Haj Khelil A, Laradi S, Miled A, et al. (2004). Clinical and molecular aspects of hemoglobinopathies in Tunisia. *Clin Chim Acta* **340**: 127–137.

Haj Khelil A, Denden S, Leban N, et al. (2010). Hemoglobinopathies in North Africa: a review. *Hemoglobin* **34**: 1–23.

Hamamy HA and Al-Allawi NA (2013). Epidemiological profile of common haemoglobinopathies in Arab countries. *J Community Genet* **4**(2): 147–167.

Hammami A, Elgazzeh M, Chalbi N, and Mansour BA (2005). [Endogamy and consanguinity in Mauritania]. *Tunis Med* **83**(1): 38–42.

Hanein S, Perrault I, Gerber S, et al. (2008). Population history and infrequent mutations: how old is a rare mutation? GUCY2D as a worked example. *Eur J Hum Genet* **16**: 115–123.

Hashem N (1978). Preventive aspects of genetic morbidity. Volumes I, II, III. In *Proceedings of First International Conference*, Cairo, Egypt, March 30 to April 4.

Healy DG, Wood NW, and Schapira AH (2008). Test for LRRK2 mutations in patients with Parkinson's disease. *Pract Neurol* **8**: 381–385.

Hentati F (2008). Muscular dystrophies in Arab countries. The First Educational Workshop in Developing Countries on Genetic Counseling, 20–28 November, Beirut, Lebanon.

Hovorka O and Ehlers E (1897). Mal de Meleda. *Arch Syph* **40**: 251–256.

Hussien FH (1971). Endogamy in Egyptian Nubia. *J Biosocial Sci* **3**: 251–257.

Ishihara L, Gibson RA, Warren L, et al. (2007). Screening for *LRRK2* G2019S and clinical comparison of Tunisian and North American Caucasian Parkinson's disease families. *Mov Disord* **22**: 55–61.

Ismail SR, Hashine MM, Mourad MI, and Abdel Kader M (1996). Inheritance of non-syndromal genetic deafness. *J Egypt Public Health Assoc* **71**(5–6): 403–438.

Itto AB, Hamzi K, Bellayou H, et al. (2013). Evolution of molecular diagnosis of Duchenne muscular dystrophy. *J Mol Neurosci* **50**(2): 314–316.

Jacoub B (1938). Tay–Sachs disease in Egypt. *J Egypt Med Assoc* **21**: 377.

Jamil MA-N (1975). *A History of the Maghrib*, 2nd edn. Cambridge University Press.

Johnson RE, Kondratick CM, Prakash S, and Prakash L (1999). hRAD30 mutations in the variant form of xeroderma pigmento-sum. *Science* **285**: 263–265.

Kefi M, Amouri R, Driss A, et al. (2003). Phenotype and sarcoglycan expression in Tunisian LGMD 2C patients sharing the same del521-T mutation. *Neuromuscul Disord* **13**: 779–787.

Kefi R, Stevanovitch A, Bouzaid E, and Béraud-Colomb E (2005). Diversité mitochondriale de la population de Taforalt (12.000 ans, Maroc): une approche génétique à l'étude du peuplement de l'Afrique du Nord. *Anthropologie* **1**: 55–64.

Kefi R, Hsouna S, Romdhane L, et al. (2015). Phylogeny and genetic structure of Tunisians and their position within Mediterranean populations. *Mitochondrial DNA* **26**: 593–604.

Kerem E, Kalman YM, Yahav Y, et al. (1995). Highly variable incidence of cystic fibrosis and different mutation distribution among different Jewish ethnic groups in Israel. *Hum Genet* **96**(2): 193–197.

Kerkeni E, Monastiri K, Saket B, et al. (2006). Association among education level, occupation status, and consanguinity in Tunisia and Croatia. *Croat Med J* **47**: 656–661.

Kerkeni E, Monastiri K, Saket B, Guediche MN, and Ben Cheikh H (2007). Interplay of socio-economic factors, consanguinity, fertility, and offspring mortality in Monastir, Tunisia. *Croat Med J* **48**: 701–707.

Khedhiri S, Chkioua L, Bouzidi H, et al. (2012). Mucopolysaccharidosis IVA within Tunisian patients: confirmation of the two novel GALNS gene mutations. *Pathol Biol (Paris)* **60**(3): 190–192.

Khemir S, Tebib N, Nasrallah F, et al. (2009). Phenylketonuria in Tunisian institutions for the mentally handicapped. *Arch Dis Child* **94**: 647–648.

Khemir S, El Asmi M, Sanhaji H, et al. (2011). Phenylketonuria is still a major cause of mental retardation in Tunisia despite the possibility of treatment. *Clin Neurol Neurosurg* **113**: 727–730.

Khemir S, Siala H, Taieb SH, et al. (2012). Screening of three Mediterranean phenylketonuria mutations in Tunisian families. *J Genet* **91**(1): 91–94.

Klein C, Lisowska-Grospierre B, LeDeist F, Fischer A, and Griscelli C (1993). Major histocompatibility complex class II deficiency: clinical manifestations, immunologic features, and outcome. *J Pediatr* **123**(6): 921–928.

Kogan A, Shinar Y, Lidar M, et al. (2001). Common MEFV mutations among Jewish ethnic groups in Israel: high frequency of carrier and phenotype III states and absence of a perceptible biological advantage for the carrier state. *Am J Med Genet* **102**: 272–276.

Kotb MA, Abdallah HK, and Kotb A (2004). Liver glycogenoses: are they a possible cause of polyneuropathy? A cross-sectional study. *J Trop Pediatr* **50**: 196–202.

Kozlowski K, Bacha L, Massen R, et al. (1988). A new type of spondylo-metaphyseal dysplasia – Algerian type: report of five cases. *Pediatr Radiol* **18**: 221–226.

Krotoski D, Namaste S, Raouf RK, et al. (2009). Conference report: second conference of the Middle East and North Africa newborn screening initiative – partnerships for sustainable newborn screening infrastructure and research opportunities. *Genet Med* **11**(9): 663–668.

Lamdouar Bouazzaoui N (1994). Consanguinité et santé publique au Maroc. *Bulletin de l'Academie Nationale de Médicine* **178**: 1013–1027.

Lamzouri A, Ratbi I, Laarabi FZ, Barkat A, and Sefiani A (2012). Low prevalence of p.G352fsdelG mutation in phenylketonuria patients from Morocco. *Genet Test Mol Biomarkers* **16**(8): 996–998.

Lee HS, Sambuughin N, Cervenakova L, et al. (1999). Ancestral origins and worldwide distribution of the PRNP 200K mutation causing familial Creutzfeldt-Jakob disease. *Am J Hum Genet* **64**: 1063–1070.

Lesage S, Leutenegger AL, Ibanez P, et al. (2005). *LRRK2* haplotype analyses in European and North African families with Parkinson disease: a common founder for the G2019S mutation dating from the 13th century. *Am J Hum Genet* **77**: 330–332.

Li Y, Pohl E, Boulouiz R, et al. (2010). Mutations in *TPRN* cause a progressive form of autosomal-recessive nonsyndromic hearing loss. *Am J Hum Genet* **86**: 479–484.

Loumi O, Ferec C, Mercier B, et al. (2008). CFTR mutations in the Algerian population. *J Cyst Fibros* **7**: 54–59.

Lucotte G (2007). High prevalences of carriers of the 35delG mutation of connexin 26 in the Mediterranean area. *Int J Pediatr Otorhinolaryngol* **71**(5): 741–746.

Maalej M, Hentati D, Slimene M, et al. (2007). Skin cancer in Tunisia: a retrospective study – 1379 cases and risk factors. *Tunis Med* **85**: 728–733.

Maktouf C, Bchir A, Louzir H, et al. (2006). Megaloblastic anemia in North Africa. *Haematologica* **91**(7): 990–991.

Marrakchi S, Audebert S, Bouadjar B, et al. (2003). Novel mutations in the gene encoding secreted lymphocyte antigen-6/urokinase-type plasminogen activator receptor-related protein-1 (SLURP-1) and description of five ancestral haplotypes in patients with Mal de Meleda. *J Invest Dermatol* **120**: 351–355.

Marrakchi Z, Marrachi S, Meziou TJ, Turki H, and Zahaf A (2006). Mal de Meleda. 16 cases. *Tunis Med* **84**: 423–426.

Masmoudi S, Elgaied-Boulila A, Kassab I, et al. (2000). Determination of the frequency of connexin26 mutations in inherited sensorineural deafness and carrier rates in the Tunisian population using DGGE. *J Med Genet* **37**(11): E39.

Masmoudi S, Tlili A, Majava M, et al. (2003). Mapping of a new autosomal recessive nonsyndromic hearing loss locus (*DFNB32*) to chromosome 1p13.3–22.1. *Eur J Hum Genet* **11**: 185–188.

Messaoud O, Ben Rekaya M, Cherif W, et al. (2010a). Genetic homogeneity of mutational spectrum of group-A xeroderma pigmentosum in Tunisian patients. *Int J Dermatol* **49**: 544–548.

Messaoud O, Ben Rekaya M, Kefi R, et al. (2010b). Identification of a primarily neurological phenotypic expression of xeroderma pigmentosum complementation group A in a Tunisian family. *Br J Dermatol* **162**(4): 883–886.

Messaoud T, Bel Haj Fredj S, Bibi A, et al. (2005). Molecular epidemiology of cystic fibrosis in Tunisia. *Ann Biol Clin* **63**: 627–630.

Mhiri C, Triki C, Fakhfakh F, et al. (1996). [Duchenne muscular dystrophy: clinico-pathologic study of 10 families from South Tunisia]. *Tunis Med* **74**(2): 85–92.

Mili A, Ben Charfeddine I, Mamaï O, et al. (2012a). Molecular and biochemical characterization of Tunisian patients with glycogen storage disease type III. *J Hum Genet* **57**(3): 170–175.

Mili A, Ben Charfeddine I, Amara A, et al. (2012b). A c.3216_3217delGA mutation in AGL gene in Tunisian patients with a glycogen storage disease type III: evidence of a founder effect. *Clin Genet* **82**(6): 534–539.

Moulard B, Genton P, Grid D, et al. (2002). Haplotype study of West European and North African Unverricht–Lundborg chromosomes: evidence for a few founder mutations. *Hum Genet* **111**: 255–262.

Mujawar Q, Naganoor R, Patil H, et al. (2009). Pycnodysostosis with unusual findings: a case report. *Cases Journal* **2**(7): 6544.

Naamane H, El Maataoui O, Ailal F, et al. (2010). The 752delG26 mutation in the *RFXANK* gene associated with major histocompatibility complex class II deficiency: evidence for a founder effect in the Moroccan population. *Eur J Pediatr* **169**(9): 1069–1074.

Nabli F, Farhat E, Kefi M, Amouri R, and Hentati F (2010). Les dystrophies musculaires en Tunisie: une expérience de 10 ans. *Journées de Neurologie de Langue Françaises*. Available at: http://archives.jnlf.fr/data/ModuleProgramme/PageSite/2010-1/Resume/7131.asp.

Navon R, Nutman J, Kopel R, et al. (1981). Hereditary heat labile hexosaminidase B: its complications for recognizing Tay–Sachs genotypes. *Am J Hum Genet* **33**: 907–915.

Nègre V, Denoyelle F, Chevallier B, et al. (1990). [Hereditary epidermolysis bullosa and laryngeal involvement]. *Arch Fr Pediatr* **47**(2): 121–124.

Nishioka K, Kefi M, Jasinska-Myga B, et al. (2010). A comparative study of *LRRK2*, PINK1 and genetically undefined familial Parkinson's disease. *J Neurol Neurosurg Psychiatry* **81**(4): 391–395.

Ornetti P, Prati C, Fery-Blanco C, et al. (2008). Pedicle stress fracture: an unusual complication of pycnodysostosis. *Clin Rheumatol* **27**(3): 385–387.

Osimani S, Husson I, Passemard S, et al. (2010). Craniosynostosis: a rare complication of pycnodysostosis. *Eur J Med Genet* **53**(2): 89–92.

Ouachani A (2001). Les glycogénoses type III: analyse clinique, explorations dynamiques et prise en charge diététique. Etude longitudinale de 17 observations. Thèse de Médecine, Faculté de Médecine de Tunis.

Ouederni M, Vincent QB, Frange P, et al. (2011). Major histocompatibility complex class II expression deficiency caused by a *RFXANK* founder mutation: a survey of 35 patients. *Blood* **118**(19): 5108–5118.

Ouragini H, Cherif F, Daoud W, et al. (2008). Haplotypic classification of dystrophic epidermolysis bullosa in Tunisian consanguineous families: implication for diagnosis. *Arch Dermatol Res* **300** (7): 365–370.

Ozkara HA and Topçu M (2004). Sphingolipidoses in Turkey. *Brain Dev* **26**(6): 363–366.

Parvari R, Moses S, Shen J, et al. (1997). A single-base deletion in the 3′-coding region of glycogen-debranching enzyme is prevalent in glycogen storage disease type IIIA in a population of North African Jewish patients. *Eur J Hum Genet* **5**: 266–270.

Piccolo F, Roberds SL, Jeanpierre M, et al. (1995). Primary adhalinopathy: a common cause of autosomal recessive muscular dystrophy of variable severity. *Nat Genet* **10**(2): 243–245.

Plaza S, Calafell F, Helal A, et al. (2003). Joining the pillars of Hercules: mtDNA sequences show multidirectional gene flow in the western Mediterranean. *Ann Hum Gene* **67**: 312–328.

Ratbi I, Hajji S, Ouldim K, et al. (2007). [The mutation 35delG of the gene of the connexin 26 is a frequent cause of autosomal-recessive non-syndromic hearing loss in Morocco]. *Arch Pediatr* **14** (5): 450–453.

Ratbi I, Génin E, Legendre M, et al. (2008). Cystic fibrosis carrier frequency and estimated prevalence of the disease in Morocco. *J Cyst Fibros* **7**: 440–443.

Reda SM, Afifi HM, and Amine MM (2009). Primary immunodeficiency diseases in Egyptian children: a single-center study. *J Clin Immunol* **29**(3): 343–351.

Reith W and Mach B (2001). The bare lymphocyte syndrome and the regulation of MHC expression. *Annu Rev Immunol* **19**: 331–373.

Rezaei N, Aghamohammadi A, Moin M, et al. (2006). Frequency and clinical manifestations of patients with primary immunodeficiency disorders in Iran: update from the Iranian Primary Immunodeficiency Registry. *J Clin Immunol* **26**(6): 519–532.

Riahi Z, Hammami H, Ouragini H, et al. (2013). Update of the spectrum of *GJB2* gene mutations in Tunisian families with autosomal recessive nonsyndromic hearing loss. *Gene* **525**(1): 1–4.

Richard P, Gaudon K, Haddad H, et al. (2008). The *CHRNE* 1293insG founder mutation is a frequent cause of congenital myasthenia in North Africa. *Neurology* **71**: 1967–1972.

Riou SE, Younsi C, and Chaabouni H (1989). Consanguinité dans la population du Nord de la Tunisie. *Tunis Med* **67**: 167–172.

Romdhane L and Abdelhak S (2011). Genetic diseases in the Tunisian population. *Am J Med Genet A* **155**: 238–267.

Romdhane L and Abdelhak S (2012). Genetic diseases in the North African populations. In Kumar D, ed., *Genomics and Health in the Developing World*. Oxford University Press.

Romdhane L, Kefi R, Azaiez H, et al. (2012). Founder mutations in Tunisia: implications for diagnosis in North Africa and Middle East. *Orphanet J Rare Dis* **7**: 52.

Romdhane L, Ben Halim N, Rejeb I, et al. (2014). Specific aspects of consanguinity: some examples from the Tunisian population. *Hum Hered* **77**(1–4): 167–174.

Romdhane L, Messaoud O, Bouyacoub Y, et al. (2016). Comorbidity in the Tunisian population. *Clin Genet* **89**(3): 312–319.

Ruzzi L, Gagnoux-Palacios L, Pinola M, et al. (1997). Homozygous mutation in the integrin alpha6 gene in junctional epidermolysis bullosa with pyloric atresia. *J Clin Invest* **99**(12): 2826–2831.

Scriver CR and Kaufman S (2001). Hyperphenylalaninaemia. In Scriver CR, Beaudet AL, Sly WS, and Valle D. eds., *The Metabolic and Molecular Bases of Inherited Disease*, 8th edition. McGraw Hill.

Senhaji MA, Abidi O, Nadifi S, et al. (2013). c.1643_1644delTG XPC mutation is more frequent in Moroccan patients with xeroderma pigmentosum. *Arch Dermatol Res* **305**(1): 53–57.

Shatla H, El-Shakankiry HM, Fateen E, Ahmed NS, and Mustafa HA (1999). Prevalence of GM2 gangliosidosis among cases with early onset seizures and psychomotor retardation. *Bull Egypt Soc Physiol Sci* **19**: 249–273.

Shawky RM, Riad MS, Osman HM, and Bahaa NM (2001). Screening for some inborn errors of amino acid metabolism which impair mental function. *Egypt J Med Hum Genet* **2**: 71–91.

Shawky RM, El-Sedfy HH, Mahmoud AO, Rashad M, and Bahaa Eldin EM (2007). Study of amino acid disorders among a high risk group of Egyptian infants and children. *Egypt J Med Hum Genet* **8**: 173–189.

Snoeckx RL, Hassan DM, Kamal NM, Van Den Bogaert K, and Van Camp G (2005). Mutation analysis of the *GJB2* (connexin 26) gene in Egypt. *Hum Mutat* **26**: 60–61.

Soufir N, Ged C, Bourillon A, et al. (2010). A prevalent mutation with founder effect in xeroderma pigmentosum group C from North Africa. *J Invest Dermatol* **130**: 1537–1542.

Stevenson AC, Johnston HA, Stewart MIP, and Golding DR (1966). Congenital malformations: a report of a study of series of consecutive births in 24 centres. *Bull WHO* **34**(Suppl.): 1–125.

Stoffman N, Magal N, Shohat T, et al. (2000). Higher than expected carrier rates for familial Mediterranean fever in various Jewish ethnic groups. *Eur J Hum Genet* **8**: 307–310.

Tabutin D, Vilquin E, and Biraben JN (2002). *L'histoire de la population de l'Afrique du Nord pendant le deuxième millénaire*. Louvain-la-Neuve.

Tadmouri GO, Al Ali MT, Al-Haj Ali S, and Al Khaja N (2006). CTGA: the database for genetic disorders in Arab populations. *Nucleic Acids Res* **34**(database issue): D602–D606.

Talbi J, Khadmaoui A, Soulaymani A, and Chafik A (2007). Etude de la consanguinité dans la population marocaine: Impact sur le profil de la santé. *Antropo* **15**: 1–11.

Tamary H, Bar-Yam R, Shalmon L, et al. (2000). Fanconi anaemia group A (*FANCA*) mutations in Israeli non-Ashkenazi Jewish patients. *Br J Haematol* **111**(1): 338–343.

Teebi A and Farag T (1997). *Genetic Disorders among Arab Populations*. Oxford University Press.

Temtamy SA, Adbel-Salam M, Hussein FA, Meguid NA, and El-Gindy E (1991). Clinical biochemical and cytogenetic studies of mental retardation in Egyptian children. *J Egypt Public Health Assoc* **65**(Suppl.): 189–199.

Temtamy SA, Ismail SR, El-Beshlawy AM, et al. (2007). Fanconi anemia: cytogenetic and clinical studies on a group of Fanconi anemia patients in Egypt. *Haema* **10**: 61–67.

Tlili A, Rebeh IB, Aifa-Hmani M, et al. (2008). *TMC1* but not *TMC2* is responsible for autosomal recessive nonsyndromic hearing impairment in Tunisian families. *Audiol Neurootol* **13**: 213–218.

Touitou I (2001). The spectrum of familial Mediterranean fever (FMF) mutations. *Eur J Hum Genet* **9**: 473–483.

Touzri RA, Mohamed Z, Khalil E, et al. (2008). [Ocular malignancies of xeroderma pigmentosum: clinical and therapeutic features]. *Ann Dermatol Venereol* **135**(2): 99–104.

Tullio-Pelet A, Salomon R, Hadj-Rabia S, et al. (2000). Mutant WD-repeat protein in triple-A syndrome. *Nat Genet* **263**: 332–335.

United Nations, Department of Economic and Social Affairs, Population Division (2010). World Population Prospects: The 2010 Revision, CD-ROM Edition.

Veten FM, Abdelhamid IO, Meiloud GM, et al. (2012). Hb S [β6(A3)Glu→Val, GAG>GTG] and β-globin gene cluster haplotype distribution in Mauritania. *Hemoglobin* **36**(4): 311–315.

Wiszniewski W, Fondaneche MC, Lambert N, et al. (2000). Founder effect for a 26-bp deletion in the RFXANK gene in North African major histocompatibility complex class II-deficient patients belonging to complementation group B. *Immunogenetics* **51**(4–5): 261–267.

Yilmaz E, Ozen S, Balci B, et al. (2001). Mutation frequency of familial Mediterranean fever and evidence for a high carrier rate in the Turkish population. *Eur J Hum Genet* **9**: 553–555.

Zaoui S and Biémont C (2002). [Frequency of consanguineous unions in the Tlemcen area (West Algeria)]. *Sante* **12**(3): 289–295.

Zeegers MPA, van Poppel F, Vlietinck R, Spruijt L, and Ostrer H (2004). Founder mutations among the Dutch. *Eur J Hum Genet* **12**: 591–600.

Zghal M, El-Fekih N, Fazaa B, et al. (2005). [Xeroderma pigmentosum: cutaneous, ocular, and neurologic abnormalities in 49 Tunisian cases]. *Tunis Med* **83**(12): 760–763.

Zghal M, Fazaa B, and Kamoun M (2006). Xeroderma pigmentosum. EMC (Elsevier SAS, Paris). *Dermatologie* **10**(A): 98–660.

Zilber N, Kahana E, and Abraham M (1991). The Libyan Creutzfeldt-Jakob disease focus in Israel: an epidemiologic evaluation. *Neurology* **41**(9): 1385–1389.

Zlotogora J (2007). Multiple mutations responsible for frequent genetic diseases in isolated populations. *Eur J Hum Genet* **15**(3): 272–278.

Zschocke J (2003). Phenylketonuria mutations in Europe. *Hum Mutat* **21**(4): 345–356.

12 Birth Defects and Genetic Disease in Sub-Saharan Africa[1]

Ambroise Wonkam

12.1 Introduction

12.1.1 Definition and Causes

A birth defect is defined as any abnormality affecting body structures or function that is present from birth, or may be diagnosed only later in life. Causes can be divided broadly into two groups: genetic and partially genetic causes originating mostly before conception, and causes developing after conception but before birth. In industrialized countries, genetic abnormalities account for about 7 percent of birth defects, mainly numerical or structural chromosomal abnormalities, e.g., Down syndrome or single genes defects, e.g., sickle cell anemia. Partial genetic birth defects are due to a combination of genes that puts the fetus at risk in the presence of specific environmental factors, and account for an estimated 20–30 percent of all birth defects, a number of which are lethal, e.g., congenital heart disease (Turnpenny and Ellard 2007). Causes of birth defects originating after conception are primarily non-genetic. These include teratogens that interfere with normal growth and development of the embryo or fetus (maternal illness, maternal infections, drugs, radiation, and environmental pollutants), mechanical forces that deform the fetus, and vascular accidents that disrupt the normal growth of organs. This category accounts for an estimated 5–10 percent of all birth defects (Turnpenny and Ellard 2007). About 50 percent of birth defects are of unknown cause. These types of defects may be due to new mutations, submicroscopic chromosome deletions, or epigenetic modification. With ongoing advances in genomic technology, including affordable individual whole-genome sequencing, analysis of copy number variants, methylation, and genome-wide association studies, the underlying causes of more birth defects will become apparent (Butler 2009; Lupski et al. 2010; Stefansson et al. 2008; Vissers et al. 2009).

[1] An earlier version of this chapter has previously been published as Wonkam A (2012) The burden of birth defects and genetic disease in sub-Saharan Africa. In Kumar D, ed., *Genomics and Health in the Developing World*. Oxford University Press. © Oxford University Press, Inc. Reproduced with permission of the Licensor through PLSclear.

12.1.2 The Toll

According to the World Health Organization (WHO), 7.6 million children are born yearly with severe genetic conditions or malformations that are predominantly genetic in origin, and 90 percent of these children are born in low- or middle-income countries (WHO 2005). When disability is considered in addition to mortality, the global toll of birth defects reflects a particularly harsh reality. Many infants burdened with serious birth defects die early in life, particularly in middle- and low-income countries, and those who survive almost always have the potential to be disabled, either as a direct or indirect result of their condition (Christianson and Modell 2004). However, genetic conditions outside the industrialized world receive little attention due to the focus on communicable diseases such as HIV and malaria, where most resources are diverted.

12.1.3 Misunderstandings

Several misunderstandings underlie the reasons why care and prevention of birth defects have been accorded relatively low priority by international donors, health agencies, and national governments (March of Dimes Global 2006). The first is that health policy-makers are generally unaware of the global toll of birth defects and associated disabilities, because data documenting the extent of the problem have been lacking. A second misunderstanding is the widespread belief that effective care and prevention of birth defects require costly, high-technology interventions that are beyond the national health budgets of low- and middle-income countries. A third misunderstanding, which follows from the previous point, is that programmatic attention to care and prevention of birth defects will necessarily draw funding away from other high-priority interventions in maternal and child health.

12.1.4 The Challenges

There is an increasing trend in developing countries, where the demographic and socioeconomic transition imposes more constraints on dealing with the double burden of infectious and non-infectious diseases. It is predicted that by 2020 non-communicable diseases, including genetic conditions, will cause seven out of every ten deaths in developing countries (Mayosi et al. 2009). Up to 85 percent of the world's more than seven billion people live in middle- and low-income countries (UNICEF 2003), many of which are characterized by poor environments and inadequate health care systems. This is a tragedy, because 70 percent of birth defects could either be prevented or the children affected offered care that would be life-saving or significantly reduce disability (Christianson and Modell 2004). To achieve the Millennium Development Goal for child survival (MDG-4), neonatal deaths need to be prevented (Martines et al. 2005). The data in the *March of Dimes Global Report on Birth Defects* show that such efforts must also recognize the global contribution of birth defects to infant and childhood mortality if they are to succeed (March of Dimes Global 2006).

In sub-Saharan Africa (SSA), where sickle cell disease (SCD) occurs at its highest frequency, with up to 224,200 new affected babies yearly, hemoglobinopathies alone represent a health burden comparable to that of communicable and other major diseases (Weatherall 2008) and HbS will have an increasing effect on public health systems (Piel et al. 2013). Care and preventive strategies must take into account the growing trend of risk factors correlated to birth defects in developing countries. This chapter is a brief review of recent literature dealing with birth defects in SSA. It gives a global view of the main diseases and their impact on populations living in SSA.

12.2 Global Burden of Birth Defects in Africa

12.2.1 Health Transition in Africa

Most low- and middle-income countries have undergone gradual improvements in key national sectors over 40 years, including economic and health sectors. These improvements have been most notable in countries of Latin America, East Asia, Middle Asia, and North Africa, and underpin a gradual, consistent shift in the major causes of population mortality and morbidity. This positive health transition is marked by the decreasing occurrence of infectious diseases and malnutrition, mirrored by declining infant and under-five-years mortality rates, and increasing population life expectancy. Unfortunately, rates of birth defects and their associated developmental disabilities have not decreased in the same populations within the same time period. In fact, a growing body of data suggests that birth defects are a far more significant contributor to infant childhood mortality and disability in low- and middle-income countries than previously estimated (IOM 2003).

In SSA, non-communicable diseases are emerging in both rural and urban areas, most prominently among poor people living in urban settings, and are resulting in increasing pressure on acute and chronic health care services. Major factors include demographic change, leading to a rise in the proportion of people older than 60 years, despite the negative effect of HIV/AIDS on life expectancy (Mayosi et al. 2009). The scale of the challenge posed by the combined and growing burden of HIV/AIDS and non-communicable diseases including genetic conditions demands an extraordinary response that few SSA countries are able to provide. Concerted action is needed to strengthen the district-based primary health care system, to integrate the care of birth defects and management of risk factors, to develop national surveillance systems, and to apply interventions of proven cost-effectiveness in the primary and secondary prevention of specific genetic conditions like SCD within populations and health services.

12.2.2 Birth Prevalence, Population Prevalence, Death, and Disability

Birth prevalence is the number of infants affected in one or a defined collective group of birth defects per 1000 live births. A few specific conditions, however, including the hemoglobin disorders, G6PD deficiency, oculocutaneous albinism, and neural tube defects have higher birth prevalence in SSA. The birth prevalence of birth defects is about 20 percent higher in SSA than in high-income countries (March of Dimes Global 2006).

The birth prevalence of post-conception birth defects due to teratogens differs widely, but is generally higher in middle- and low-income countries. In SSA, there tend to be few, if any, regulations governing the use of some of these substances, and health services are not directed toward identifying and controlling exposure (Penchaszadeh 2002). An undetermined number, but undoubtedly in the hundreds of thousands, of infants are born annually with serious birth defects of teratogenic origin.

Population prevalence refers to the number of affected individuals per 1000 in a defined population. For birth defects, this number is usually far lower than birth prevalence, especially in middle- and low-income countries, because serious birth defects shorten life. In SSA, the difference between birth prevalence and population prevalence is so marked that studies may identify few or no living individuals with serious birth defects. In rural South Africa in the early and mid-1990s, for example, the birth prevalence of Down syndrome was reported to be 2.09 per 1000 live births and the population prevalence to be only 0.75 per 1000 children aged 2–9 years (Christianson et al. 2002; Venter et al. 1995).

Similarly, after the first description of SCD in a Grenadian dental student in Chicago in 1910, it was not until the late 1940s that reports began to appear from Africa itself (Serjeant 2005). The prevalence of SCD in the newborn population was 3 percent and 2 percent in Nigeria and Cameroon, respectively (Odunvbun et al. 2008; Weatherall 2001), but the population prevalence is likely less. While median survival for patients with sickle cell anemia (HbSS) in North America, Europe, and the Caribbean is now estimated to be 45–55 years (Platt et al. 1994; Wierenga et al. 2001), only 7 percent of Nigerian adults and 2 percent of Cameroonian adults compared with 46 percent of US adults with SCD were >35 years (Aliyu et al. 2008; Wonkam 2009). Approximately two-thirds of the children (60.9 percent) were under five years of age in Gabon (Koko et al. 1998). Moreover, 90 percent of children with SCD die before the diagnosis can be made (Williams et al. 2009).

As effective care becomes available, population prevalence rises gradually to approximate birth prevalence. In the 1970s, 50 percent of sickle cell anemia patients died by the age of five years in Nigeria (Tattah and Ekere 1975). The recognition of SCD, and near-total extinction of the people's belief in reincarnation as the cause of SCD (Onwubalili 1983), the provision of well-organized holistic care, including counseling, preventive health, and nutritional education, prompt treatment of illness and free supplies of vitamin supplements, malarial prophylactics and other necessary medication, significantly reduced illness and deaths and improved the quality of life of people living with HbSS in Nigeria. The mortality rate fell from 20.6 percent in 1988 to 0.6 percent in 1995 ($p < 0.0001$); the number of hospital admissions fell from 350 (119 percent) in 1988 to 30 (4 percent) in 1995 ($p < 0.0001$); the number of patients transfused with blood fell from 260 (90 percent) in 1988 to 25 (2 percent) in 1995 ($p < 0.00001$) (Akinyanju et al. 2005).

By conservative estimates, some 3.3 million deaths each year are associated with birth defects. This estimate includes approximately 50 percent who die in low-income countries, including SSA. It is also estimated that without appropriate care another 3.2 million children born with a serious birth defect are disabled each year (Christianson and Modell 2004).

12.3 Determinants Influencing Birth Defects in Africa

12.3.1 Malaria

Compared to noncarriers, healthy carriers of recessive genes for sickle cell anemia, thalassemia, and G6PD deficiency have a well-documented survival advantage against the lethal effects of malaria. As a result, carriers are more likely to reach reproductive age. Consequently, the birth prevalence of thalassemia, SCD, and G6PD deficiency is high in sub-Saharan Africa (Aidoo et al. 2002; Enevold et al. 2008; Föller et al. 2009; Vento et al. 2006; Williams et al. 2005).

12.3.2 Consanguineous Marriage

Currently, couples related as second-cousins or closer ($F \geq 0.0156$) and their progeny account for an estimated 10.4 percent of the global population (Bittles and Black 2010). The highest rates of consanguineous marriage occur in North Africa, where 25–80 percent of unions involve related family members (Abdel-Meguid et al. 2000; Ben Arab et al. 2004; Bittles et al. 1991). The practice is also accepted in parts of SSA, and even couples who regard themselves as unrelated may exhibit high levels of homozygosity because marriage within clan, tribe, caste, or biraderi boundaries has been a long-established tradition (Bittles and Black 2010; Hampshire and Smith 2001; Saka et al. 2009). Parental consanguinity increases the birth prevalence of autosomal recessive birth defects. The risk for neonatal and childhood death, intellectual disability, and serious birth defects is almost doubled for first-cousin unions (Bittles and Black 2010; Mansour et al. 2009; Nouira et al. 2008).

12.3.3 Advanced Maternal Age and Chromosomal Anomalies

The increasing risk of chromosomal abnormalities, particularly Down syndrome, with advancing maternal age is well recognized. The percentage of women of advanced maternal age (greater than 35 years) delivering infants is high in middle- and low-income countries and sub-Saharan Africa, which do not have universal screening, prenatal diagnosis, and associated services (average 11–18 percent; see Mokhtar and Abdel-Fattah 2001; Obed et al. 1995; Oloyede 2008; Orji and Ndububa 2004; Tebeu et al. 2007; Watcham et al. 2007).

12.3.4 Migration

Sickle cell disease, which was previously rare in South Africa, is being introduced through an influx of migrants from West and Central Africa (Beighton and Botha

1986; Wonkam et al. 2012). Other well-known examples of the introduction of single-gene disorders into the African continent or country by migration include porphyria and Fanconi anemia, which appears in high frequency in the Afrikaner population in South Africa (Jenkins 1990; Parker et al. 2008; Tipping et al. 2001).

Migration also occurs as a result of increased urbanization, as people move from rural settings and traditional lifestyles to towns and cities. The rapid growth of cities led to an increase in urban populations of common disorders with a genetic predisposition, such as insulin-dependent diabetes mellitus (Mayosi et al. 2009). Etiologic factors involved include not only genetic predisposition, but also lifestyle and dietary changes (e.g., high-fat, high-caloric diets and lack of exercise leading to obesity, increased risk of diabetes, and coronary artery disease) and exposure to occupational, recreational, and environmental toxins and pollutants, particularly alcohol and smoke (WHO and the World Alliance for the Prevention of Birth Defects 1999).

12.3.5 National Level of Health Care and Poverty

The proportion of gross domestic product spent on research and development is extremely low in SSA (Hardy et al. 2008). Most SSA countries lack clinical and diagnostic facilities for genetic medicine and have limited capacity for research to address African genetic problems (Wonkam et al. 2006). Furthermore, human genetics research in SSA has often failed to address the ethical obligation to build capacity. For example, we identified 50 studies on PubMed published since 1989 that used Cameroonian DNA samples. Only seven publications (14 percent) included a Cameroonian collaborative center, and only 14 (28 percent) included a Cameroonian author (Wonkam et al. 2011).

As services become available, though, the condition can be alleviated for an increasing proportion of children at birth. For example, structural birth defects including congenital heart defects, cleft lip and palate, and clubfoot can be corrected with pediatric surgery. Children with functional problems, including thalassemia, sickle cell disorders, and congenital hypothyroidism can also survive with appropriate treatment. In addition, many children with chronic disabilities due to birth defects such as Down syndrome or neural tube defects can be treated so they will live longer, more productive lives (Christianson and Modell 2004).

12.4 Distribution of Specific Birth Defects in Sub-Saharan Africa

The categories include birth defects of genetic or partially genetic origin, including single-gene defects, chromosomal disorders, multifactorial disorders, and birth defects caused by teratogens.

12.4.1 Birth Defects of Genetic or Partially Genetic Origin

Single-Gene Defects: Common Recessive Disorders

These include three major disorders in SSA: the hemoglobin disorders, sickle cell anemia and thalassemia, and oculocutaneous albinism.

Sickle Cell Anemia

The WHO has estimated that at least 5 percent of the population are carriers for one or other of the most serious forms, the alpha- and beta-thalassemias and the structural variant hemoglobins S, C, and E, which are found at polymorphic frequencies in many countries. All of these hemoglobinopathies are believed to provide protection against malaria (Clegg and Weatherall 1999). It is estimated that more than 300,000 children are born each year with a severe inherited disorder of hemoglobin, and that approximately 80 percent of these births occur in low- or middle-income countries (Piel et al. 2013; Weatherall 2010).

Sickle cell disease is caused by a point mutation leading to a single amino acid substitution in the beta subunit of hemoglobin (Hb), the principal oxygen transporter in red blood cells. In SSA, SCD occurs at its highest frequency; the prevalence of sickle cell trait is estimated to be 5–40 percent (Weatherall 2001). It has been estimated that SCD results in the annual loss of several millions of disability-adjusted life years, particularly in the developing world (Weatherall 2006). Sickle cell disease is responsible for high burdens on families (Wonkam et al. 2013a) and high mortality (Makani et al. 2011a) in many African countries. A notable feature of SCD is the frequent occurrence of pain and acute clinical episodes, which are generally attributable to vasoocclusive crises (Castro et al. 1994; Platt et al. 1991). As a consequence of these complications, SCD patients have increased mortality compared with control populations (Platt et al. 1994). Sickle cell disease has recently been recognized as a problem of major public health significance by the WHO. Despite the fact that >70 percent of sufferers live in Africa, expenditure on the related care and research in the continent is negligible, and most advances in the understanding and management of this condition have been based on research conducted in the north. African countries need to focus research and interventions on areas that will lead to the maximum impact (Makani et al. 2007). Typically, clinical severity of SCD is extremely variable, from death in childhood to being relatively well even until old age. The reasons for this heterogeneity are not fully understood. Abnormal transcranial Doppler velocity (TCD) and pulmonary hypertension, which are diagnosed by Doppler echocardiography, are proxies of stroke episode risk and cause of death, respectively. Transcranial Doppler velocity and pulmonary hypertension are reliable clues, using a non-invasive tool, for the identification of these complications before end-stage progression occurs (Gladwin and Vichinsky 2008; Nichols et al. 2001). Interindividual variation in fetal hemoglobin (HbF) levels is one of the main modifiers that contribute to the clinical heterogeneity observed in SCD patients. Higher expression of HbF in adulthood ameliorates morbidity and mortality in SCD. It is now clear that common HbF variation is a quantitative genetic trait shaped by common polymorphisms. Genetic variation in three major loci account for a relatively large proportion (20–50 percent) of the phenotypic variation in HbF levels: (1) a single-base substitution (T/C) at position -158 of the Gγ globin gene, termed *Xmn*I Gγ site (Nagel and Platt 2001); (2) the *HMIP* locus (*HBS1L-MYB intergenic polymorphism*) on chromosome 6q (Labie et al. 1985); (3) the oncogene *BCL11A* on chromosome 2 (Menzel and Thein 2009; Uda et al. 2008). It is expected that the identification of genetic polymorphisms that modulate HbF levels will shed light on the molecular mechanisms that control HbF expression, and on the etiology of the clinical heterogeneity observed in general in SCD patients. In addition, it could open perspectives into new therapeutic approaches for HbF reactivation, but an immediate application would be an improved prediction of one's ability to produce HbF, which, in turn, may improve prediction of disease severity. Fortunately, a few African researchers have recently published works that are relevant to African SCD patients (Makani et al. 2011b; Wonkam and Hurst 2014).

The mortality rate associated with SCD has remained high despite the use of appropriate interventions to manage the various forms of crisis. The organisms causing bacteremia in African children with sickle cell anemia are the same as those in developed countries (*Streptococcus pneumoniae* [44/108 isolates; 41 percent], non-typhi *Salmonella* species [19/108; 18 percent], *Haemophilus influenzae* type b [13/108; 12 percent], *Acinetobacter* species [7/108; 7 percent], and *Escherichia coli* [7/108; 7 percent] (Serjeant 2005; Williams et al. 2009)). Introduction of conjugate vaccines against *S. pneumoniae* and *H. influenzae* into the childhood immunization schedules of African countries could substantially affect survival of children with sickle cell anemia. Methods for screening SCD are now well established and inexpensive, and it is vital that they are applied to define the magnitude of the problem that will be posed by these conditions in the future, and effective enrollment of affected neonates in a comprehensive follow-up program adapted to the socioeconomic conditions, health structures, and cultural background of the African setting (De Montalembert and Tshilolo 2007). In developed countries, newborn screening has resulted in a reduction in the SCD mortality rate from 16 percent to <1 percent (Komba et al. 2009; Streetly et al. 2009). Similarly, in the Republic of Benin, following neonatal screening, counseling, and follow up, the under-five-years mortality rate was 15.5 per 10,000, a figure that is ten times lower than the general rate recorded (Rahimy et al. 2009).

The life expectancy of patients with SCD has improved in the USA and Europe thanks to the use of penicillin prophylaxis, appropriate immunizations, neonatal screening, implementation of a quality transfusional policy, hydroxyurea therapy, detection and treatment of cerebral vasculopathy, recognition of situations that can benefit from allogenic marrow transplantation, and improvements in bone marrow transplantation techniques. The cost of almost all these techniques is far beyond the means of health care systems in Africa, where they can hardly be used (De Montalembert and Tshilolo 2007). Hematopoietic stem cell transplantations (SCT) is now an established treatment modality with definitive indications for many hematological disorders. However, this line of treatment requires tremendous resources, and it becomes increasingly difficult for transplanters practicing in the developing world to reconcile the difference between what is possible and what is available. On the basis of 18 years of experience and more than 1300 transplants that were performed in Egypt, this could guide how to address special issues that are important for hematopoietic SCT practices in developing countries (Mahmoud et al. 2008). A great deal can be done to help the situation by developing further North–South and South–South partnerships for disseminating better practice (Weatherall 2008).

Interestingly, up to 95 percent of Cameroonian parents with an SCD-affected child indicated they would request prenatal genetic diagnosis (PND) for SCD, and 65 percent would opt for abortion of a sickle cell anemia-affected pregnancy. The effective initiation of PND for SCD (amniocentesis and fetal DNA analysis) has laid the foundation for the first medical genetic service in Cameroon, and demonstrates that the practice of medical genetics is possible in an SSA country (Wonkam et al. 2010). The use of SCD prenatal diagnosis as the point of entry of genetic medicine was also reported in Nigeria (Akinyanju et al. 1999). There could be a need for widespread screening and counseling programs directed at developing PND expertise where this is not available (Weatherall 2005). A quantitative sociological survey of 110 doctors, 130 parents, and 89 adult SCD patients in Cameroon showed that the majority accepted the prenatal genetic diagnosis for SCD (78.7 percent, 89.8 percent, and 89.2 percent, respectively). Parents (62.5 percent) were more in favor of termination of SCD-affected pregnancy than doctors and adult patients (36.1 percent and 40.9 percent acceptance, respectively). Parents and patients who found medical abortion acceptable cited fear of having a SCD-affected child (98.1 and 88.9 percent) and the poor quality of the affected child's health (92.6 percent and 81.5 percent). The data could indicate the urgency of policy action that places emphasis on: premarital screening, early detection and care of SCD, socioeconomic measures to assist SCD-affected families, and appropriateness of considering maternal distress due to fetal anomalies in medical abortion laws. The findings signal potential value-based conflicts, and can inform concrete future policy actions as OMICS biotechnologies are increasingly employed in African countries (Wonkam et al. 2013b, 2014).

Thalassemias

The carrier frequency for beta-thalassemia in SSA ranges from 1 percent to 10 percent, rarely greater, while that for the milder forms of alpha-thalassemia is much higher, ranging from 10 percent to 20 percent in parts of SSA (Brabin et al. 2004; Weatherall 2001). The alpha-thalassemia genotypes described in SSA (–alpha [3.7]/alpha alpha and –alpha [3.7]/–alpha [3.7]) could not lead to hydrops fetalis as observed in Southeast Asia. Similarly, the nature of mild beta mutations in SSA is not usually associated with children with a beta-thalassemia major that will require transfusion. Nevertheless, a varying proportion of HbS-beta-thalassemia and SC forms of SCD do exist, and studies are needed to determine their frequency in SSA (Weatherall 2001).

Oculocutaneous Albinism

Oculocutaneous albinism (OCA) is a genetically inherited autosomal recessive condition; OCA2, tyrosine-positive albinism, is the most prevalent type found throughout SSA. Due to the lack of melanin, people with albinism are more susceptible to the harmful effects of ultraviolet radiation exposure. This population must deal with issues such as photophobia, nystagmus, decreased visual acuity, extreme sun sensitivity, and skin cancer (Asuquo et al. 2009; Hong et al. 2006). As a result, oculocutaneous albinism is associated with early death (Adegbidi et al. 2007; George et al. 2005). In Tanzania and Nigeria, for example, only 10 percent of those affected survive beyond 30 years of age (Kromberg 1992; Kromberg et al. 1989; Onuigbo 2006). An overall estimate of albinism prevalence ranges from 1/5000 to 1/15,000 (Hong et al. 2006). From epidemiologic data on OCA available for South Africa, Zimbabwe, Tanzania, Cameroon, and Nigeria, population prevalence as high as 1 in 1000 has been reported for selected populations in Zimbabwe and other specific ethnic groups in southern Africa. One clan, the Vhatavhatsindi in South Africa, had a significantly higher frequency of 1 in 832 (Lund et al. 2007).

Oculocutaneous albinism is a highly visible birth defect in dark-skinned people. As a consequence, people with albinism also face social discrimination as a result of their difference in appearance. Several myths account for this event. Albinos are believed to possess good and evil magical powers (Imperato and Imperato 2006; Lund 2005; McBride and Leppard 2002). On the white magic side, some organs are believed to confer luck, health, and prosperity. For this reason albinos in Tanzania and Burundi were still, in 2008, prey to ritual murders and mutilations to obtain various body parts such as arms, legs, and genitals for preparation of amulets (Aquaron et al. 2009). The WHO is currently investigating the issues concerning this vulnerable population. Disseminating accurate medical information explaining the genetic basis of albinism will be necessary to eliminate ignorance and superstition.

Children with albinism living away from home in rural areas appear to have inadequate sun protection strategies (Lund and Taylor 2008). Changes in health policy could address these deficiencies by providing more detailed health care information, giving advice on appropriate styles of hat to wear, and guidance on how to assess commercial sun-protection products. The existence of tens of thousands of people living with albinism in Africa reiterates the need for increased awareness of, and public health interventions for, albinism in order to better address the medical, psychological, and social needs of this vulnerable population.

Cystic Fibrosis

Mutations of the cystic fibrosis (CF) transmembrane conductance regulator (CFTR) result in defective chloride secretion and excessive sodium reabsorption, and a depletion of the airway surface liquid causing impaired mucociliary clearance. The ongoing infection and inflammation leads to persistent symptoms and irreversible lung damage (Mogayzel and Flume 2010). Cystic fibrosis is the most common single-gene birth defect in Caucasian populations, with a birth prevalence of approximately 1 in 2000 live births. The defect had been considered rare in other populations, but reports of birth prevalence of 1 in 2560 and 1 in 2608 live births in Jordan and Egypt, respectively, challenge that assumption. In addition, studies in migrants from the Middle East and Africa to Europe are revealing different CFTR mutations in patients (Ratbi et al. 2008; Spinelli et al. 2009).

The assumption that CF is considered a rare disease in black Africans seems also to be incorrect. Previous studies indicate that African Americans have their own set of "common" CF mutations that originate from the native African population. One novel mutation, 3120+1G>A, occurred with a frequency of 12.3 percent and 46 percent in the black South African population and among African Americans, indicating a common ethnic origin of the mutation in different African populations (Dörk et al. 1998; Macek et al. 1997). Since this mutation accounts for between 15 and 65 percent of CF chromosomes in South African blacks, a corrected CF carrier frequency would be between 1 in 14 and 1 in 59. Hence, the incidence of CF would be predicted to be between 1 in 784 and 1 in 13,924 births in this population (Padoa et al. 1999). Four other mutations have been detected, resulting in the identification of a total of 62.5 percent of mutations in black South Africans (Goldman et al. 2001). Screening for 16 "common Caucasian" mutations identified 52 percent of CF alleles in African Americans, while screening for eight "common African" mutations accounted for an additional 23 percent. The combined detection rate of 75 percent was comparable to the sensitivity of mutation analysis in Caucasian CF patients (Macek et al. 1997). There are several possible reasons why CF in African populations is not being detected. Some of these are misdiagnosis as chronic pulmonary infection, malnutrition, tuberculosis, infantile diarrhea, failure to thrive, or a high infant mortality rate (Feuillet-Fieux et al. 2004).

Rare Single-Gene Defects

The birth prevalence of rare single-gene defects is generally 1 in 10,000 or higher. However, in some clusters of people in Africa, gene frequencies of rare single-gene defects are greater than expected due to the founder effect, genetic drift, and geographic or cultural isolation (WHO and the World Alliance for the Prevention of Birth Defects 1999). Typical examples include porphyria in South Africa. In spite of their infrequency, certain rare single-gene disorders for which treatment is available, for example hemophilia, are important in Africa because of the high costs associated with the care of affected individuals. Collectively, these disorders also add significantly to the infant and child health burden.

Genetic Predisposition to Hemolysis and Neonatal Jaundice: G6PD Deficiency

G6PD deficiency, an X-linked recessive disorder, is the most common enzyme deficiency disease in humans. An estimated 7.5 percent of the world's population carries a G6PD-deficiency gene. Hundreds of variants of the disorder with differing clinical severity have been described, with approximately 90 percent of those affected being male. Individuals may be asymptomatic or may experience varying degrees of anemia caused by rapid breakdown of red blood cells. Acute crises can be induced by infection or the ingestion of drugs, or certain foods, such as fava beans. In newborns with severe G6PD deficiency, neonatal jaundice due to the rapid breakdown of red blood cells may result in damage to the brain and auditory nerves, a condition called *kernicterus*. Kernicterus can lead to intellectual disability, cerebral palsy, deafness, and death.

More than five million infants are born annually with G6PD deficiency of varying severity, mainly in tropical SSA, the eastern Mediterranean, and North Africa

(Luzzatto et al. 1995). G6PD deficiency in Nigerian males from the western region of the country affects between 20 and 26 percent of the population (Egesie et al. 2008). A similar prevalence was found in Sierra Leone, with 11.3 percent in male subjects (Jalloh et al. 2008). In a Nigerian study, 30.9 percent of the jaundiced neonates were G6PD-deficient, compared with 13.3 percent of controls (Sodeinde et al. 1995). G6PD deficiency (40 percent) was the major etiological factor in kernicterus (Ahmed et al. 1995; Owa 1989). More data are needed on the adverse effects of G6PD deficiency in African populations.

Rhesus-Negativity

When a rhesus-negative woman has a rhesus-positive partner, the fetus is often rhesus-positive. In such cases, fetal red cells may pass into the mother's circulation and stimulate the production of antibodies that cross the placenta and destroy fetal red cells, causing severe anemia. An affected fetus may be stillborn or become severely jaundiced soon after birth, with the risk of severe mental and neurological disability. Rhesus-negativity values in blood donors range from 4 percent in Guinée (Loua et al. 2007) to 6 percent in Kenya (Mwangi 1999), and to 7.26 percent in Nigeria (Nwauche and Ejele 2004). Avoidable fetal and neonatal death from rhesus hemolytic disease will continue in any country that does not have a systematic prevention program. This problem seems to have received relatively little attention in SSA.

Chromosomal Disorders: Down Syndrome

Birth prevalence of trisomy 21 may be as high as 2 per 1000 live births in Africa (WHO 1996). Early infant or childhood death from congenital heart disease and infection is common among infants and children with Down syndrome (DS) in SSA (Ali 2009; Faye et al. 2004; Lawrenson et al. 2006). As a result, the population prevalence of chromosomal disorders is low (Adeyokunnu 1982; Christianson and Modell 2004). Still, there is reason to suspect an underestimation, inasmuch as in previous studies it was suggested that there may be clinical difficulties in making the diagnosis in African neonates (Christianson 1996). These findings suggest that if mothers themselves cannot see the differences between their DS children and unaffected children, clinical diagnosis based on physical stigmata may be difficult. Furthermore, acceptance of the diagnosis may be delayed until late milestones can be observed in the affected infants (Christianson and Kromberg 1996). The difficulties experienced by medical and nursing staff in diagnosing DS in black neonates extends into infancy and childhood, despite the fact that the clinical features of black DS infants and children do not differ from those seen in DS patients in other ethnic groups (Christianson 1997).

Prenatal screening for DS is not routine in many antenatal clinics in SSA (Oloyede 2008). It has been demonstrated that professionals and women in SSA are favorably disposed to first-trimester prenatal diagnosis services where available (Salako et al. 2006; Wonkam et al. 2006). In North Africa, many countries, including Egypt and Tunisia, have established experience in biochemical or echographic screening followed by invasive procedures (amniocentesis) for DS (Baghagho et al. 2004; Chelli et al. 2008; Mokhtar and Abdel-Fattah 2002). In SSA, however, only a few centers in South Africa have a long experience in screening diagnosis and treatment of children with DS (Dangor and Ross 2006; Venter et al. 1995). We report a single experience from Cameroon of available reports on prenatal genetic diagnosis options for DS in SSA (Wonkam et al. 2010).

It is important to improve professional and public awareness of DS prenatal screening options. Many more personnel and centers should be encouraged to provide first-trimester screening, and the NT scan may be the most feasible. Collaboration with African countries with long experience of prenatal diagnosis (PND) may help to cost-effectively build capacity where it is lacking and offer this service to the public. Quantitative fluorescence polymerase chain reaction (QF-PCR) appears as an interesting option in the context, since no culture of amniocytes will be required. When initiating prenatal genetic diagnosis in Cameroon, we successfully established such collaboration with Geneva University Hospitals.

Multifactorial Disorders

Key multifactorial disorders include congenital heart defects, neural tube defects, and cleft lip with or without cleft palate.

Congenital Heart Defects

Congenital heart defects (CHD) are the most common form of birth defects, occurring in 4–8 per 1000 live births. The majority of CHDs, about 90 percent, have a multifactorial cause. An estimated 1,040,800 infants are born each year with a multifactorial CHD (Turnpenny and Ellard 2007). The remaining CHDs are associated with chromosome abnormalities (5–8 percent), single-gene defects (3–5 percent), and teratogens (2–3 percent) (Seashore and Wappner 1996). With epidemiologic transition, CHDs rather than rheumatoid heart defects are dominant in some African settings, and their relative frequencies are similar to reports elsewhere – ventricular septal defects, oval fossa type atrial septal defects, Fallot's tetralogy, and patent arterial duct together accounted for 77.8 percent of the structural heart defects in Africa (Ali 2009; Edwin et al. 2010; Ejim et al. 2009; Okoromah et al. 2008).

Christiaan Barnard (1922–2002) introduced heart transplantation in humans for the first time, on the African continent (Toledo-Pereyra 2010). Despite this historical fact, the wide range of children with diverse native CHDs is a reflection of non-availability of definitive surgical facilities in some African countries. Humanitarian organizations are very active on the African continent as part of an outreach project that sends teams of specialist doctors and health professionals to perform open heart surgery on children and adults with congenital heart abnormalities (Lindblom 2009; Tippett et al. 2008). Regional and international collaboration could be mutually beneficial, inasmuch as studies demonstrate feasibility of surgery for adolescents and adults with congenitally malformed hearts in Sudan, Kenya, Nigeria, and Cameroon, with a good outcome (Edwin et al. 2010; Sulafa and Karani 2007; Tantchou Tchoumi and Butera 2009). Nevertheless, due to poverty and illiteracy of parents, the postsurgical follow-up of patients is challenging (Sulafa and Karani 2007; Tantchou Tchoumi and Butera 2009). In addition, the mean age at referral to a pediatric cardiologist is estimated to be 16.9 months in Kenya (Awori et al. 2007). This suggests that a significant number of patients may miss the opportunity to have optimal surgical intervention, indicating a need for policies of early diagnosis, close monitoring, and timely intervention (Sani et al. 2007).

With the establishment of tertiary health care institutions in many SSA countries, availability of echocardiographic facilities, as well as increasing numbers of pediatric cardiologists, more of these cases are likely to be seen in the future. There is an urgent need for the government to establish a well-equipped cardiothoracic surgical center to cater to these patients, either free or at highly subsidized rates.

Single-gene congenital heart defects are starting to be recognized and characterized, especially in South Africa, with molecular characterization of long QT-syndrome (Brink et al. 2005), arrhythmogenic right ventricular cardiomyopathy (Matolweni et al. 2006), or Marfan syndrome (Child et al. 2007).

Neural Tube Defects

Neural tube defects (NTDs), including spina bifida, anencephaly, and encephalocele, occur when part of the neural tube, which forms the spine, spinal cord, skull, and brain, fails to close between 21 and 28 days after conception – before women realize they are pregnant. A paleopathological study reports a meningocele in an ancient Egyptian infant mummy from the XI dynasty (2100–1955 BC; see Boano et al. 2009). Many children affected by NTDs have multiple lifelong disabilities, including varying degrees of lower limb paralysis, bowel and bladder incontinence, hydrocephalus, and intellectual and learning disabilities. Each year spina bifida and anencephaly, the two most common forms of NTDs, occur in an estimated 300,000 newborns worldwide (Shin et al. 2010). Although there is scarcity of information, NTDs are not rare and the burden of disease in SSA has been estimated at 1.99 cases per 1000 births in Cameroon (Njamnshi et al. 2008). The cranial lesions are mainly congenital primary hydrocephalus in 44 percent of cases, encephalocele in 28 percent, and microcephaly in 17 percent. The spinal lesions were spinal dysraphism in more than 90 percent, all in the lumbosacral region (Adeleye and Olowookere 2009; Ba et al. 2007; Taye and Bedru 2009). Cervical spina bifida is rarely reported in SSA (Andronikou et al. 2006).

Some African countries are developing considerable experience in surgical treatment, like endoscopic third ventriculostomy or choroid plexus cauterization (ETV/CPC) in Uganda or South Africa (Warf 2005; Warf and Campbell 2008; Warf et al. 2009). Still, delay in presentation even with associated malnutrition and local sepsis seems not to significantly affect the outcome of repair. Intracranial sepsis is the main cause of death in NTD patients who are not operated on in time in Nigeria and Cameroon (Alatise et al. 2006; de Paul Djientcheu et al. 2008; Idowu and Apemiye 2008).

Prenatal management and outcome at birth are limited by poverty and cultural beliefs. Prevention is possible and may be better than palliative care in developing countries. A call is made for the establishment of nationwide efforts to fully clarify the epidemiology of this silent epidemic and develop appropriate preventive strategies such as folic acid awareness among women of reproductive age, and preconception folic acid supplements (Sayed et al. 2006; van't Veer et al. 2008).

Cleft Lip with or without Cleft Palate

Cleft lip, with or without cleft palate, is a congenital malformation etiologically different from isolated cleft palate. It comprises a partial or complete cleft of the upper lip, with or without a cleft of the alveolar ridge (gum) or the palate. Multifactorial inheritance accounts for the majority of cleft lip, with or without cleft palate, with birth prevalence ranging from 0.3 per 1000 live births in African American populations, to 1 per 1000 in Caucasians. A number of potentially causative genes are under investigation, though certain drugs and maternal smoking are considered to play a role (WHO 2002a). Cleft lip with and without cleft palate is also associated with other syndromes caused by chromosomal abnormalities, single-gene defects, and teratogens.

Cleft lip and palate is a significant congenital malformation in SSA, with a variable birth prevalence based on hospital records on the continent from 0.3, 0.6, and 0.9 per 1000 live births in Zambia, Zimbabwe, and Sudan, respectively (Elliott et al. 2008; Pham and Tollefson 2007; Suleiman et al. 2005). There seems to be a higher incidence of familial tendency and associated congenital malformations than that reported elsewhere. In a large Kenyan study there was a male predominance, and most clefts were on the left side. A retrospective and prospective study featured a positive family history in 3.5 percent and 30.9 percent, respectively. Associated congenital malformations were 8.2 percent for the retrospective study and 25 percent for the prospective study (Spritz et al. 2007; Wanjeri and Wachira 2009). A seasonal variation was observed in Zambia that could be explained, at least in part, by environmental factors affecting the development of cleft lip with or without cleft palate. But the genetic background also affects the distribution, since there was a considerable variation between the groups of orofacial clefts occurring in the Caucasian versus South African black patients with cleft deformities (Bütow et al. 2007). Bantu tribes have lower risk of orofacial clefts than Nilotic/paranilotic tribes in Kenya (Spritz et al. 2007). The awareness of oral and maxillofacial surgery by both the public and medical specialties has led to rapid development and expansion of the oral and maxillofacial surgery specialty, with management of diverse and complex problems within a well-defined anatomical region. There is an indication of low awareness of the specialty by the public and professionals in Nigeria (Adewole and Akinwande 2007). Access to treatment is likely the major determinant of regional disparity in clefts in Zambia and Zimbabwe (Elliott et al. 2008; Pham and Tollefson 2007). Humanitarian plastic surgery training missions have been essential for the development of plastic surgery in some SSA countries (Saboye 1999). Financial, logistical, and training support to cleft teams has led to more surgeries for children with cleft in Ghana and Uganda (Donkor et al. 2007; Hodges and Hodges 2000).

There is an associated substantial clinical and psychosocial impact on patients with cleft lip/palate. Compliance with primary surgical treatment in people with cleft lip and palate is a well-known problem, especially in developing countries fraught with poverty and ignorance. The West African subregion has a high dropout rate (20 percent) after lip repair (Onah et al. 2008). Cultural and religious factors seem to have an important role in the explanations, labels, and treatment that follows the birth of a child with cleft lip/palate in Nigeria (Olasoji et al. 2007). Studies have emphasized the need for greater collaboration and sharing of information between modern medical practitioners and traditional healers in Nigeria and South Africa (Dagher and Ross 2004; Olasoji et al. 2007; Ross 2007). By viewing cultural differences on a continuum, following the asset-based approach, applying knowledge based on contextually relevant research, and recognizing family uniqueness, families may be empowered to participate fully in the team approach to support their children with cleft lip/palate and craniofacial anomalies in attaining their full potential in the SSA context (Louw et al. 2006).

A larger population-based study is warranted to further clarify these various epidemiological findings, to provide a basis for further prevention and therapeutic approaches to orofacial clefts in SSA.

12.4.2 Individual Teratogen-Associated Birth Defects

Birth defects due to teratogens are among the more readily preventable. Pregnancies in middle- and low-income countries, by comparison to high-income countries, are more likely to be at risk from potential teratogens for several reasons, including increased frequency of intrauterine infection, poor maternal nutrition, low socioeconomic and educational levels, lack of environmental protection policies, and poorly regulated access to medication (Penchaszadeh 2002). Broadly, three of the five causal groups of teratogens have the most significance: (1) congenital infections, (2) maternal illness and altered maternal metabolism, and (3) recreational and therapeutic drugs.

12.4.3 Congenital Infection

The TORCH organisms – toxoplasmosis, other (syphilis, varicella-zoster, human parvovirus B19), rubella, cytomegalovirus (CMV), and herpes – account for the most common infections associated with birth defects (Stegman and Carey 2002).

Toxoplasmosis

Toxoplasma gondii's importance for humans refers mainly to primary infection during pregnancy, resulting in abortion/stillbirth or congenital toxoplasmosis. Foci of high prevalence exist in Latin America, parts of Eastern/Central Europe, the Middle East, parts of Southeast Asia, and Africa (Pappas et al. 2009). The results indicated that the prevalence of *T. gondii* among pregnant women was relatively high in the rural (57.6 percent) compared to urban (46.5 percent) areas. The IgG positivity was correlated, in general, with age, as it was higher in the older age groups in Egypt (El-Gozamy et al. 2009). Nevertheless, there are significant variations in IgG seroprevalence in pregnant women in SSA. It was reported to be 25.3 percent in Burkina Faso (Simpore et al. 2006), 35 percent in Tanzania (Doehring et al. 1995), 40 percent in Senegal (Ndir et al. 2004), 50.6 percent in Morocco (El Mansouri et al. 2007), 53.6 percent in Benin (Rodier et al. 1995), 60 percent in Ethiopia and Côte d'Ivoire (Adou-Bryn et al. 2004; Negash et al. 2008), 71.2 percent in Gabon (Nabias et al. 1998), 75.4 percent in Nigeria (Onadeko et al. 1996), and 77.1 percent in Cameroon (Ndumbe et al. 1992). Despite the high IgG seroprevalence, country-specific awareness and screening is needed, since cases of congenital toxoplasmosis do occur in Africa (Ben Abdallah et al. 2009).

Congenital Syphilis

Syphilis is a common sexually transmitted infection. Over 11 million new infections occur each year worldwide. The continued occurrence of congenital syphilis is an indictment of the inadequate antenatal care services and poor quality of programs to control sexually transmitted infections. More than one million infants are born with congenital syphilis each year. In SSA alone, previous studies showed 6–16 percent of pregnant women have active syphilis (Gichangi et al. 2004; Jenniskens et al. 1995; Kamanga 1995; Ndumbe et al. 1992; Romoren and Rahman 2006; Walker and Walker 2007). The burden of congenital syphilis is also increasing along with AIDS pandemics (Mullick et al. 2004; Saloojee et al. 2004; Walker and Walker 2002).

Unless the mother is treated with penicillin, as many as 40 percent of embryos or fetuses with syphilis die before or shortly after birth. Those who survive are at risk for brain damage, blindness, hearing loss, and bone and tooth problems if they are not treated with an antibiotic shortly after birth (Fitzgerald et al. 1998). Although screening and appropriate treatment of infected pregnant women can prevent fetal and maternal complications, traditional screening algorithms requiring multiple tests have proven to be difficult to implement in resource-poor settings because of the lack of investigative ability and the pressure on health facilities to discharge infants early (Lech 2003; Saloojee et al. 2004; Walker and Walker 2002). Diagnosis of congenital syphilis is problematic, since more than half of all infants are asymptomatic, and signs in symptomatic infants may be subtle and nonspecific. Newer diagnostic tests such as enzyme immunoassays, polymerase chain reaction, and immunoblotting have made diagnosis more sensitive and specific, but are largely unavailable in the settings where they are most needed (Saloojee et al. 2004). Universal prenatal syphilis screening using rapid point-of-care tests will improve both maternal and infant outcomes and is cost-effective (Rydzak and Goldie 2008).

In SSA, with high maternal syphilis prevalence, important goals for patients, health care workers, and health authorities are earlier antenatal care attendance, rapid on-site testing, improved partner treatment, and a repeat test late in pregnancy to manage incident cases. Using the decision analysis to estimate the incremental cost-effectiveness of two on-site antenatal syphilis screening strategies to avert congenital infections – qualitative rapid plasma reagin (on-site RPR) and treponemal immunochromatographic strip assay (on-site ICS) – compared to the current practice (off-site RPR/*Treponema pallidum* hemagglutination [TPHA]), on-site antenatal screening with ICS was a cost-effective approach to reducing the incidence of congenital syphilis in a South African setting (Blandford et al. 2007; Bronzan et al. 2007). It was shown to be feasible in Bolivia, Kenya, South Africa, and Mozambique to introduce rapid syphilis testing in settings without laboratory services at a small incremental cost per woman screened (Deperthes et al. 2004; Levin et al. 2007). In settings with laboratories, the cost of ICS is similar to that of RPR. Nevertheless, in a recent study in Nigeria, the seroprevalence rate of syphilis was extremely low (12/18,101). The peak age-specific incidence of 0.02 percent was in the 20–24 years age group. This calls for a review of the policy of routine antenatal serology screening for syphilis according to the setting (Bukar et al. 2009).

Varicellla-Zoster Virus and Herpes Simplex Virus

Primary infection with varicella-zoster virus (VZV; chickenpox) can pose risks to the fetus if a mother is infected during pregnancy. A susceptible pregnant woman who is exposed to an infected household member has a 90 percent risk of contracting the illness. Maternal infection can lead to congenital varicella syndrome, a group of birth defects that can include defects of muscle and bone, malformed and paralyzed limbs, a smaller-than-normal head, blindness, seizures, and intellectual disability. This syndrome affects about 2 percent of babies whose mothers are infected with varicella during the first 20 weeks of pregnancy, but is rare if infection occurs after 20 weeks (March of Dimes Global 2006). The birth prevalence of VZV in SSA countries is unknown.

Herpes simplex virus type 2 (HSV-2) can cause genital infection and can be passed to the fetus by the mother in utero, during birth, and postnatally. Herpes simplex virus infection in the newborn can present as isolated skin, eye, or mouth infection, or as disseminated disease with CNS involvement that can result in microcephaly, intellectual disability, and seizures. Mortality rates as high as 25 percent have been reported for disseminated disease. Morbidity is also high. While 95 percent of newborns with isolated infection are normal at two years of age, only 60 percent of newborns who survive disseminated disease are normal at that age (Whitley et al. 1991). Unfortunately, many women are asymptomatic, with no evidence of genital lesions at the time of primary infection, complicating control efforts. Maternal HSV seroprevalences of greater than 50 percent have been documented in specific populations in Central America, Africa, and Asia (Ghebrekidan et al. 1999; Kaur et al. 1999; Nahmias et al. 1990; Oberle et al. 1989).

In recent decades the alarming spread of HIV infection in Africa, and associated frequent reactivation of herpes infections, is leaving less room for neglect (Ghebrekidan et al. 1999). In Côte d'Ivoire, the herpes viral isolation and the direct immunofluorescence (IF) assay of the genital secretions were found to have a prevalence of 12.7 percent among the pregnant women who tested positive (the majority of whom were from modest socioeconomic origins and young); 10 percent had antecedents of genital herpes, 2.7 percent showed an asymptomatic excretion, and 36.8 percent had an acute episode during their pregnancy, including a primary infection in the twentieth week. The scarcity of data that exist in regard to herpes virus infections in many African regions could partly be due to the mild nature of their primary infections and the lack of means for their proper diagnosis (Duff et al. 2009).

Human Parvovirus

Human parvovirus (HPV) B19 causes fifth disease (erythema infectosum), a mild viral illness usually seen in children. When the infection occurs in pregnant women, it acts as a teratogen and can cause fetal death. The virus disrupts the fetus' ability to produce red blood cells, leading to anemia, heart failure, and up to a 9 percent risk of miscarriage or stillbirth. Fetal deaths are more likely when a pregnant woman contracts the infection in the first 20 weeks of pregnancy. An estimated 1 per 3333 newborns are infected in high-income countries (Markenson and Yancey 1998). In South Africa, in a prospective study of 1967 pregnant women who were routinely screened for recent human B19 parvovirus infection, 64 (3.3 percent) were identified as being IgM positive. No adverse effects were documented by ultrasound in any of the fetuses. The outcome of pregnancy was favorable in 95.1 percent of these women, with no evidence of hydrops fetalis or any congenital abnormalities. Two neonates (3.4 percent) were small for gestational age, and there was one abortion. Samples of blood obtained from 20 neonates born to women with evidence of recent infection were B19 parvovirus IgM negative. Recent infection with human B19 parvovirus in pregnancy constitutes a low risk for the development of adverse fetal effects; hence, routine antenatal screening is not warranted (Guidozzi et al. 1994).

Congenital Rubella Syndrome

Rubella poses a serious threat to the fetus if the mother contracts it during the first 16 weeks of pregnancy. The annual incidence of rubella worldwide is difficult to ascertain because the disease is often mild and overlooked, or, when it is more severe, is often misdiagnosed as measles. About 25 percent of babies whose mothers contract rubella during the first trimester of pregnancy are born with one or more birth defects, which, together, are referred to as *congenital rubella syndrome*. These birth defects include blindness, hearing impairment, heart defects, intellectual disability, and, less frequently, movement disorders. The infection occasionally causes miscarriage and stillbirth. Children with CRS also are at increased risk of diabetes, which may develop during childhood or adulthood (March of Dimes Global 2002). Although the situation is improving, more than 100,000 children are born annually with congenital rubella syndrome (WHO 2000b). In high-income countries, successful rubella immunization programs have virtually eliminated CRS.

In the African region only 2 percent of countries had rubella vaccine programs; nevertheless, the few available data indicate a high immunization rate. Within the Western Cape a total of 95.3 percent of women in the 15–24 years age group, 97.5 percent in the 25–34 years group, and 98 percent in the 35–45 years age group were immune to rubella (Corcoran and Hardie 2005); an 83.9 percent immunization rate was reported in Cameroon (Ndumbe et al. 1992). Any vaccination program has to take into account the natural immunization rate in a given country.

Cytomegalovirus

Cytomegalovirus (CMV) infection is the most common congenital infection in high-income countries. Intrauterine infections occur in mothers who have developed a first-time CMV infection during pregnancy. These women have a 30–40 percent risk of passing the virus to their fetus (March of Dimes Global 2001). Most of the time, the infected infant shows no symptoms at birth, but 14 percent of asymptomatic newborns with infection develop serious disabilities over the next several years. These include neurological, growth, and developmental problems; sight or hearing problems; and dental abnormalities. About 10 percent of infected newborns have symptomatic CMV disease. Of these, most who survive suffer from progressive deafness and/or intellectual disability. An estimated 4 per 10,000 newborns are affected in middle- and low-income countries (Christianson and Modell 2004). The overall prevalence of congenital CMV was 5.71 percent among neonates in intensive care units in Egypt, with higher frequency in premature and low-birth-weight neonates (Morgan et al. 2003). Up to 7.3 percent of 68 infants with congenital anomalies tested positive for CMV IgM antibodies by microassay ELISA technique in an Egyptian cohort (Shaaban et al. 1995). Cytomegalovirus was found to be the most prevalent infection in both groups of women (19.2 percent) and in the babies (24.2 percent), indicating the importance of this virus in intrauterine infection, in a community study in South Africa (Bos et al. 1995). A 15-month longitudinal study of CMV infection in 178 Gambian mothers and their babies was undertaken; 25 (14 percent) of the babies were congenitally infected despite the fact that 87 percent of their mothers were antibody-positive to the virus (Bello and Whittle 1991). The few data available seem to indicate that CMV could play a considerable role in congenital anomalies in SSA.

It is anticipated that with HIV, the prevalence of the TORCH organisms related to birth defects will not decrease unless serious preventive actions are taken. This will require more research for a better understanding of the problem with a country-specific approach, and public health action for screening and preventions.

12.4.4 Maternal Illness and Altered Maternal Metabolism

Maternal Diabetes Mellitus

Pregnancy in type 1 and type 2 diabetes mellitus (DM) is associated with an increased rate of adverse outcomes for both mother and fetus. Insulin-dependent DM (type 1; IDDM) affects 0.5 percent of pregnancies in high-income countries (Khoury et al. 1989), and increases the risk of serious birth defects up to three-fold. The risk of fetal abnormalities due to maternal IDDM can be reduced by good periconception diabetic control (Kinsley 2007).

The number of cases of diabetes worldwide has increased significantly in the last decade. Characteristically, the incidence of gestational diabetes (GDM) reflects the incidence of type 2 diabetes mellitus (T2DM) in the background population, which is a warning that a rapid increase in the incidence is to be expected, concomitant with the already-observed increase in the incidence of T2DM. Although the majority of all deliveries worldwide take place in the so-called developing world, little is known about the prevalence of diabetes in pregnancy in rural areas of East Africa (Zeck and McIntyre 2008).

Gestational diabetes mellitus complicates about 7.7 percent of pregnancies in Morocco (Bouhsain et al. 2009); 5.4 percent in Nigeria (Adegbola and Ajayi 2008); 8.8 percent in Limpopo, South Africa (Mamabolo et al. 2007); and 11 percent in Mozambique (Challis et al. 2002). The prevalence rate of GDM was found to be 3.7 percent in rural Ethiopia (Seyoum et al. 1999).

The perinatal mortality was 6.1 percent, 10 percent, and 12.5 percent in South Africa, Tanzania, and Nigeria, respectively (Huddle et al. 1993; Ozumba et al. 2004). Major congenital anomalies in pregnancy in diabetic women were 1.7 percent and 2 percent in South Africa and Tanzania, respectively (Huddle et al. 1993; Lutale et al. 1991). Maternal diabetes is a risk factor for congenital heart diseases in Alexandria, Egypt (Bassili et al. 2000). A study in Libya showed that 13.8 percent of infants of mothers with IDDM had easily recognizable birth defects, compared to 3 percent in the non-diabetic maternal population (Malik and Mir 1992). Maternal hyperglycemia is an important factor affecting maternal well-being and neonatal morbidity and mortality in Sudan (Abdelgadir et al. 2003).

The institution of a specialized service for pregnant diabetic women from Soweto for a sustained period of 20 years proved feasible and was of significant benefit. The interventions, centered around a diabetes nurse educator, were relatively simple and easy to implement, even though late presentation/referral of patients remains a problem (Huddle 2005). Programs and policies have to be established, including organization of the health system to provide care, medicines, and other tools necessary for management of diabetes in pregnancy, consideration of accessibility and affordability of care, education for health care workers, and education of pregnant and non-pregnant women of reproductive age.

Maternal Epilepsy

Children of epileptic mothers have a greater risk for birth defects than is seen in the general population. Risks of major malformations, growth retardation, and hypoplasia of the midface and fingers, known as *anticonvulsant embryopathy*, increases in infants exposed to anticonvulsant drugs in utero, including sodium valproate and phenytoin.

One-quarter of the 40 million individuals estimated to have epilepsy in the developing world live in Africa (Onwuekwe et al. 2009). Data on the incidence of and prognosis for epilepsy in SSA are scarce, but prevalence data show that epilepsy is two or three times more common than in industrialized countries in non-tropical areas. Door-to-door surveys and neurological assessment showed a prevalence of 11.2/1000 in Tanzania (Winkler et al. 2009), 12.5/1000 in rural Zambia (Birbeck and Kalichi 2004), 14.2/1000 in Senegal (Ndoye et al. 2005), 15.7/1000 in Togo (Balogou et al. 2007), 15.9/1000 in Benin (Debrock et al. 2000), 29.5 percent in Ethiopia (Almu et al. 2006), 41/1000 in rural Kenya (Edwards et al. 2008), 64/1000 in rural Ethiopia (Tekle-Haimanot et al. 1997), and up to 105/1000 in a rural area in Cameroon

(Prischich et al. 2008). Active convulsive epilepsy was estimated to be 4.5/1000 in rural Kenya (Edwards et al. 2008) and 7.4/1000 in Tanzania (Dent et al. 2005). The high prevalence of epilepsy and low life expectancy indicate that incidence is high. Reasons for this type of frequency could be ascribed to malnutrition, trauma, genetic conditions, antenatal and prenatal care, diseases resulting in febrile convulsions, and/or cerebral infections, and may even result in death in some diseased individuals. Relative contributions of each cause of epilepsy are difficult to determine. Only a few case–control studies have been done in SSA. Infections, particularly cysticercosis in its endemic areas, cause most cases of epilepsy. The implementation of studies to accurately determine the causes of epilepsy in SSA is urgently required. Such studies will help to lower the incidence of epilepsy in this region and better understand the etiology of epilepsy in other areas (Preux and Druet-Cabanac 2005).

Despite the high frequency of epilepsy in SSA, there is only anecdotal data concerning birth defects due to antiepileptic drugs (Lekwuwa et al. 1995). In developing countries, the fetus of an epileptic mother is at even greater risk because antiepileptic therapy is less likely to be well controlled, multiple-drug therapy is more common, and cheaper drugs, more likely to cause fetal damage, are used in place of more expensive, less teratogenic anticonvulsant drugs (Shorvon and Farmer 1988). The general fertility rate for reproductive-aged epileptic women for the three-year period 2003–2006 preceding the study was 46 live births per 1000 women-years, decreased by two-thirds in Kenya (Kariuki et al. 2008). The social burden of the disease largely outweighed the medical burden in women living with epilepsy in SSA (Birbeck et al. 2008). Phenobarbitone was the most frequently prescribed drug (Preux et al. 2000), and phenytoin therapy is not unusual (Valodia et al. 1999). Fetal valproate syndromes have been reported in sibling pairs in South Africa (Christianson et al. 1994).

Maternal Malnutrition

Recent estimates indicate that globally over two billion people are at risk for vitamin A, iodine, and/or iron deficiency, in spite of recent efforts in the prevention and control of these deficiencies. The prevalence is especially high in Southeast Asia and SSA, and pregnant women and young children are at greatest risk. Other micronutrient deficiencies of public health concern include zinc, folate, and the B vitamins. However, there are limited data on the actual prevalence of these deficiencies. Finally, in many settings, more than one micronutrient deficiency exists, suggesting the need for simple approaches that evaluate and address multiple micronutrient malnutrition (Ramakrishnan 2002).

Folic Acid

Overwhelming evidence links an increased intake of synthetic folic acid (the bio-available form of folate) to a reduced risk of NTDs. Because NTDs result from early failure of neural tube closure, which occurs before pregnancy can be confirmed, research indicates that at least half of the cases of NTDs can be prevented if women consume folic acid before conception and during early pregnancy (CDC 2005). These and other findings lent support to the 1992 recommendation by the US Public Health Service and the 1998 recommendation of the Institute of Medicine of the US National Academy of Sciences that all women of childbearing age should consume 0.4 mg (400 μg) of synthetic folic acid daily, in addition to folate provided by a varied diet, in order to reduce their risk of an NTD-affected pregnancy (IOM et al. 1998).

A 1997 demographic and health survey in Mozambique showed that 47 percent of girls 15–19 years old living in Manica province (west-central Mozambique) were pregnant or had already had a child. Supplementation periods can have an important impact on girls' hematological status and birth defect prevention (Horjus et al. 2005). There was a high prevalence of anemia, and folate deficiency varies among different populations in SSA. Up to 57.7 percent of pregnant women were reported to have anemia and folate deficiency in Sudan (Abdelrahim et al. 2009). Studies also evidenced folate deficiency in Nigerian pregnant women (Kamau-Mbuthia and Elmadfa 2007; Vanderjagt et al. 2007); folate deficiency, defined by red blood cell folate of less than 100 μg/l, was observed in 6 percent of pregnant women in Congo (Mashako et al. 1991). However, only 2 percent had low plasma folate (<6.8 nmol/l) in Ethiopia (Gibson et al. 2008), and in a cross-sectional study of 150 pregnant women in Blantyre, Malawi, 23 percent were iron-deficient with no evidence of folate deficiencies (van den Broek and Letsky 2000).

In SSA, effective use of iron/folic acid during pregnancy was estimated without intervention to be 40 percent, but reached 86 percent when improving women's access to supplements in Senegal (Seck and Jackson 2009). A lack of awareness and a lack of supplies and materials were the main barriers to the use of iron/folic acid during pregnancy in Cameroon (Tita et al. 2005). Maternal international multiple micronutrient preparations (UNIMMAPs), rather than iron and folic acid alone, modestly but significantly increased fetal growth in Senegal, Niger, Guinea-Bissau, Zimbabwe, and Tanzania (Friis et al. 2004; Kaestel et al. 2005; Makola et al. 2003; Roberfroid et al. 2008; Zagré et al. 2007).

If a woman has already had a baby with an NTD, studies have shown that a larger daily dose of folic acid (4000 μg or 4 mg), beginning at least one month before pregnancy and throughout the first trimester, reduces the risk of having another affected pregnancy by about 70 percent. Women with diabetes, epilepsy being treated with sodium valproate, and, possibly, obesity also are at increased risk of having a baby with an NTD and should discuss the possibility of taking a higher level of folic acid with their primary care provider (March of Dimes Global 2004). An increasing body of evidence suggests that folic acid may help prevent other major malformations, including congenital heart defects, as well as coronary heart disease, certain forms of cancer, and possibly dementia (Canfield et al. 2005; McIlroy et al. 2002).

A decrease in NTD rates post-fortification was reported in black populations in South Africa (Sayed et al. 2008). Nutrient education for dietary diversification, and nutrient supplementation, should be explored as interventions for improving nutrient intake among pregnant women. The economic benefit flowing from the prevention of NTDs greatly exceeds the costs of implementing folic acid fortification.

Iodine

Iodine is part of the thyroid hormones, which contribute to brain development in the fetus and regulate human metabolism; iodine deficiency is the most frequent cause of preventable mental retardation. Related disorders include lethargy, physical disabilities, goiter, stillbirth, and neonatal death. UNICEF considers iodine deficiency the single most important cause during pregnancies of brain damage and intellectual disability, with most cases caused before birth (UNICEF 1998). Despite these efforts, though, approximately 1.6 billion people in more than 100 countries live in areas where iodine is not available in sufficient amounts; those most at risk include about one-third of China's population. It is also a severe problem in some areas in SSA (Abuye et al. 2008; Boulet 1997; Cherinet and Kelbessa 2000). In 1998, 60,000 babies were born worldwide with severe congenital hypothyroidism and an estimated 28 million pregnancies were still at risk from maternal iodine deficiency (UNICEF 1998). About 50 million people worldwide live with intellectual, motor, and hearing disabilities due to iodine deficiency disorders (IDDs) (UNICEF 2000).

The gross prevalence of IDDs among schoolchildren was 53.3 percent in Ethiopia (Cherinet and Kelbessa 2000). The virtual absence of iodized salt in Ethiopian households shows that currently there is no salt iodization program in the country (Abuye et al. 2008). A study in Sudan suggests that, in areas with mild iodine deficiency, neonates may be at the limit of decompensation as evidenced by their enhanced TSH and T3 levels, as well as increased T4 compared to their mothers (Eltom et al. 2001). Similarly, the incidence of neonatal chemical hypothyroidism (NCH) in Saki (Nigeria) was 14.7 per 1000 babies. The results suggest environmental iodine deficiency and relative chemical hypothyroidism of mothers and neonates in Saki when compared to mothers and neonates from non-iodine-deficient areas (Ojule and Osotimehin 1998). In school-aged children in South Africa, surveys demonstrated goiter prevalence in the 20–29.9 percent range, and a median urinary iodine level in the 2–4.9 μg/dl range, indicating iodine deficiency of moderate severity. Iodized salt was not available in any of the three community shops studied (Benedé et al. 1997). Study in a tertiary care infant welfare clinic in Ibadan, southwestern Nigeria, a geographical area recognized to be outside Nigeria's endemic goiter belt (goiter prevalence <5.0 percent), suggests that the breast-fed child of an iodine-replete mother resident in a non-iodine-deficient area may be iodine-deficient and in need of iodine supplementation. Deficiency affects most of the Zairian population and requires public health measures on a larger scale than previously estimated. Combined iodine and selenium deficiency affects Equateur, Haut-Zaire, and Kivu, where endemic myxoedematous cretinism occurs, but also in Shaba, where it was not previously described. Besides combined iodine and selenium deficiency, another factor (thiocyanate?) must be taken into account to explain the peculiarly elevated prevalence of endemic myxoedematous cretinism in Central Africa (Ngo et al. 1997). Besides low levels of iodine intake, cassava consumption and living at high altitudes were responsible for the observed variation and severity in goiter rates. These observations have important implications for planning IDD control programs in Africa (Akanji et al. 1996).

Progress has been made due to affordable, cost-effective technology and alliances between government, the private sector, and international agencies. Almost all but Ethiopia, of the 53 countries of Africa that were identified as having IDD problems, had implemented IDD control programs by 1995. The programs emphasized the long-term use of iodized salt. By February 1996, over 50 percent of the salt consumed in Africa was estimated to be iodized. It also appears that the proportion of households using iodized salt is increasing. Fortification of household salt appears to be an appropriate method of controlling iodine deficiency in Benin (Melse-Boonstra et al. 1998). Preliminary evidence suggests that the prevalence of goiter in three districts of Kenya declined rapidly after the introduction of legislation authorizing the use of iodized salt. Success is not just limited to declines in IDD (Kavishe 1997). In South Africa, mandatory iodization of salt virtually eradicated iodine deficiency within one year in South African schoolchildren, but the goiter rate in these children did not decline (Jooste et al. 2000; Kalk 1998). Countrywide, around 30 percent of low socioeconomic status households obtained their salt from unconventional sources, compared to less than 5 percent in high socioeconomic status households, emphasizing the vulnerability of low socioeconomic groups to the use of non-iodized salt. Similarly, a short prospective study was conducted to assess thyroid status in healthy full-term newborns (n = 90) of a large maternity hospital in Marrakech (Morocco) as part of the validation of a national salt iodization program. High TSH (>5 mU/l) was detected in 89 percent of infants tested; however, iodine deficiency seems to remain a widespread problem in this population and justifies a large-scale survey of iodine status (Zaida et al. 2006).

Intervention strategies should mobilize all players involved in unconventional marketing channels of household salt to provide only iodized salt to consumers, as required by law (Jooste 2005). A follow-up of proper implementation of the program is required to reach the realistic goal of eradicating IDD in the next century.

Recreational and Therapeutic Drugs

Fetal Alcohol Spectrum Disorder

Fetal alcohol spectrum disorder (FASD) encompasses a range of effects that can occur in an individual whose mother drank alcohol during pregnancy (SAMHSA 2005). These effects, which can range from mild to severe, include physical, behavioral, mental, and learning disabilities with lifelong implications for affected individuals (Hoyme et al. 2005). The only cause of FASD is alcohol use during pregnancy, and no threshold for safety is known. Multidisciplinary approaches – including a dysmorphology examination, developmental and neuropsychological testing, and a structured maternal interview to gather data about prenatal drinking practices and other demographic and family information, as recommended in published guidelines – are used for diagnosis (Hoyme et al. 2005; Peadon et al. 2008). The facial anomalies associated with fetal alcohol syndrome (FAS), some of which are also present in individuals with less severe forms of the broader category of FASD, are typically identified with the aid of linear distance measurements taken between facial landmarks (smooth philtrum, small palpebral fissures, and thin upper lip; see Douglas and Mutsvangwa 2010).

The prevalence of FAS appears to be much higher in certain other regions of the world. In the wine-growing Western Cape Province of South Africa, more than one-quarter of pregnant women abuse alcohol during pregnancy and more than 4 percent of six- to seven-year-old school children had FAS (the highest rate ever reported for a stable community) (Viljoen et al. 2001). Comparable studies in urban Johannesburg found 2.7 percent of children had FAS (Croxford and Viljoen 1999; May et al. 2000; Viljoen et al. 2001). This finding has raised concerns about the prevalence of FAS in other middle- and low-income countries where alcohol is available and used by women of reproductive age (Rosenthal et al. 2005).

Therapeutic Drugs

A number of therapeutic drugs have been implicated as teratogens, and their adverse effects linked to the timing of use during pregnancy, dose, genetic susceptibility, and other factors. Jolted into action by the thalidomide tragedy, developed Western countries began to establish pharmacovigilance systems for identifying and responding to adverse drug reactions and events about 40 years ago. These systems focus on side-effects, adverse reactions, and drug interactions. In developing countries, especially in Africa, the scope for pharmacovigilance needs to be broader (despite the additional challenges this brings) because of growing problems with substandard and counterfeit drugs, and the need to have an early warning signal system for the development of antimicrobial resistance to the "new essential drugs" that are barely beyond the clinical trial stage in Africa; e.g., artemisinin-combination therapy (ACT) for malaria, and anti-retrovirals (ARVs) for HIV/AIDS. Zambia learned important lessons from its own initial experiences in attempting to use ACT as a pathfinder for pharmacovigilance, as well as its experience with other drug information systems (Huff-Rousselle et al. 2007).

Thalidomide causes severe limb-reduction birth defects and was formerly used to combat morning sickness, but is also used to treat leprosy, macular degeneration in AIDS, and psoriasis, and also inhibits monocyte tumor necrosis factor (TNF)-α production and co-stimulates T cells. It was tested for immune modulation in patients with HIV infection and tuberculosis (TB) in a placebo-controlled study. In the presence of thalidomide, antigen-specific immune responses *in vitro* and in patients with HIV/TB were enhanced (Baccaglini et al. 2007; Bekker et al. 2000). There is a need to monitor such limited anecdotal reports of benefit existing for thalidomide or other agents on HIV patients (Marais et al. 2009). Indeed, implications for birth defects specifically in pregnant women affected by HIV should not be underestimated.

Anticonvulsant drugs are associated with major malformations, including NTDs, microcephaly, intellectual disability, growth restriction, and malformations of the face and fingers.

Anticoagulants have been linked to nasal hypoplasia, stippling of bones, optic atrophy, microcephaly, growth and intellectual disability, and fetal and neonatal hemorrhage (Adab et al. 2001; Holmes et al. 2001). These medications are causes for concern in SSA, where use is less regulated and obstetric or fetal outcomes are unmonitored; the data are needed.

Unlike the Western world, valvular disease ranks among the major cardiovascular afflictions in Africa. Acute rheumatic fever and chronic rheumatic valvular disease in their most virulent form are still commonly encountered and impose a huge burden on limited health care resources (Essop and Nkomo 2005). Pregnant patients with mechanical valves require careful attention to ensure maternal survival and prevent fetal warfarin embryopathy. Prolonged subcutaneous heparin and frequent monitoring of the partial thromboplastin time are sometimes impractical in this setting. The pregnancy rate in open heart surgical patients at Kenyatta National Hospital was approximately one-quarter of that in the general Kenyan population. The risks of maternal and fetal complications with heparin seem to be greater than the likely drawbacks associated with warfarin use in these patients (Ogendo 1999). The mean fetal birth weight was low (2.54 ± 0.98 kg). In South Africa, a study on maternal and fetal sequelae of anticoagulation during pregnancy in patients with mechanical heart valve prostheses revealed nine abortions (18 percent), seven stillbirths (14 percent), two neonatal deaths (4 percent), and two instances of warfarin embryopathy (4 percent) (Sareli et al. 1989).

12.5 Conclusion and Perspectives

As revealed and emphasized by the present review, as African countries go through an epidemiological transition, with a reduction in childhood and infant mortality due to improved public health measures, babies who would have previously died of these diseases before they were recognized are now surviving to present for diagnosis and treatment. Hence, they are presenting an increasing global health burden. In addition, some specific factors like malaria, migration, consanguineous marriages, maternal age, and the poor level of national health systems contribute to the bulk of the burden of birth defects in SSA.

In the SSA context, there is an urgent need to develop projects to gather more epidemiological data on birth defects, the stage in development at which each country needs to introduce medical genetic services, the relative roles of care and prevention in public health policies, the appropriate application of knowledge and technology to these purposes, and the financial implications. WHO, in partnership with a March of Dimes affiliate, the World Alliance of Organizations for the Prevention of Birth Defects, identified the need to integrate medical genetic services into primary health care and to further integrate these with secondary and tertiary medical genetic services and with reproductive, obstetric, pediatric, and other medical and social services, and also considered how this could be achieved within the limits of local circumstances, customs, available manpower, and material resources (WHO and the World Alliance for the Prevention of Birth Defects 1999). The objectives are to care for those with a disorder in order to reduce suffering and to improve health by prevention.

When raised in the context of developing health policy for many middle- and low-income countries, the need for prevention is usually given priority over the requirement for patient care. In fact, the bulk of care and prevention of birth defects is most appropriately carried out routinely in primary and secondary care settings (Christianson and Modell 2004; WHO and the World Alliance for the Prevention of Birth Defects 1999). In low-income countries the most effective interventions for prevention include family planning, optimizing women's diets, managing maternal health problems, and avoiding maternal infections. These are feasible and affordable in financially constrained environments and have a cascade of beneficial effects in addition to preventing birth defects. Because risk factors for birth defects (including advanced maternal age, poverty, maternal medical complications, infection, poor nutrition, smoking, and alcohol and drug use) are common to other adverse maternal and child health outcomes, interventions to reduce birth defects will contribute overall to women's, maternal, neonatal, and child health. Care and prevention of birth defects should be considered an integral and cost-effective arm of public health programs directed at saving the lives of and reducing disability among women, newborns, and children. In fact, some interventions, like fortification of food with micronutrients, benefit the entire population (Christianson and Modell 2004).

Our review suggests that the goals of prenatal care in SSA should be revisited in order to implement a program based on real scientific evidence. Specific purposes of consultations would be: (1) to screen birth defect risk factors which, when recognized, lead to specific actions, including: determination of advanced maternal age, nucal translucency screening, TORCH screening, micronutrient inquiry and supplementation when necessary (iodine, folate), history of therapeutic or recreational drugs (anticonvulsant and alcohol); (2) to prevent and/or detect (and treat) specific complications of pregnancy including infection (venereal disease, HIV), anemia and trace element deficiencies, and GDM; (3) to provide counseling, support, and information for pregnant women and their families (including the partner) concerning SCD screening (prenatal and neonatal) and, if possible, prenatal genetic diagnosis and reproductive options. These potentially effective actions aimed at preventable birth defects can only have a real public health impact if implemented within an organized maternal health system with a functional network of delivery units, if truly quality care is given, and if the relationships between health care providers and the population are based on mutual respect. Sub-Saharan African women use prenatal care extensively when it is accessible; this opportunity must be used to implement evidence-based actions with appropriate and realistic goals of birth defect prevention.

Moreover, the predicted increase in non-communicable diseases including birth defects and disabilities could be economically burdensome for the basic health care infrastructure of countries that lack resources to address this emerging disease burden. Biomedical research could stimulate development of health care and biomedical infrastructure. If this development is sustainable, it provides an opportunity to alleviate the burden of both communicable and non-communicable diseases through diagnosis, prevention, and treatment (Peprah and Wonkam 2013). Improved capabilities of African institutions to undertake research will be realized if ministries of health implement at least three key recommendations of the World Health Organization. First, capacity building will be achieved most effectively through the development of academic public research and industrial partnerships between developed and developing countries and between developing countries themselves, encouraged by tax or other incentives, and through extensive networking within regions where there is evolving expertise in biotechnology. Second, to be able to utilize the vast quantities of genomic data that are being generated, African countries must develop a critical mass of expertise in bioinformatics. Finally, all African countries need to evolve appropriate national frameworks to consider the ethical implications of genomics research and its applications in their own unique social, cultural, economic, and religious contexts; empirical data from a broad spectrum of stakeholders and the public are needed for the development of effective genomic policies and programs of disease that affect the continent (Wonkam and Mayosi 2014).

Because of their uneven distribution in high-frequency populations, reflecting their complex population genetics, the true magnitude of burden of hemoglobinopathies alone is still unknown. In many SSA countries there are few facilities for the diagnosis and management of genetic and partially genetic birth defects, and there are limited data about their frequency, clinical course, or mortality. Without this information it will be impossible to persuade governments of the increasing importance of these diseases. The situation will only be improved by concerted action on the part of the international community of the richer countries, together with input from the major international health organizations and funding agencies (Weatherall 2010).

References

Abdelgadir M, Elbagir M, Eltom A, Eltom M, and Berne C (2003). Factors affecting perinatal morbidity and mortality in pregnancies complicated by diabetes mellitus in Sudan. *Diabetes Res Clin Pract* **60**(1): 41–47.

Abdel-Meguid N, Zaki MS, and Hammad SA (2000). Premarital genetic investigations: effect of genetic counselling. *East Mediterr Health J* **6**(4): 652–660.

Abdelrahim II, Adam GK, Mohmmed AA, et al. (2009). Anaemia, folate and vitamin B12 deficiency among pregnant women in an area of unstable malaria transmission in eastern Sudan. *Trans R Soc Trop Med Hyg* **103**(5): 493–496.

Abuye C, Berhane Y, and Ersumo T (2008). The role of changing diet and altitude on goitre prevalence in five regional states in Ethiopia. *East Afr J Public Health* **5**(3): 163–168.

Adab N, Jacoby A, Smith D, and Chadwick D (2001). Additional educational needs in children born to mothers with epilepsy. *J Neurol Neurosurg Psychiatry* **70**(1): 15–21.

Adegbidi H, Yedomon H, Atadokpede F, et al. (2007). Skin cancers at the National University Hospital of Cotonou from 1985 to 2004. *Int J Dermatol.* **46**(Suppl. 1): 26–29.

Adegbola O and Ajayi GO (2008). Screening for gestational diabetes mellitus in Nigerian pregnant women using fifty-gram oral glucose challenge test. *West Afr J Med* **27**(3): 139–143.

Adeleye AO and Olowookere KG (2009). Central nervous system congenital anomalies: A prospective neurosurgical observational study from Nigeria. *Congenit Anom (Kyoto)* **49**(4): 258–261.

Adewole RA and Akinwande JA (2007). Public and professional perception of oral and maxillofacial surgery (a pilot study). *Nig Q J Hosp Med* **17**(1): 8–12.

Adeyokunnu AA (1982). The incidence of Down's syndrome in Nigeria. *J Med Genet* **19**(4): 277–279.

Adou-Bryn KD, Ouhon J, Nemer J, Yapo CG, and Assoumou A (2004). Serological survey of acquired toxoplasmosis in women of child-bearing age in Yopougon (Abidjan, Côte d'Ivoire). *Bull Soc Pathol Exot* **97**(5): 345–348.

Ahmed H, Yukubu AM, and Hendrickse RG (1995). Neonatal jaundice in Zaria, Nigeria: a second prospective study. *West Afr J Med* **14**(1): 15–23.

Aidoo M, Terlouw DJ, Kolczak MS, et al. (2002). Protective effects of the sickle cell gene against malaria morbidity and mortality. *Lancet* **359**(9314): 1311–1312.

Akanji AO, Mainasara AS, and Akinlade KS (1996). Urinary iodine excretion in mothers and their breast-fed children in relation to other childhood nutritional parameters. *Eur J Clin Nutr* **50**(3): 187–191.

Akinyanju OO, Disu RF, Akinde JA, et al. (1999). Initiation of prenatal diagnosis of sickle-cell anaemia in Africa. *Prenat Diagn* **19**(4): 299–304.

Akinyanju OO, Otaigbe AI, and Ibidapo MO (2005). Outcome of holistic care in Nigerian patients with sickle cell anaemia. *Clin Lab Haematol* **27**(3): 195–199.

Alatise OI, Adeolu AA, Komolafe EO, Adejuyigbe O, and Sowande OA (2006). Pattern and factors affecting management outcome of spina bifida cystica in Ile-Ife, Nigeria. *Pediatr Neurosurg* **42** (5): 277–283.

Ali SK (2009). Cardiac abnormalities of Sudanese patients with Down's syndrome and their short-term outcome. *Cardiovasc J Afr* **20**(2): 112–115.

Aliyu ZY, Gordeuk V, Sachdev V, et al. (2008). Prevalence and risk factors for pulmonary artery systolic hypertension among sickle cell disease patients in Nigeria. *Am J Hematol* **83**(6): 485–490.

Almu S, Tadesse Z, Cooper P, and Hackett R (2006). The prevalence of epilepsy in the Zay Society, Ethiopia: an area of high prevalence. *Seizure* **15**(3): 211–213.

Andronikou S, Wieselthaler N, and Fieggen AG (2006). Cervical spina bifida cystica: MRI differentiation of the subtypes in children. *Childs Nerv Syst* **22**(4): 379–384.

Aquaron R, Djatou M, and Kamdem L (2009). Sociocultural aspects of albinism in sub-Saharan Africa: mutilations and ritual murders committed in East Africa (Burundi and Tanzania). *Med Trop* **69**(5): 449–453. [French]

Asuquo ME, Ngim O, Ebughe G, and Bassey EE (2009). Skin cancers amongst four Nigerian albinos. *Int J Dermatol* **48**(6): 636–638.

Awori MN, Ogendo SW, Gitome SW, Ong'uti SK, and Obonyo NG (2007). Management pathway for congenital heart disease at Kenyatta National Hospital, Nairobi. *East Afr Med J* **84**(7): 312–317.

Ba MC, Ly Ba A, Hossini A, et al. (2007). [The occipital encephaloceles: report of 16 cases]. *Mali Med* **22**(2): 54–57.

Baccaglini L, Atkinson JC, Patton LL, et al. (2007). Management of oral lesions in HIV-positive patients. *Oral Surg Oral Med Oral Pathol Oral Radiol Endod* **103**(Suppl. S50): 1–23.

Baghagho EE, Kharboush IF, El-Kaffash DM, et al. (2004). Maternal serum alpha fetoprotein among pregnant females in Alexandria. *J Egypt Public Health Assoc* **79**(1–2): 59–81.

Balogou AA, Grunitzky EK, Belo M, et al. (2007). Management of epilepsy patients in Batamariba district, Togo. *Acta Neurol Scand* **116**(4): 211–216.

Bassili A, Mokhtar SA, Dabous NI, et al. (2000). Risk factors for congenital heart diseases in Alexandria, Egypt. *Eur J Epidemiol* **16**(9): 805–814.

Beighton P and Botha MC (1986). Inherited disorders in the black population of southern Africa. Part I: Historical and demographic background; genetic haematological conditions. *S Afr Med J* **69** (4): 247–249.

Bekker LG, Haslett P, Maartens G, Steyn L, and Kaplan G (2000). Thalidomide-induced antigen-specific immune stimulation in patients with human immunodeficiency virus type 1 and tuberculosis. *J Infect Dis* **181**(3): 954–965.

Bello C and Whittle H (1991). Cytomegalovirus infection in Gambian mothers and their babies. *J Clin Pathol* **44**(5): 366–369.

Ben Abdallah R, Aoun K, Siala E, et al. (2009). Congenital toxoplasmosis: clinical and biological analysis of 11 cases in Tunisia. *Arch Pediatr* **16**(2): 118–121.

Ben Arab S, Masmoudi S, Beltaief N, Hachicha S, and Ayadi H (2004). Consanguinity and endogamy in Northern Tunisia and its impact on non-syndromic deafness. *Genet Epidemiol* **27**(1): 74–79.

Benedé JG, Oelofse A, van Stuijvenberg ME, et al. (1997). Endemic goitre in a rural community of KwaZulu-Natal. *S Afr Med J* **87**(3): 310–313.

Birbeck GL and Kalichi EM (2004). Epilepsy prevalence in rural Zambia: a door-to-door survey. *Trop Med Int Health* **9**(1): 92–95.

Birbeck GL, Chomba E, Atadzhanov M, Mbewe E, and Haworth A (2008). Women's experiences living with epilepsy in Zambia. *Am J Trop Med Hyg* **79**(2): 168–172.

Bittles AH and Black ML (2010). Evolution in health and medicine Sackler colloquium: consanguinity, human evolution, and complex diseases. *PNAS* **107**(Suppl. 1): 1779–1786.

Bittles AH and Mason WM, Greene J, and Rao NA (1991). Reproductive behavior and health in consanguineous marriages. *Science* **252**(5007): 789–794.

Blandford JM, Gift TL, Vasaikar S, et al. (2007). Cost-effectiveness of on-site antenatal screening to prevent congenital syphilis in rural eastern Cape Province, Republic of South Africa. *Sex Transm Dis* **34**(Suppl. 7): S61–S66.

Boano R, Fulcheri E, Martina MC, et al. (2009). Neural tube defect in a 4000-year-old Egyptian infant mummy: a case of meningocele from the museum of anthropology and ethnography of Turin (Italy). *Eur J Paediatr Neurol* **13**(6): 481–487.

Bos P, Steele D, Alexander J (1995). Prevalence of antibodies to rubella, herpes simplex 2 and cytomegalovirus in pregnant women and in neonates at Ga-Rankuwa. *Cent Afr J Med* **41**(1): 14–17.

Bouhsain S, Dami A, Elannaz H, et al. (2009). [A critical study of the screening practices of gestational diabetes of a service of gynecology and obstetrics]. *Ann Biol Clin (Paris)* **67**(2): 159–162.

Boulet M (1997). Micronutrient deficiencies: reports from the field – Africa. *Glob Impacts* **13**.

Brabin BJ, Prinsen-Geerligs PD, Verhoeff FH, et al. (2004). Haematological profiles of the people of rural southern Malawi: an overview. *Ann Trop Med Parasitol* **98**(1): 71–83.

Brink PA, Crotti L, Corfield V, et al. (2005). Phenotypic variability and unusual clinical severity of congenital long-QT syndrome in a founder population. *Circulation* **112**(17): 2602–2610.

Bronzan RN, Mwesigwa-Kayongo DC, Narkunas D, et al. (2007). On-site rapid antenatal syphilis screening with an immunochromatographic strip improves case detection and treatment in rural South African clinics. *Sex Transm Dis* **34**(Suppl. 7): S55–S60.

Bukar M, Audu BM, Takai UI, Ajayi BB, and Kullima AA (2009). Is routine antenatal screening for syphilis in Nigeria still justified clinically and economically? *Saudi Med J* **30**(10): 1311–1315.

Butler MG (2009). Genomic imprinting disorders in humans: a mini-review. *J Assist Reprod Genet* **26**(9–10): 477–486.

Bütow KW, van Wyk PJ, and Zwahlen RA (2007). Differences in the clinical appearances of white versus black patients with facial cleft deformities: a retrospective study of a South African clinic. *SADJ* **62**(7): 298, 300–304.

Canfield MA, Collins JS, Botto LD, et al. (2005). Changes in the birth prevalence of selected birth defects after grain fortification with folic acid in the United States: findings from a multi-state population-based study. *Birth Defects Res A* **73**: 679–689.

Castro O, Brambilla DJ, Thorington B, et al. (1994). The acute chest syndrome in sickle cell disease: incidence and risk factors. The Cooperative Study of Sickle Cell Disease. *Blood* **84**(2): 643–649.

CDC (2005). *Morbidity and Mortality Weekly Report* **64**(1).

Challis K, Melo A, Bugalho A, Jeppsson JO, and Bergström S (2002). Gestational diabetes mellitus and fetal death in Mozambique: an incident case-referent study. *Acta Obstet Gynecol Scand* **81** (6): 560–563.

Chelli D, Dimassi K, Chaabouni M, et al. (2008). [Prenatal diagnosis of trisomy 21: the Tunisian experience]. *Santé* **18**(4): 199–203.

Cherinet A and Kelbessa U (2000). Determinants of iodine deficiency in school children in different regions of Ethiopia. *East Afr Med J* **77**(3): 133–137.

Child A, Comeglio P, Arno G, and Beighton P (2007). Marfan syndrome in South Africa: a molecular genetic approach to diagnosis. *S Afr Med J* **97**(9): 845–847.

Christianson AL (1996). Down syndrome in sub-Saharan Africa. *J Med Genet* **33**(2): 89–92.

Christianson AL (1997). Down syndrome in black South African infants and children: clinical features and delayed diagnosis. *S Afr Med J* **87**(8): 992–995.

Christianson AL and Kromberg JG (1996). Maternal non-recognition of Down syndrome in black South African infants. *Clin Genet* **49**(3): 141–144.

Christianson AL and Modell B (2004). Medical genetics in developing countries. *Annu Rev Genomics Hum Genet* **5**: 219–265.

Christianson AL, Chesler N, and Kromberg JG (1994). Fetal valproate syndrome: clinical and neuro-developmental features in two sibling pairs. *Dev Med Child Neurol* **36**(4): 361–369.

Christianson AL, Zwane ME, Manga P, et al. (2002). Children with intellectual disability in rural South Africa: prevalence and associated disabilities. *J Intellect Disabil Res* **46**: 179–186.

Clegg JB and Weatherall DJ (1999). Thalassemia and malaria: new insights into an old problem. *Proc Assoc Am Physicians* **111**(4): 278–282.

Corcoran C and Hardie DR (2005). Seroprevalence of rubella antibodies among antenatal patients in the Western Cape. *S Afr Med J* **95**(9): 688–690.

Croxford J and Viljoen D (1999). Alcohol consumption by pregnant women in the Western Cape. *S Afr Med J* **89**: 962–965.

Dagher D and Ross E (2004). Approaches of South African traditional healers regarding the treatment of cleft lip and palate. *Cleft Palate Craniofac J* **41**(5): 461–469.

Dangor T and Ross E (2006). Western and traditional medicine: cultural beliefs and practices of South African Muslims with regard to Down syndrome. *S Afr J Commun Disord* **53**: 27–38.

De Montalembert M and Tshilolo L (2007). [Is therapeutic progress in the management of sickle cell disease applicable in sub-Saharan Africa?]. *Med Trop (Mars)* **67**(6): 612–616.

de Paul Djientcheu V, Njamnshi AK, Wonkam A, et al. (2008). Management of neural tube defects in a Sub-Saharan African country: the situation in Yaounde, Cameroon. *J Neurol Sci* **275**(1–2): 29–32.

Debrock C, Preux PM, Houinato D, et al. (2000). Estimation of the prevalence of epilepsy in the Benin region of Zinvié using the capture-recapture method. *Int J Epidemiol* **29**(2): 330–335.

Dent W, Helbok R, Matuja WB, Scheunemann S, and Schmutzhard E (2005). Prevalence of active epilepsy in a rural area in South Tanzania: a door-to-door survey. *Epilepsia* **46**(12): 1963–1969.

Deperthes BD, Meheus A, O'Reilly K, and Broutet N (2004). Maternal and congenital syphilis programmes: case studies in Bolivia, Kenya and South Africa. *Bull World Health Organ* **82**(6): 410–416.

Doehring E, Reiter-Owona I, Bauer O, et al. (1995). *Toxoplasma gondii* antibodies in pregnant women and their newborns in Dar es Salaam, Tanzania. *Am J Trop Med Hyg* **52**(6): 546–548.

Donkor P, Bankas DO, Agbenorku P, Plange-Rhule G, and Ansah SK (2007). Cleft lip and palate surgery in Kumasi, Ghana: 2001–2005. *J Craniofac Surg* **18**(6): 1376–1379.

Dörk T, El-Harith EH, Stuhrmann M, et al. (1998). Evidence for a common ethnic origin of cystic fibrosis mutation 3120+1G–>A in diverse populations. *Am J Hum Genet* **63**(2): 656–662.

Douglas TS and Mutsvangwa TE (2010). A review of facial image analysis for delineation of the facial phenotype associated with fetal alcohol syndrome. *Am J Med Genet A* **152A**(2): 528–536.

Duff P, Barth WH Jr., and Post MD (2009). Case records of the Massachusetts General Hospital: case 4-2009 – a 39-year-old pregnant woman with fever after a trip to Africa. *N Engl J Med* **360**(5): 508–516.

Edwards T, Scott AG, Munyoki G, et al. (2008). Active convulsive epilepsy in a rural district of Kenya: a study of prevalence and possible risk factors. *Lancet Neurol* **7**(1): 50–56.

Edwin F, Sereboe LA, Tettey MM, et al. (2010). Experience from a single centre concerning the surgical spectrum and outcome of adolescents and adults with congenitally malformed hearts in West Africa. *Cardiol Young* **4**: 1–6.

Egesie OJ, Joseph DE, Isiguzoro I, and Egesie UG (2008). Glucose-6-phosphate dehydrogenase (G6PD) activity and deficiency in a population of Nigerian males resident in Jos. *Niger J Physiol Sci* **23**(1–2): 9–11.

Ejim EC, Ike SO, Anisiuba BC, Onwubere BJ, and Ikeh VO (2009). Ventricular septal defects at the University of Nigeria Teaching Hospital, Enugu: a review of echocardiogram records. *Trans R Soc Trop Med Hyg* **103**(2): 159–161.

El Mansouri B, Rhajaoui M, Sebti F, et al. (2007). [Seroprevalence of toxoplasmosis in pregnant women in Rabat, Morocco]. *Bull Soc Pathol Exot* **100**(4): 289–290.

El-Gozamy BR, Mohamed SA, and Mansour HA (2009). Toxoplasmosis among pregnant women in Qualyobia Governorate, Egypt. *J Egypt Soc Parasitol* **39**(2): 389–401.

Elliott RF, Jovic G, and Beveridge M (2008). Seasonal variation and regional distribution of cleft lip and palate in Zambia. *Cleft Palate Craniofac J* **45**(5): 533–538.

Eltom A, Eltom M, Idris M, and Gebre-Medhin M (2001). Thyroid function in the newborn in relation to maternal thyroid status during labour in a mild iodine deficiency endemic area in Sudan. *Clin Endocrinol (Oxf)* **55**(4): 485–490.

Enevold A, Lusingu JP, Mmbando B, et al. (2008). Reduced risk of uncomplicated malaria episodes in children with alpha+-thalassemia in northeastern Tanzania. *Am J Trop Med Hyg* **78**(5): 714–720.

Essop MR and Nkomo VT (2005). Rheumatic and nonrheumatic valvular heart disease: epidemiology, management, and prevention in Africa. *Circulation* **112**(23): 3584–3591.

Faye M, Hennequin M, Yam AA, and Ba I (2004). [Evaluation of oral health and access to care in Senegalese children with Down syndrome: preliminary study]. *Dakar Med* **49**(1): 64–69.

Feuillet-Fieux MN, Ferrec M, Gigarel N, et al. (2004). Novel CFTR mutations in black cystic fibrosis patients. *Clin Genet* **65**(4): 284–287.

Fitzgerald DW, Behets FM, Lucet C, and Roberfroid D (1998). Prevalence, burden, and control of syphilis in Haiti's rural Artibonite region. *Int J Infect Dis* **2**(3): 127–131.

Föller M, Bobbala D, Koka S, et al. (2009). Suicide for survival: death of infected erythrocytes as a host mechanism to survive malaria. *Cell Physiol Biochem* **24**(3–4): 133–140.

Friis H, Gomo E, Nyazema N, et al. (2004). Effect of multimicronutrient supplementation on gestational length and birth size: a randomized, placebo-controlled, double-blind effectiveness trial in Zimbabwe. *Am J Clin Nutr* **80**(1): 178–184.

George AO, Ogunbiyi AO, Daramola OO, and Campbell OB (2005). Albinism among Nigerians with malignant melanoma. *Trop Doct* **35**(1): 55–56.

Ghebrekidan H, Rudén U, Cox S, Wahren B, and Grandien M (1999). Prevalence of herpes simplex virus types 1 and 2, cytomegalovirus, and varicella-zoster virus infections in Eritrea. *J Clin Virol* **12**(1): 53–64.

Gibson RS, Abebe Y, Stabler S, et al. (2008). Zinc, gravida, infection, and iron, but not vitamin B-12 or folate status, predict hemoglobin during pregnancy in Southern Ethiopia. *J Nutr* **138**(3): 581–586.

Gichangi P, Renterghem LV, Karanja J, et al. (2004). Congenital syphilis in a Nairobi maternity hospital. *East Afr Med J* **81**(11): 589–593.

Gladwin MT and Vichinsky E (2008). Pulmonary complications of sickle cell disease. *N Engl J Med* **359**(21): 2254–2265.

Goldman A, Labrum R, Claustres M, et al. (2001). The molecular basis of cystic fibrosis in South Africa. *Clin Genet* 59(1): 37–41.

Guidozzi F, Ballot D, and Rothberg AD (1994). Human B19 parvovirus infection in an obstetric population: a prospective study determining fetal outcome. *J Reprod Med* 39(1): 36–38.

Hampshire KR and Smith MT (2001). Consanguineous marriage among the Fulani. *Hum Biol* 73(4): 597–603.

Hardy BJ, Séguin B, Ramesar R, Singer PA, and Daar AS (2008). South Africa: from species cradle to genomic applications. *Nat Rev Genet* 9(Suppl. 1): S19–S23.

Hodges AM and Hodges SC (2000). A rural cleft project in Uganda. *Br J Plast Surg* 53(1): 7–11.

Holmes LB, Harvey EA, Coull BA, et al. (2001). The teratogenicity of anticonvulsant drugs. *N Engl J Med* 344(15): 1132–1138.

Hong ES, Zeeb H, and Repacholi MH (2006). Albinism in Africa as a public health issue. *BMC Public Health* 6: 212.

Horjus P, Aguayo VM, Roley JA, Pene MC, and Meershoek SP (2005). School-based iron and folic acid supplementation for adolescent girls: findings from Manica Province, Mozambique. *Food Nutr Bull* 26(3): 281–286.

Hoyme HE, May PA, Kalberg WO, et al. (2005). Pediatrics: a practical clinical approach to diagnosis of fetal alcohol spectrum disorders – clarification of the 1996 Institute of Medicine criteria. *Pediatrics* 115(1): 39–47.

Huddle KR (2005). Audit of the outcome of pregnancy in diabetic women in Soweto, South Africa, 1992–2002. *S Afr Med J* 95(10): 789–794.

Huddle KR, England M, and Nagar A (1993). Outcome of pregnancy in diabetic women in Soweto, South Africa 1983–1992. *Diabet Med* 10(3): 290–294.

Huff-Rousselle M, Simooya O, Kabwe V, et al. (2007). Pharmacovigilance and new essential drugs in Africa: Zambia draws lessons from its own experiences and beyond. *Glob Public Health* 2(2): 184–203.

Idowu OE and Apemiye RA (2008). Outcome of myelomeningocoele repair in sub-Saharan Africa: the Nigerian experience. *Acta Neurochir (Wien)* 150(9): 911–913.

Imperato GH and Imperato PJ (2006). Beliefs and practices concerning twins, hermaphrodites, and albinos among the Bamana and Maninka of Mali. *J Community Health* 31(3): 198–224.

IOM (Institute of Medicine) (2003). *Reducing Birth Defects. Meeting the Challenge in the Developing World.* National Academy Press.

IOM. Howson CP, Kennedy ET, and Horwitz A, eds. (1998). *Prevention of Micronutrient Deficiencies: Tools for Policymakers and Public Health Workers.* National Academy Press.

Jalloh A, Jalloh M, Gamanga I, et al. (2008). G6PD deficiency assessment in Freetown, Sierra Leone, reveals further insight into the molecular heterogeneity of G6PD A–. *J Hum Genet* 53(7): 675–679.

Jenkins T (1990). Medical genetics in South Africa. *J Med Genet* 27: 760–779.

Jenniskens F, Obwaka E, Kirisuah S, et al. (1995). Syphilis control in pregnancy: decentralization of screening facilities to primary care level, a demonstration project in Nairobi, Kenya. *Int J Gynaecol Obstet* 48(Suppl.): S121–S128.

Jooste PL (2005). Sources of household salt in South Africa. *Asia Pac J Clin Nutr* 14(2): 159–162.

Jooste PL, Weight MJ, and Lombard CJ (2000). Short-term effectiveness of mandatory iodization of table salt, at an elevated iodine concentration, on the iodine and goiter status of schoolchildren with endemic goiter. *Am J Clin Nutr* 71(1): 75–80.

Kaestel P, Michaelsen KF, Aaby P, and Friis H (2005). Effects of prenatal multimicronutrient supplements on birth weight and perinatal mortality: a randomised, controlled trial in Guinea-Bissau. *Eur J Clin Nutr* 59(9): 1081–1089.

Kalk WJ (1998). Iodine deficiency disorders in South Africa. *S Afr Med J* 88(3): 352–354.

Kamanga J (1995). Maternal and congenital syphilis in Zambia. *Afr Health* 18(1): 15–17.

Kamau-Mbuthia E and Elmadfa I (2007). Diet quality of pregnant women attending an antenatal clinic in Nakuru, Kenya. *Ann Nutr Metab* **51**(4): 324–330.

Kariuki JG, Joshi MD, Adam AM, Kwasa TO, and Machoki M (2008). Fertility rate of epileptic women at Kenyatta National Hospital. *East Afr Med J* **85**(7): 341–346.

Kaur R, Gupta N, Nair D, Kakkar M, and Mathur MD (1999). Screening for TORCH infections in pregnant women: a report from Delhi. *Southeast Asian J Trop Med Public Health* **30**(2): 284–286.

Kavishe F (1997). Salt iodization programmes in Africa. *SCN News* **15**: 10.

Khoury MJ, Becerra JE, Coredo JF, and Erickson JD (1989). Clinical-epidemiological assessment of pattern of birth defects associated with human teratogen: application to diabetic embryopathy. *Pediatrics* **84**: 658–665.

Kinsley B (2007). Achieving better outcomes in pregnancies complicated by type 1 and type 2 diabetes mellitus. *Clin Ther* **29**(Suppl. D): S153–S160.

Koko J, Dufillot D, M'Ba-Meyo J, Gahouma D, and Kani F (1998). Mortality of children with sickle cell disease in a pediatric department in Central Africa. *Arch Pediatr* **5**(9): 965–969.

Komba AN, Makani J, Sadarangani M, et al. (2009). Malaria as a cause of morbidity and mortality in children with homozygous sickle cell disease on the coast of Kenya. *Clin Infect Dis* **49**(2): 216–222.

Kromberg JGR (1992). Albinism in the South African Negro: IV: attitudes and the death myth. *Birth Defects* **28**: 159–166.

Kromberg JGR, Castle D, Zwane EM, and Jenkins T (1989). Albinism and skin cancer in southern Africa. *Clinical Genetics* **36**: 43–52.

Labie D, Pagnier J, Lapoumeroulie C, et al. (1985). Common haplotype dependency of high G gamma-globin gene expression and high Hb F levels in beta-thalassemia and sickle cell anemia patients. *PNAS* **82**: 2111–2114.

Lawrenson JB, Kalis NN, Pribut H, et al. (2006). Why are some South African children with Down syndrome not being offered cardiac surgery? *S Afr Med J* **96**(9 Pt 2): 914–919.

Lech MM (2003). Non-effective partner notification system: a missed opportunity for the reduction of sexually transmitted infections in sub-Saharan Africa. *Med Wieku Rozwoj* **7**(4 Pt 1): 503–509.

Lekwuwa GU, Adewole IF, and Thompson MO (1995). Antiepileptic drugs and teratogenicity in Nigerians. *Trans R Soc Trop Med Hyg* **89**(2): 227.

Levin CE, Steele M, Atherly D, et al. (2007). Analysis of the operational costs of using rapid syphilis tests for the detection of maternal syphilis in Bolivia and Mozambique. *Sex Transm Dis* **34**(Suppl. 7): S47–S54.

Lindblom D (2009). [Heart surgery in Sudan: experiences from a highly specialized cardiac surgical centre managed by an Italian humanitarian organization]. *Lakartidningen* **106**(38): 2370–2373.

Loua A, Lamah MR, Haba NY, and Camara M (2007). Frequency of blood groups ABO and rhesus D in the Guinean population. *Transfus Clin Biol* **14**(5): 435–439.

Louw B, Shibambu M, and Roemer K (2006). Facilitating cleft palate team participation of culturally diverse families in South Africa. *Cleft Palate Craniofac J* **43**(1): 47–54.

Lund PM (2005). Oculocutaneous albinism in southern Africa: population structure, health and genetic care. *Ann Hum Biol* **32**(2): 168–173.

Lund PM and Taylor JS (2008). Lack of adequate sun protection for children with oculocutaneous albinism in South Africa. *BMC Public Health* **8**: 225.

Lund PM, Maluleke TG, Gaigher I, and Gaigher MJ (2007). Oculocutaneous albinism in a rural community of South Africa: a population genetic study. *Ann Hum Biol* **34**(4): 493–497.

Lupski JR, Reid JG, and Gonzaga-Jauregui C, et al. (2010). Whole-genome sequencing in a patient with Charcot-Marie-Tooth neuropathy. *N Engl J Med* **362**(13): 1181–1191.

Lutale JK, Justesen A, Lema RS, Swai AB, and McLarty DG (1991). Outcome of pregnancy in diabetic patients in Dar es Salaam, Tanzania. *Diabet Med* **8**(9): 881–884.

Luzzatto L, Mehta A, and Meloni T (1995). Haemoglobinuria and haptoglobin in G6PD deficiency. *Br J Haematol* **91**(2): 511–512.

Macek M Jr., Mackova A, Hamosh A, et al. (1997). Identification of common cystic fibrosis mutations in African-Americans with cystic fibrosis increases the detection rate to 75%. *Am J Hum Genet* **60**(5): 1122–1127.

Mahmoud H, El-Haddad A, Fahmy O, et al. (2008). Hematopoietic stem cell transplantation in Egypt. *Bone Marrow Transplant* **42**(Suppl. 1): S76–S80.

Makani J, Williams TN, and Marsh K (2007). Sickle cell disease in Africa: burden and research priorities. *Ann Trop Med Parasitol* **101**(1): 3–14.

Makani J, Menzel S, Nkya S, et al. (2011a). Genetics of fetal hemoglobin in Tanzanian and British patients with sickle cell anemia. *Blood* **117**(4): 1390–1392.

Makani J, Cox SE, Soka D, et al. (2011b). Mortality in sickle cell anemia in Africa: a prospective cohort study in Tanzania. *PLoS One* **6**(2): e14699.

Makola D, Ash DM, Tatala SR, et al. (2003). A micronutrient-fortified beverage prevents iron deficiency, reduces anemia and improves the hemoglobin concentration of pregnant Tanzanian women. *J Nutr* **133**(5): 1339–1346.

Malik SJ and Mir NA (1992). Perinatal mortality in high risk pregnancy: a prospective study of preventable factors. *Asia Oceania J Obstet Gynaecol* **18**(1): 45–48.

Mamabolo RL, Alberts M, Levitt NS, Delemarre-van de Waal HA, and Steyn NP (2007). Prevalence of gestational diabetes mellitus and the effect of weight on measures of insulin secretion and insulin resistance in third-trimester pregnant rural women residing in the Central Region of Limpopo Province, South Africa. *Diabet Med* **24**(3): 233–239.

Mansour H, Klei L, Wood J, et al. (2009). Consanguinity associated with increased risk for bipolar I disorder in Egypt. *Am J Med Genet B Neuropsychiatr Genet* **150B**(6): 879–885.

Marais S, Wilkinson RJ, Pepper DJ, and Meintjes G (2009). Management of patients with the immune reconstitution inflammatory syndrome. *Curr HIV/AIDS Rep* **6**(3): 162–171.

March of Dimes Global (2001). *Cytomegalovirus in Pregnancy: Fact Sheet.* Dimes Birth Defects Foundation.

March of Dimes Global (2002). *Sexually Transmitted Infections in Pregnancy: Fact Sheet.* Dimes Birth Defects Foundation.

March of Dimes Global (2004). *Folic Acid: Fact Sheet.* Dimes Birth Defects Foundation.

March of Dimes Global (2006). *The March of Dimes Global Report on Birth Defects: The Hidden Toll of Dying and Disabled Children.* Dimes Birth Defects Foundation.

Markenson G and Yancey M (1998). Parvovirus B19 infections in pregnancy. *Obstet Gynecol* **91**: 125–128.

Martines J, Paul VK, Bhutta ZA, et al. (2005). Lancet Neonatal Survival Steering Team. Neonatal survival: a call for action. *Lancet* **365**(9465): 1189–1197.

Mashako L, Preziosi P, Nsibu C, et al. (1991). Iron and folate status in Zairian mothers and their newborns. *Ann Nutr Metab* **35**(6): 309–314.

Matolweni LO, Bardien S, Rebello G, et al. (2006). Arrhythmogenic right ventricular cardiomyopathy type 6 (ARVC6), support for the locus assignment, narrowing of the critical region and mutation screening of three candidate genes. *BMC Med Genet* **7**: 29.

May PA, Brooke L, Gossage JP, et al. (2000). Epidemiology of fetal alcohol syndrome in a South African community in the Western Cape Province. *Am J Public Health* **90**(12): 1905–1912.

Mayosi BM, Flisher AJ, Lalloo UG, et al. (2009). The burden of non-communicable diseases in South Africa. *Lancet* **374**(9693): 934–947.

McBride SR and Leppard BJ (2002). Attitudes and beliefs of an albino population toward sun avoidance: advice and services provided by an outreach albino clinic in Tanzania. *Arch Dermatol* **138**(5): 629–632.

McIlroy SP, Dynan KB, Lawson JT, Patterson CC, and Passmore AP (2002). Moderately elevated plasma homocysteine, methylenetetrahydrofolate reductase genotype, and risk for stroke, vascular dementia, and Alzheimer disease in Northern Ireland. *Stroke* **33**(10): 2351–2356.

Melse-Boonstra A, Rozendaal M, Rexwinkel H, et al. (1998). Determination of discretionary salt intake in rural Guatemala and Benin to determine the iodine fortification of salt required to control iodine deficiency disorders: studies using lithium-labeled salt. *Am J Clin Nutr* **68**(3): 636–641.

Menzel S and Thein SL (2009). Genetic architecture of hemoglobin F control. *Curr Opin Hematol* **16**: 179–186.

Mogayzel PJ Jr. and Flume PA (2010). Update in cystic fibrosis 2009. *Am J Respir Crit Care Med* **181** (6): 539–544.

Mokhtar MM and Abdel-Fattah MM (2001). Consanguinity and advanced maternal age as risk factors for reproductive losses in Alexandria, Egypt. *Eur J Epidemiol* **17**(6): 559–565.

Mokhtar MM and Abdel-Fattah MM (2002). Major birth defects among infants with Down's syndrome in Alexandria, Egypt (1995–2000). *J Trop Pediatr* **48**(4): 247–249.

Morgan MA, el-Ghany SM, Khalifa NA, Sherif A, and Rasslan LR (2003). Prevalence of cytomegalovirus (CMV) infection among neonatal intensive care unit (NICU) and healthcare workers. *Egypt J Immunol* **10**(2): 1–8.

Mullick S, Broutet N, Htun Y, Temmerman M, and Ndowa F (2004). Controlling congenital syphilis in the era of HIV/AIDS. *Bull World Health Organ* **82**(6): 431–432.

Mwangi J (1999). Blood group distribution in an urban population of patient targeted blood donors. *East Afr Med J* **76**(11): 615–618.

Nabias R, Ngouamizokou A, Migot-Nabias F, Mbou-Moutsimbi RA, and Lansoud-Soukate J (1998). [Serological investigation of toxoplasmosis in patients of the M.I.P. center of Franceville (Gabon)]. *Bull Soc Pathol Exot* **91**(4): 318–320.

Nagel RL and Platt OS (2001). General pathophysiology of sickle cell anemia. In Steinberg MH, Forget BG, Higgs DR, and Nagel RL, eds., *Disorders of Hemoglobin: Genetics, Pathophysiology, and Clinical Management*. Cambridge University Press, pp. 494–526.

Nahmias AJ, Lee FK, and Beckman-Nahmias S (1990). Sero-epidemiological and -sociological patterns of herpes simplex virus infection in the world. *Scand J Infect Dis Suppl* **69**: 19–36.

Ndir I, Gaye A, Faye B, Gaye O, and Ndir O (2004). [Seroprevalence of toxoplasmosis among women having spontaneous abortion and pregnant women following in a center of health up-town in Dakar]. *Dakar Med* **49**(1): 5–9.

Ndoye NF, Sow AD, Diop AG, et al. (2005). Prevalence of epilepsy: its treatment gap and knowledge, attitude and practice of its population in sub-urban Senegal – an ILAE/IBE/WHO study. *Seizure* **14**(2): 106–111.

Ndumbe PM, Andela A, Nkemnkeng-Asong J, Watonsi E, and Nyambi P (1992). Prevalence of infections affecting the child among pregnant women in Yaounde, Cameroon. *Med Microbiol Immunol* **181**(3): 127–130.

Negash T, Tilahun G, and Medhin G (2008). Seroprevalence of *Toxoplasma gondii* in Nazaret town, Ethiopia. *East Afr J Public Health* **5**(3): 211–214.

Ngo DB, Dikassa L, Okitolonda W, et al. (1997). Selenium status in pregnant women of a rural population (Zaire) in relationship to iodine deficiency. *Trop Med Int Health* **2**(6): 572–581.

Nichols FT, Jones AM, and Adams RJ (2001). Stroke prevention in sickle cell disease (STOP) study guidelines for transcranial Doppler testing. *J Neuroimaging* **11**(4): 354–362.

Njamnshi AK, Djientcheu Vde P, Lekoubou A, et al. (2008). Neural tube defects are rare among black Americans but not in sub-Saharan black Africans: the case of Yaounde – Cameroon. *J Neurol Sci* **270**(1–2): 13–17.

Nouira S, Kamoun I, Ouragini H, et al. (2008). Clinical and genetic investigation of atrial septal defect with atrioventricular conduction defect in a large consanguineous Tunisian family. *Arch Med Res* **39**(4): 429–433.

Nwauche CA and Ejele OA (2004). ABO and rhesus antigens in a cosmopolitan Nigeria population. *Niger J Med* **13**(3): 263–266.

Obed SA, Armah JO, and Wilson JB (1995). Advanced maternal age and pregnancy. *West Afr J Med* **14**(4): 198–201.

Oberle MW, Rosero-Bixby L, Lee FK, et al. (1989). Herpes simplex virus type 2 antibodies: high prevalence in monogamous women in Costa Rica. *Am J Trop Med Hyg* **41**(2): 224–229.

Odunvbun ME, Okolo AA, and Rahimy CM (2008). Newborn screening for sickle cell disease in a Nigerian hospital. *Public Health* **122**(10): 1111–1116.

Ogendo SW (1999). Pregnancy in open heart surgical patients at Kenyatta National Hospital. *East Afr Med J* **76**(1): 19–22.

Ojule AC and Osotimehin BO (1998). Maternal and neonatal thyroid status in Saki, Nigeria. *Afr J Med Med Sci* **27**(1–2): 57–61.

Okoromah CA, Ekure EN, Ojo OO, Animasahun BA, and Bastos MI (2008). Structural heart disease in children in Lagos: profile, problems and prospects. *Niger Postgrad Med J* **15**(2): 82–88.

Olasoji HO, Ugboko VI, and Arotiba GT (2007). Cultural and religious components in Nigerian parents' perceptions of the aetiology of cleft lip and palate: implications for treatment and rehabilitation. *Br J Oral Maxillofac Surg* **45**(4): 302–305.

Oloyede OA (2008). Down syndrome screening in Nigeria. *Int J Gynaecol Obstet* **100**(1): 88–89.

Onadeko MO, Joynson DH, Payne RA, and Francis J (1996). The prevalence of toxoplasma antibodies in pregnant Nigerian women and the occurrence of stillbirth and congenital malformation. *Afr J Med Med Sci* **25**(4): 331–334.

Onah II, Opara KO, Olaitan PB, and Ogbonnaya IS (2008). Cleft lip and palate repair: the experience from two West African sub-regional centres. *J Plast Reconstr Aesthet Surg* **61**(8): 879–882.

Onuigbo WI (2006). Epidemiology of skin cancer arisen from the burn scars in Nigerian Ibos. *Burns* **32**(5): 602–604.

Onwubalili JK (1983). Sickle-cell anaemia: an explanation for the ancient myth of reincarnation in Nigeria. *Lancet* **2**(8348): 503–505.

Onwuekwe IO, Onodugo OD, Ezeala-Adikaibe B, et al. (2009). Pattern and presentation of epilepsy in Nigerian Africans: a study of trends in the southeast. *Trans R Soc Trop Med Hyg* **103**(8): 785–789.

Orji EO and Ndububa VI (2004). Obstetric performance of women aged over forty years. *East Afr Med J* **81**(3): 139–141.

Owa JA (1989). Relationship between exposure to icterogenic agents, glucose-6-phosphate dehydrogenase deficiency and neonatal jaundice in Nigeria. *Acta Paediatr Scand* **78**(6): 848–852.

Ozumba BC, Obi SN, and Oli JM (2004). Diabetes mellitus in pregnancy in an African population. *Int J Gynaecol Obstet* **84**(2): 114–119.

Padoa C, Goldman A, Jenkins T, and Ramsay M (1999). Cystic fibrosis carrier frequencies in populations of African origin. *J Med Genet* **36**(1): 41–44.

Pappas G, Roussos N, and Falagas ME (2009). Toxoplasmosis snapshots: global status of *Toxoplasma gondii* seroprevalence and implications for pregnancy and congenital toxoplasmosis. *Int J Parasitol* **39**(12): 1385–1394.

Parker M, Corrigall AV, Hift RJ, and Meissner PN (2008). Molecular characterization of erythropoietic protoporphyria in South Africa. *Br J Dermatol* **159**(1): 182–191.

Peadon E, Fremantle E, Bower C, and Elliott EJ (2008). International survey of diagnostic services for children with fetal alcohol spectrum disorders. *BMC Pediatr* **8**: 12.

Penchaszadeh VB (2002). Preventing congenital anomalies in developing countries. *Community Genet* **5**: 61–69.

Peprah E and Wonkam A (2013). Biomedical research, a tool to address the health issues that affect African populations. *Global Health* **9**: 50.

Pham AM and Tollefson TT (2007). Cleft deformities in Zimbabwe, Africa: socioeconomic factors, epidemiology, and surgical reconstruction. *Arch Facial Plast Surg* 9(6): 385–391.
Piel FB, Patil AP, Howes RE, et al. (2013). Global epidemiology of sickle haemoglobin in neonates: a contemporary geostatistical model-based map and population estimates. *Lancet* 381(9861): 142–151.
Platt OS, Thorington BD, Brambilla DJ, et al. (1991). Pain in sickle cell disease: rates and risk factors. *N Engl J Med* 325(1): 11–16.
Platt OS, Brambilla BJ, Rosse WF, et al. (1994). Mortality in sickle cell disease: life expectancy and risk factors for early death. *N Engl J Med* 330: 1639–1644.
Preux PM and Druet-Cabanac M (2005). Epidemiology and aetiology of epilepsy in sub-Saharan Africa. *Lancet Neurol* 4(1): 21–31.
Preux PM, Tiemagni F, Fodzo L, et al. (2000). Antiepileptic therapies in the Mifi Province in Cameroon. *Epilepsia* 41(4): 432–439.
Prischich F, De Rinaldis M, Bruno F, et al. (2008). High prevalence of epilepsy in a village in the Littoral Province of Cameroon. *Epilepsy Res* 82(2–3): 200–210.
Rahimy MC, Gangbo A, Ahouignan G, and Alihonou E (2009). Newborn screening for sickle cell disease in the Republic of Benin. *J Clin Pathol* 62(1): 46–48.
Ramakrishnan U (2002). Prevalence of micronutrient malnutrition worldwide. *Nutr Rev* 60(5 Pt 2): S46–S52.
Ratbi I, Génin E, Legendre M, et al. (2008). Cystic fibrosis carrier frequency and estimated prevalence of the disease in Morocco. *J Cyst Fibros* 7(5): 440–443.
Roberfroid D, Huybregts L, Lanou H, et al. (2008). MISAME Study Group. Effects of maternal multiple micronutrient supplementation on fetal growth: a double-blind randomized controlled trial in rural Burkina Faso. *Am J Clin Nutr* 88(5): 1330–1340.
Rodier MH, Berthonneau J, Bourgoin A, et al. (1995). Seroprevalences of Toxoplasma, malaria, rubella, cytomegalovirus, HIV and treponemal infections among pregnant women in Cotonou, Republic of Benin. *Acta Trop* 59(4): 271–277.
Romoren M and Rahman M (2006). Syphilis screening in the antenatal care: a cross-sectional study from Botswana. *BMC Int Health Hum Rights* 6: 8.
Rosenthal J, Christianson A, and Cordero J (2005). Fetal alcohol syndrome prevention in South Africa and other low-resource countries. *Am J Public Health* 95(7): 1099–1101.
Ross E (2007). A tale of two systems: beliefs and practices of South African Muslim and Hindu traditional healers regarding cleft lip and palate. *Cleft Palate Craniofac J* 44(6): 642–648.
Rydzak CE and Goldie SJ (2008). Cost-effectiveness of rapid point-of-care prenatal syphilis screening in sub-Saharan Africa. *Sex Transm Dis* 35(9): 775–784.
Saboye J (1999). Plastic surgery training missions in developing countries. A 10-year experience at missions in Mali. *Ann Chir Plast Esthet* 44(1): 35–40.
Saka B, Mouhari-Touré A, Kombaté K, et al. (2009). [Scattered papules in three Togolese children from a consanguineous marriage: epidermodysplasia verruciformis]. *Med Trop* 69(3): 293–294.
Salako AA, Iyaniwura CA, Jeminusi OA, and Sofowora R (2006). Sexual behaviour, contraception and fertility among in-school adolescents in Ikenne Local Government, south-western Nigeria. *Niger J Clin Pract* 9(1): 26–36.
Saloojee H, Velaphi S, Goga Y, et al. (2004). The prevention and management of congenital syphilis: an overview and recommendations. *Bull World Health Organ* 82(6): 424–430.
SAMHSA (Substance Abuse and Mental Health Services Administration) (2005). *The Language of Fetal Alcohol Spectrum Disorders*. Center for Substance Abuse Prevention.
Sani MU, Mukhtar-Yola M, and Karaye KM (2007). Spectrum of congenital heart disease in a tropical environment: an echocardiography study. *J Natl Med Assoc* 99(6): 665–669.
Sareli P, England MJ, Berk MR, et al. (1989). Maternal and fetal sequelae of anticoagulation during pregnancy in patients with mechanical heart valve prostheses. *Am J Cardiol* 63(20): 1462–1465.

Sayed AR, Nixon J, and Bourne DE (2006). Folic acid awareness among women of reproductive age in Cape Town. *S Afr Med J* **96**(1): 12.

Sayed AR, Bourne D, Pattinson R, Nixon J, and Henderson B (2008). Decline in the prevalence of neural tube defects following folic acid fortification and its cost-benefit in South Africa. *Birth Defects Res A Clin Mol Teratol* **82**(4): 211–216.

Seashore MR and Wappner RS (1996). *Genetics in Primary Care & Clinical Medicine*. Prentice-Hall.

Seck BC and Jackson RT (2009). Providing iron/folic acid tablets free of charge improves compliance in pregnant women in Senegal. *Trans R Soc Trop Med Hyg* **103**(5): 485–492.

Serjeant GR (2005). Mortality from sickle cell disease in Africa. *BMJ* **330**(7489): 432–433.

Seyoum B, Kiros K, Haileselase T, and Leole A (1999). Prevalence of gestational diabetes mellitus in rural pregnant mothers in northern Ethiopia. *Diabetes Res Clin Pract* **46**(3): 247–251.

Shaaban AA, Abdel-Raoof TA, and Mohamed MA (1995). Role of cytomegalovirus in inducing congenital anomalies. *J Egypt Public Health Assoc* **70**(3–4): 343–355.

Shin M, Besser LM, Siffel C, et al. (2010). Prevalence of spina bifida among children and adolescents in 10 regions in the United States. *Pediatrics* **126**(2): 274–279.

Shorvon SD and Farmer PJ (1988). Epilepsy in developing countries: a review of epidemiological, sociocultural, and treatment aspects. *Epilepsia* **29**(Suppl. 1): S36–S54.

Simpore J, Savadogo A, Ilboudo D, et al. (2006). *Toxoplasma gondii*, HCV, and HBV seroprevalence and co-infection among HIV-positive and -negative pregnant women in Burkina Faso. *J Med Virol* **78**(6): 730–733.

Sodeinde O, Chan MC, Maxwell SM, Familusi JB, and Hendrickse RG (1995). Neonatal jaundice, aflatoxins and naphthols: report of a study in Ibadan, Nigeria. *Ann Trop Paediatr* **15**(2): 107–113.

Spinelli E, Seia M, Melotti P, et al. (2009). CFTR mutation in an Arab patient: clinical and functional features of 875+1G->A/875+1G->A genotype. *J Cyst Fibros* **8**(4): 282–284.

Spritz RA, Arnold TD, Buonocore S, et al. (2007). Distribution of orofacial clefts and frequent occurrence of an unusual cleft variant in the Rift Valley of Kenya. *Cleft Palate Craniofac J* **44**(4): 374–377.

Stefansson H, Rujescu D, Cichon S, et al. (2008). Large recurrent microdeletions associated with schizophrenia. *Nature* **455**(7210): 232–236.

Stegmann BJ and Carey JC (2002). TORCH infections. *Curr Women Health Reports* **2**: 253–258.

Streetly A, Latinovic R, Hall K, and Henthorn J (2009). Implementation of universal newborn bloodspot screening for sickle cell disease and other clinically significant haemoglobinopathies in England: screening results for 2005–7. *J Clin Pathol* **62**(1): 26–30.

Sulafa KM and Karani Z (2007). Diagnosis, management and outcome of heart disease in Sudanese patients. *East Afr Med J* **84**(9): 434–440.

Suleiman AM, Hamzah ST, Abusalab MA, and Samaan KT (2005). Prevalence of cleft lip and palate in a hospital-based population in the Sudan. *Int J Paediatr Dent* **15**(3): 185–189.

Tantchou Tchoumi JC and Butera G (2009). Rheumatic valvulopathies occurrence, pattern and follow-up in rural area: the experience of the Shisong Hospital, Cameroon. *Bull Soc Pathol Exot* **102**(3): 155–158.

Tattah EB and Ekere MC (1975). Death patterns in sickle cell anemia. *JAMA* **233**(8): 889–890.

Taye K and Bedru A (2009). Pattern of neural tube defects at Tikur Anbessa Hospital, Addis Ababa, Ethiopia. *Ethiop Med J* **47**(1): 71–76.

Tebeu PM, Kouam L, and Nghonguia MF, et al. (2007). [Historical cohort study of delivery in women of forty years of age and older]. *Rev Med Liege* **62**(7–8): 509–514.

Tekle-Haimanot R, Forsgren L, and Ekstedt J (1997). Incidence of epilepsy in rural central Ethiopia. *Epilepsia* **38**(5): 541–546.

Tippett B, Wilson J, Shepherd J, Cutler R, and Webster P (2008). Operation open heart – Rwanda. *Heart Lung Circ* **17**(Suppl. 4): S82–S83.

Tipping AJ, Pearson T, Morgan NV, et al. (2001). Molecular and genealogical evidence for a founder effect in Fanconi anemia families of the Afrikaner population of South Africa. *PNAS* **98**(10): 5734–5739.

Tita AT, Selwyn BJ, Waller DK, Kapadia AS, and Dongmo S (2005). Evidence-based reproductive health care in Cameroon: population-based study of awareness, use and barriers. *Bull World Health Organ* **83**(12): 895–903.

Toledo-Pereyra LH (2010). Heart transplantation. *J Invest Surg* **23**(1): 1–5.

Turnpenny P and Ellard S (2007). *Emery's Elements of Medical Genetics*, 13th edn. Churchill Livingstone.

Uda M, Galanello R, Sanna S, et al. (2008). Genome-wide association study shows BCL11A associated with persistent fetal hemoglobin and amelioration of the phenotype of beta-thalassemia. *PNAS* **105**(5): 1620–1625.

UNICEF (1998). *The State of the World's Children*. UNICEF.

UNICEF (2000). *The State of the World's Children*. UNICEF.

UNICEF (2003). *The State of the World's Children*. UNICEF.

Valodia PN, Seymour MA, Kies BM, and Folb PI (1999). Optimization of phenytoin therapy in adults with epilepsy in the Western Cape, South Africa. *J Clin Pharm Ther* **24**(5): 381–385.

van den Broek NR and Letsky EA (2000). Etiology of anemia in pregnancy in south Malawi. *Am J Clin Nutr* **72**(Suppl. 1): S247–S256.

Vanderjagt DJ, Brock HS, Melah GS, et al. (2007). Nutritional factors associated with anaemia in pregnant women in northern Nigeria. *J Health Popul Nutr* **25**(1): 75–81.

van't Veer T, Meester H, Poenaru D, et al. (2008). Quality of life for families with spina bifida in Kenya. *Trop Doct* **38**(3): 160–162.

Venter PA, Christianson AL, Hutamo CM, Makhura MP, and Gericke GS (1995). Congenital anomalies in rural black South African neonates: a silent epidemic? *S Afr Med J* **85**(1): 15–20.

Vento S, Cainelli F, and Cesario F (2006). Infections and thalassaemia. *Lancet Infect Dis* **6**(4): 226–233.

Viljoen DL, Carr LG, Foroud TM, et al. (2001). Alcohol dehydrogenase-2*2 allele is associated with decreased prevalence of fetal alcohol syndrome in the mixed-ancestry population of the Western Cape Province, South Africa. *Alcohol Clin Exp Res* **25**(12): 1719–1722.

Vissers LE, de Vries BB, and Veltman JA (2009). Genomic microarrays in mental retardation: from CNV to gene, from research to diagnosis. *J Med Genet* **47**: 289–297.

Walker DG and Walker GJ (2002). Forgotten but not gone: the continuing scourge of congenital syphilis. *Lancet Infect Dis* **2**(7): 432–436.

Walker GJ and Walker DG (2007). Congenital syphilis: a continuing but neglected problem. *Semin Fetal Neonatal Med* **12**(3): 198–206.

Wanjeri JK and Wachira JM (2009). Cleft lip and palate: a descriptive comparative, retrospective, and prospective study of patients with cleft deformities managed at 2 hospitals in Kenya. *J Craniofac Surg* **20**(5): 1352–1355.

Warf BC (2005). Comparison of endoscopic third ventriculostomy alone and combined with choroid plexus cauterization in infants younger than 1 year of age: a prospective study in 550 African children. *J Neurosurg* **103**(Suppl. 6): 475–481.

Warf BC and Campbell JW (2008). Combined endoscopic third ventriculostomy and choroid plexus cauterization as primary treatment of hydrocephalus for infants with myelomeningocele: long-term results of a prospective intent-to-treat study in 115 East African infants. *J Neurosurg Pediatr* **2**(5): 310–316.

Warf B, Ondoma S, Kulkarni A, et al. (2009). Neurocognitive outcome and ventricular volume in children with myelomeningocele treated for hydrocephalus in Uganda. *J Neurosurg Pediatr* **4**(6): 564–570.

Watcham SJ, Schön S, and Christianson AL (2007). Neglect in the care of pregnant South African women of advanced maternal age. *S Afr Med J* **97**(11): 1064, 1068–1069.

Weatherall DJ (2001). Phenotype–genotype relationships in monogenic disease: lessons from the thalassaemias. *Nat Rev Genet* 2(4): 245–255.

Weatherall DJ (2005). The global problem of genetic disease. *Ann Hum Biol* 32(2): 117–122.

Weatherall DJ (2006). Inherited disorders of hemoglobin. In Jamison DT, Breman JG, Measham AR, et al., eds., *Disease Control Priorities in Developing Countries*, 2nd edn. Oxford University Press, pp. 663–680.

Weatherall DJ (2008). Hemoglobinopathies worldwide: present and future. *Curr Mol Med* 8(7): 592–599.

Weatherall DJ (2010). The inherited diseases of hemoglobin are an emerging global health burden. *Blood* 115(22): 4331–4336.

Whitley R, Arvin A, Prober C, et al. (1991). Predictors of morbidity and mortality in neonates with herpes simplex virus infections. The National Institute of Allergy and Infectious Diseases Collaborative Antiviral Study Group. *N Engl J Med* 324(7): 450–454.

WHO (1996). *Control of Hereditary Diseases*. WHO Technical Report Series 865. WHO.

WHO (2002a). *Global Strategies to Reduce the Health Care Burden of Craniofacial Anomalies*. WHO.

WHO (2002b). *Report of a Meeting on Preventing Congenital Rubella Syndrome: Immunization Strategies, Surveillance Needs*. WHO.

WHO (2005). *Control of Genetic Diseases. Report 2005*. Available at: http://apps.who.int/gb/ebwha/pdf_files/EB116/B116_3-en.pdf.

WHO and the World Alliance for the Prevention of Birth Defects (1999). *Services for the Prevention and Management of Genetic Disorders and Birth Defects in Developing Countries*. WHO.

Wierenga KJJ, Hambleton IR, and Lewis NA (2001). Survival estimates for patients with homozygous sickle-cell disease in Jamaica: a clinic-based population study. *Lancet* 357: 680–683.

Williams TN, Mwangi TW, Wambua S, et al. (2005). Negative epistasis between the malaria-protective effects of alpha+-thalassemia and the sickle cell trait. *Nat Genet* 37(11): 1253–1257.

Williams TN, Uyoga S, Macharia A, et al. (2009). Bacteraemia in Kenyan children with sickle-cell anaemia: a retrospective cohort and case–control study. *Lancet* 374(9698): 1364–1370.

Winkler AS, Kerschbaumsteiner K, Stelzhammer B, et al. (2009). Prevalence, incidence, and clinical characteristics of epilepsy: a community-based door-to-door study in northern Tanzania. *Epilepsia* 50(10): 2310–2313.

Wonkam A (2009). Phenotype, burden and perspectives of sickle cell anaemia studies in sub-Saharan Africa: the case of Cameroon. In *ICBD Proceedings*, October 4, 2009, New Delhi, India.

Wonkam A and Hurst S (2014). A call for policy action in Sub-Saharan Africa to rethink diagnostics for pregnancy affected by sickle cell disease: differential views of medical doctors, parents and adult patients predict value conflicts in Cameroon. *OMICS*. DOI: 10.1089/omi.2013.0167.

Wonkam A and Mayosi BM (2014). Genomic medicine in Africa: promise, problems and prospects. *Genome Medicine* 6: 11.

Wonkam A, Njamnshi AK, and Angwafo III FF (2006). Knowledge and attitudes concerning medical genetics amongst physicians and medical students in Cameroon (sub-Saharan Africa). *Genet Med* 8: 331–338.

Wonkam A, Muna RR, Rotimi C, and Newport M (2010). Capacity-building in human genetics for developing countries: initiatives and perspectives in sub-Saharan Africa. *Public Health Genomics* 13(7–8): 492–494.

Wonkam A, Kenfack MA, Muna WF, and Ouwe-Missi-Oukem-Boyer O (2011). Ethics of human genetic studies in sub-Saharan Africa: the case of Cameroon through a bibliometric analysis. *Dev World Bioeth* 11(3): 120–127.

Wonkam A, Ponde C, and Nicholson N, et al. (2012). The burden of sickle cell disease in Cape Town. *S Afr Med J* 102(9): 752–754.

Wonkam A, Mba CZ, Mbanya D, et al. (2013a). Psychosocial burden of sickle cell disease on parents with an affected child in Cameroon. *J Genet Couns*. DOI: 10.1007/s10897-013-9630-2.

Wonkam A, de Vries J, Royal CD, Ramesar R, and Angwafo FF 3rd (2013b). Would you terminate a pregnancy affected by sickle cell disease? Analysis of views of patients in Cameroon. *J Med Ethics*. DOI: 10.1136/medethics-2013-101392.

Wonkam A, Ngo Bitoungui VJ, Vorster AA, et al. (2014). Association of variants at BCL11A and HBS1L-MYB with hemoglobin F and hospitalization rates among sickle cell patients in Cameroon. *Plos One*. DOI: 10.1371/journal.pone.0092506.

Zagré NM, Desplats G, Adou P, Mamadoultaibou A, and Aguayo VM (2007). Prenatal multiple micronutrient supplementation has greater impact on birthweight than supplementation with iron and folic acid: a cluster-randomized, double-blind, controlled programmatic study in rural Niger. *Food Nutr Bull* **28**(3): 317–327.

Zaida F, Sedki A, Laroche D, Lekouch N, and Bouglé D (2006). [High cord blood TSH in Morocco: iatrogenic hypothyroidism?]. *Ann Endocrinol (Paris)* **67**(1): 39–41. In French.

Zeck W and McIntyre HD (2008). Gestational diabetes in rural East Africa: a call to action. *J Womens Health (Larchmt)* **17**(3): 403–411.

13 Neurogenetic Disorders in Africa: Hereditary Spastic Paraplegia

A Case Study

Liena E. O. Elsayed, Ammar E. M. Ahmed, and Giovanni Stevanin

Many regions of Africa, the continent where the sun is always shining, are gradually becoming overwhelmed by genetic disorders, especially where consanguineous marriages are strongly favored over many generations. In this chapter we navigate through the genetic and phenotypic features observed in some African populations, using hereditary spastic paraplegia (HSP) as a model of neurogenetic disorders. Published data and unpublished studies on HSP are starting to disclose criteria that distinguish these populations due to an interesting mixture of environmental factors, traditions, and certain characteristics of African genomes.

Hereditary spastic paraplegias are a group of neurodegenerative disorders with the principal defining clinical features of insidiously progressive weakness and spasticity of the lower extremities. The prevalence of autosomal dominant (AD) HSP ranges between 0.5 and 5.5/100,000 and autosomal recessive (AR) HSP from 0.3 to 5.3/100,000 (Ruano et al. 2014). Hereditary spastic paraplegia shows remarkable genetic and clinical heterogeneity. There are mutations in more than 60 genes that can be transmitted via all modes of inheritance, including AD, AR, X-linked (XL), mitochondrial/maternal (Mito), and mixed inheritance (AD/AR). The clinical presentation can be pure or complex if additional neurological or non-neurological clinical features such as ataxia, optic atrophy, extrapyramidal features, cataract, and more, are absent or present (Figure 13.1).

The surge of information over the last decade that followed the completion of the human genome project has revolutionized genetic studies in many fields of medicine. Neurogenetics is not an exception. The advent of next-generation sequencing (NGS) techniques has contributed the greatest amount to this genetic revolution. However, the amount of literature on genetics, especially on the genetic basis of neurological diseases in Africa, does not match the size of the increasing problem of genetic (in general) and neurogenetic disorders (in particular). The published data vary from country to country and between various disorders. Overall, the paucity of data may be attributed to the lack of trained staff and well-equipped up-to-date neurogenetics research units in most African countries.

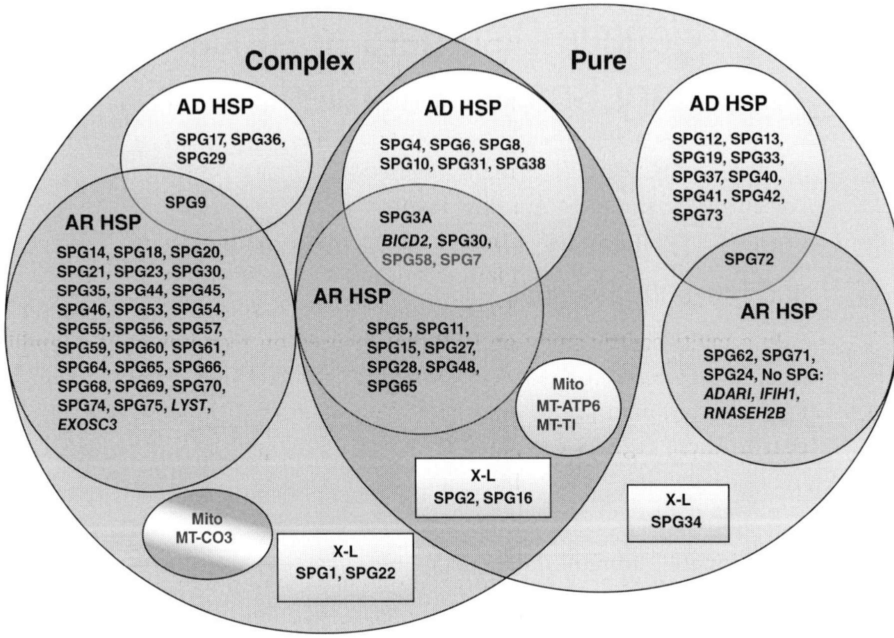

Figure 13.1 Correlation between the modes of inheritance and the clinical phenotype (pure or complex). Various patterns of inheritance: AD, AR, mitochondrial, XL, and mixed inheritance.

Hereditary spastic paraplegia has been studied relatively well in Tunisia, also with a few studies in Morocco, Algeria, and Egypt. In Sudan, it has remained almost completely unexplored from a genetic point of view (as is the case in many other sub-Saharan African countries) despite the immense amount of clinical work-up that has been conducted. At present, there is accumulating data about genetics of HSP in Sudanese patients, with a large familial cohort being studied (Elsayed et al. 2016b). This cohort will be utilized throughout this chapter to highlight the potential for conducting neurogenetic research in developing countries in the setting of limited resources and minimal experience in the field.

Sudan geographically links the Arab Peninsula, in the east, to the western part of Africa. It also represents the northern region of sub-Saharan Africa and is just to the south of the Mediterranean basin. It is inhabited by ethnically and linguistically distinct populations (Dobon et al. 2015; Salih 2010). Its location in East Africa, which is known for the high genetic diversity among its population, makes it a strategic region for genetic studies (Dobon et al. 2015). Phylogenetic analysis has shown that East Africans possess more ancestral lineages in comparison to various continental populations, placing them at the root of the human evolutionary tree (Elhassan et al. 2014).

In addition, being the main pathway for trading and pilgrim caravans with the consequent settling of travelers at some point in recent history has resulted in an interesting mixture of ethnic groups with an immense genetic heterogeneity, unique to the Sudanese population.

13.1 Consanguinity

The consanguinity rate is high among certain local African communities. First-degree consanguineous parents are frequently noted in communities with Arabic origins, where religion and tradition allow and often promote these marriages. In some African tribes, including in South Sudan, close consanguineous marriages are considered taboo, although they favor marriages within the same tribe. In Tunisia, consanguinity rate ranges from 61 percent in an HSP cohort to a maximum of 78 percent in a cohort of Pierre-Marie's heredo-ataxia, despite a much lower rate (25 percent) in the general Tunisian population (Ben Hamida et al. 1986).

In a multi-centric study on HSP that focused on recruitment of a familial cohort from Middle Eastern countries, and which included three African countries (Libya, Egypt, and Morocco), consanguinity rate was estimated collectively around 50 percent in these regions compared to a rate of <1 percent in the USA and Western Europe (Novarino et al. 2014).

Although the consanguinity is estimated to occur in around 50 percent of the general Sudanese population (Bittles and Black 2010; Romeo and Bittles 2014), a much higher rate of 90 percent (the highest reported in Africa) is observed in a Sudanese HSP cohort of extended and nuclear families (Elsayed et al. 2016b). In these families, even when non-consanguinity is claimed, it is seldom completely true, as parents may not be related but consanguinity can exist in both maternal/paternal branches separately. Interestingly, when consanguinity is not reported, a distant tribal consanguineous ancestral relation can often exist.

13.2 Mode of Inheritance

The predominant mode of inheritance of HSP is AR. The high prevalence of AR forms is mostly the result of high inbreeding. Nevertheless, founder mutations cannot be ruled out in these small groups, which splintered off from the original population with recurrent consanguineous marriages in subsequent generations. In a study of HSP in Sfax (Tunisia), the mode of transmission was almost exclusively via AR inheritance (97 percent) (Boukhris et al. 2009). In the recent study in Sudan, frank AR inheritance is almost two-thirds (66 percent), and can reach up to 90 percent if we include the sporadic cases (24 percent). Rarely, different branches of the same family show different patterns of transmission. It is noteworthy that in consanguineous families with AD HSP, it is often difficult to exclude pseudo-AD transmission patterns and digenic inheritance.

13.3 Tendency for Highly Complex Clinical Presentation

Autosomal recessive HSP has a tendency for complicated clinical presentations. This fact is highlighted in the few HSP studies conducted in African populations. In a Tunisian cohort, complicated HSP forms were more prevalent (69 percent) than the pure forms (Boukhris et al. 2009). An even higher rate (90 percent) of the studied

Sudanese HSP families showed phenotypes complicated in the majority of families with at least two additional clinical signs (in addition to the pyramidal features), which makes high levels of complexity the norm. Complicated HSP with minimal complexity (one additional complicating feature only) was found in 5 percent and non-complicated (pure) HSP constituted only 10 percent of the clinical phenotypes of HSP in these families. On the contrary, in a cohort of Algerian and Portuguese with AR inheritance, pure HSP presentation (58.7 percent) slightly predominated over the complex phenotype, which was observed in 41.3 percent of the families (Coutinho et al. 1999), which matches the pattern of AD HSP despite the fact that all families showed AR inheritance. Of note, this study was done before MRI was included in the routine clinical diagnosis and therefore pure forms might have been overestimated.

13.4 Expanding Phenotypes of Genetic Disorders

Examples have been shown in which specific HSP phenotypes have been expanded by studying additional patients. This is the case for SPG11 in a Tunisian cohort (Boukhris et al. 2009). Sudanese families are not an exception and Sudanese SPG11 patients present with slightly different phenotypes, clinical course, and age at onset. Data from Sudan have also broadened the phenotypic spectrum associated with other known HSP genes, such as SPG57. The second SPG57 family reported worldwide was Sudanese and presented a phenotype distinct from the one reported in the first SPG57 kindred (Elsayed et al. 2016b). This phenomenon is also observed in other non-HSP neurogenetic disorders: DNAJC6 (juvenile onset Parkinson disease; JOPD) (Elsayed et al. 2016a); SIL1 (Marinesco Sjogren syndrome; MSS); and ARG1 (hyperargininemia) (unpublished data).

13.5 One Family, Multiple Genetic Disorders

The Sudanese study shows some other unique features that merit a closer look in order to analyze the underlying cause.

Multiple coexisting clinical and/or genetically distinct phenotypes are the rule in many families (Figure 13.2). This further complicates the situation (Elsayed et al. 2016b). This rare phenomenon apparently occurs in communities with extremely elevated consanguinity levels, though unfortunately sufficient literature is not yet available.

This phenomenon of intra-familial heterogeneity significantly impacts the analysis of genetic data, which has to be approached with caution, especially when dealing with NGS data. It is particularly difficult to analyze data when the candidate genes in these families are novel or when the family (or branch of the family) under question is sporadic (Elsayed et al. 2016b).

13.6 Novel Genetic Entities

Despite the scanty number of studies conducted, many HSP genes were initially identified in some African populations, especially in the last decade.

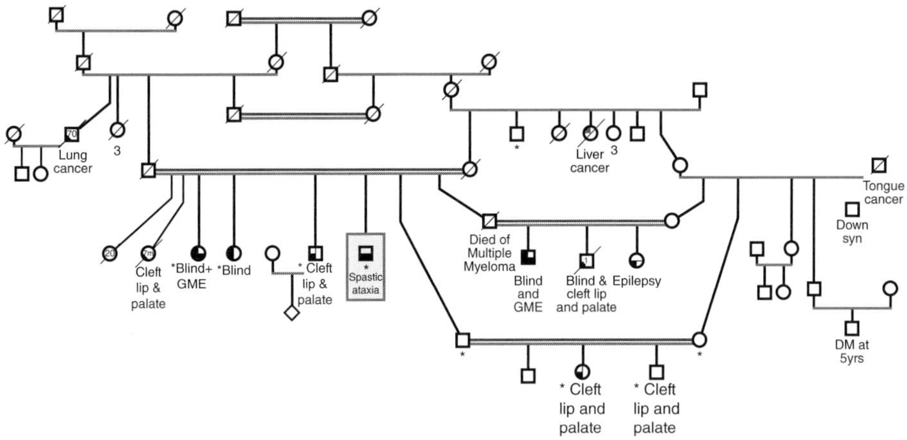

Figure 13.2 Pedigree of family F21 illustrating the high level of consanguinity. Pedigree symbols: * sampled individuals; phenotype symbols: half-fill black color: spastic ataxia individual, other filling patterns correspond to various clinical phenotypes in the family. Others are standard medical pedigree symbols.

A non-exhaustive list includes some of the most common AR HSP genes: *SPG11*, *SPG15*, *SPG5*, *SPG46*, *SPG30* (Hanein et al. 2008; Klebe et al. 2015; Martin et al. 2013; Novarino et al. 2014; Stevanin et al. 2007). The relative ease of finding a homozygous variant with resulting AR inheritance through analysis of NGS data is one of the main factors that may have contributed to these results. The particular nature of the African genome offers its specific effect through expansion of both genotypic and phenotypic manifestations of both known and newly identified genes. However, this is not always easy, especially when the AR disease is due to compound heterozygous variants, and is even harder when some regions of the exome are not well covered (e.g. in the whole-exome sequencing), or when one variant is located in a non-coding region. When there is too much consanguinity, the high possibility of multiple genetic disorders in the same family and the elevated probability of oligogenic causation make the situation rather chaotic.

13.7 Neurogenetic Diseases, a Continuum of Disorders, a Concept Further Highlighted

In the cohort from Sudan, we can easily spot that it further highlights the concept of neurogenetic disorders being a continuum due to the multiple overlaps with other neurological disorders in which spastic paraplegia/quadriplegia is present.

This concept of neurodegeneration being a continuum of disorders has been illustrated through the literature and history of HSP by two facts:

1. Many clinico-genetic HSP entities are associated with other neurological signs that can be the major clinical features of overlapping diseases. For example, hereditary ataxias (HAs) represent the other extreme of the spectrum of

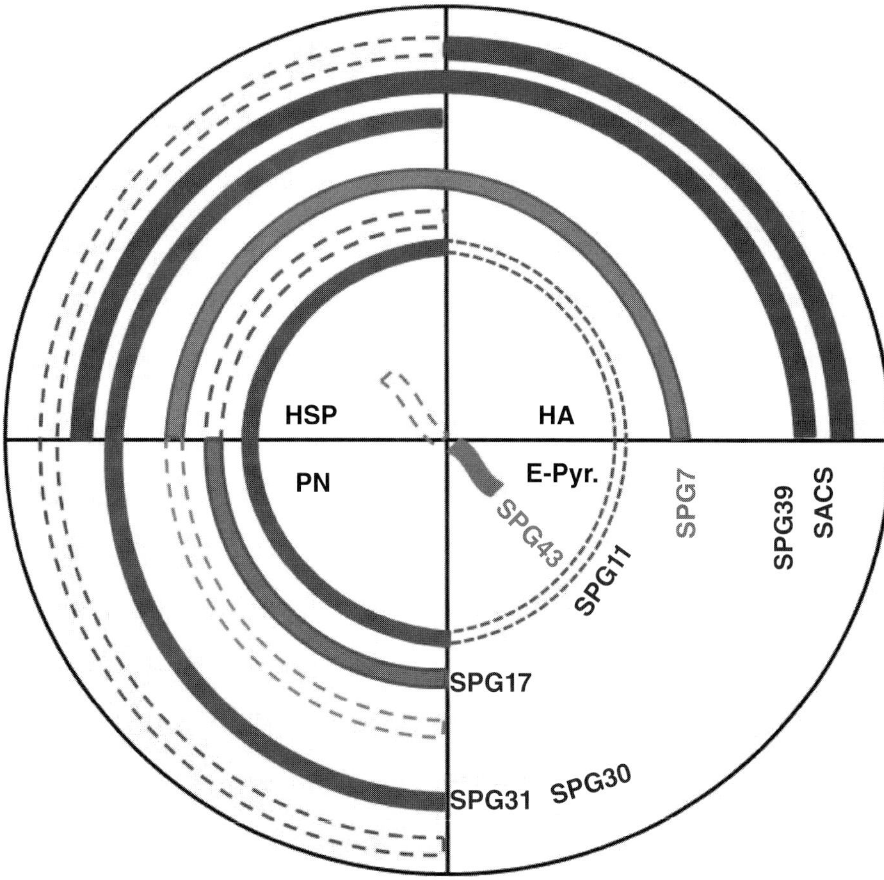

Figure 13.3 Schematic illustration of the phenomenon of genes presenting with closely related allelic phenotypes throughout the spectrum of the neurodegenerative disorders. HSP: hereditary spastic paraplegia; HA: hereditary ataxia; PN: peripheral neuropathy; and E-Pyr: extrapyramidal syndrome. Opaque curves stand for clinical presentation as pure category (HSP, HA, PN, E-Pyr). Interrupted lines imply that the gene manifests the clinical category only in complex form.

spinocerebellar neurodegenerative disorders. Hereditary ataxias may sometimes be difficult to differentiate clinically from HSPs as cerebellar manifestations are present in more than 30 clinico-genetic SPG entities.

2. Mutations in multiple SPG genes can account for HSP but also for allelic phenotypes. This is the case for motor neuron disease, where *ALS2* is a major cause of amyotrophic lateral sclerosis but few HSP families have also been reported as mutated in this gene, including in Sudan. The same applies for other neurogenetic disorders. Despite the long list of genes that manifest this phenomenon, only some examples of genes detected in Sudanese families that have emphasized this concept are highlighted in Figure 13.3.

13.8 Challenges, Milestones, and Particularities of the Neurogenetic Study in Developing Countries: Sudan as an Example

13.8.1 Cohort Building

In developing countries, the first step is cohort building. This comprises the building of a network of specialists, a referral system, a trained neurogenetic team, and an optimized clinical examination/sampling method.

Continuous Optimization of the Phenotyping Process

Questionnaire

One of the first steps is to adopt a clinical chart adapted to local needs, given the nature of the cohort, more often of pediatric origin.

Video Documentation

A second step is to optimize a video documentation system to create a video archive for the major and minor clinical signs that are encountered during recruitment. This is especially important for diagnosis establishment by the second viewer in order to eliminate alternate disorders. This has many advantages, including increasing local expertise and training of new members to ensure they are more efficient and accurate when they perform clinical assessments.

National Collaborations with Experts in Related Fields

Another critical step is to create a network for further collaborations with experts in the fields of neuroradiology, ophthalmology, and neurophysiology and psychiatry. In particular, the four main problems that are encountered for MRI in African countries, particularly in Sudan, making MRI unavailable are: (1) financial restrictions; (2) anesthesia staff unavailable in certain imaging centers, which can be critical for young children or developmentally disabled patients; (3) patients far from equipped hospitals and unable to travel; and (4) poor conservation of images/films. Central databases and help from social insurance/donations are then required.

13.8.2 Diagnostic Capacities

Most important is building a diagnostic approach through cooperation with European clinical neurogeneticist experts and to transfer this knowledge into practice in Sudan.

The formation of a neurogenetic team in Sudan was a great milestone and a real leap in neurogenetics in Sudan. Teaming up with international cooperative colleagues had the best effect on the yield. This reinforces the notion that the best results are achieved via team work.

13.8.3 Budget Management

Research project finance is always an obstacle in developing countries. This includes the expenses for sampling sessions (gloves, sterilizers, needles, etc.) as well as providing travel/transportation funds to poor families to allow them to visit the sampling center (and to the team members in case the reverse [home visits] occurs) in the peripheral regions of the country. It also includes the costs of sample shipment and customs, as well as the costs for investigations for the poor, uninsured patients (imaging, electrophysiological studies, and specialist consultation [ophthalmologists, audiologists]). While the latter can be partially covered by building collaborations, the former has to be provided from the researcher's personal wages and familial support.

There is no specific financial support for international collaborative projects involving African countries. This has consequences for the strategy, which should focus on the most informative families with high chances of success.

13.8.4 WES Analysis

In addition to the difficulties that are usually encountered in analyzing WES data for prioritization of variants and genes, WES data analysis has to be adapted to the particular nature of the African families. As previously stated, one of the most important criteria that has to be considered is the probability of occurrence of two clinical entities in the same family. This probability increases when analyzing patients belonging to two branches of the same family, or even when analyzing data of affected siblings. In fact, a causative mutation rarely segregates in more than one branch of extended families even in the presence of related phenotypic presentation. The lack of satisfactory reference genomic data about most African populations (Ibrahim and Musa 2016), even in countries where genetic studies are ongoing, imposes certain challenges when trying to analyze NGS data (including WES data). It was quite evident on analysis of data from Sudan or North African countries that variants encountered were unique to the analyzed population.

Despite all the challenges that faced the Sudanese HSP study, the success rate of the WES approach was 57 percent. The cause was a new putative gene in 36 percent, whereas in 21 percent it was a known gene. These rates could be compared to the 42 percent and 33 percent success rates for novel and known genes, respectively, of the study conducted by Novarino et al. (2014), which interestingly included families from the three African countries discussed earlier (Morocco, Egypt, and Libya).

13.9 Take Home Message

It is now obvious that an African consortium is a necessity to enhance data collection in a collaborative way in order to locate the many missing parts of genetic studies in Africa. This will be an enormous help, not only to Africans, but to the international scientific community to solve the dilemma of genetic diseases in general and neurogenetic disorders in particular.

References

Ben Hamida M, Chaabouni H, Madani S, et al. (1986). Genetic study of spinocerebellar hereditary degenerations in Tunisia: role of consanguinity in their occurrence. *J Genet Hum* **34**: 267–274.

Bittles AH and Black ML (2010). Consanguinity, human evolution, and complex diseases. *PNAS* **107**: 1779–1786.

Boukhris A, Stevanin G, Feki I, et al. (2009). Tunisian hereditary spastic paraplegias: clinical variability supported by genetic heterogeneity. *Clin Genet* **75**: 527–536.

Coutinho P, Barros J, Zemmouri R, et al. (1999). Clinical heterogeneity of autosomal recessive spastic paraplegias: analysis of 106 patients in 46 families. *Arch Neurol* **56**: 943–949.

Dobon B, Hassan HY, Laayouni H, et al. (2015). The genetics of East African populations: a Nilo-Saharan component in the African genetic landscape. *Scientific Reports* **5**: 9996.

Elhassan N, Gebremeskel EI, Elnour MA, et al. (2014). The episode of genetic drift defining the migration of humans out of Africa is derived from a large East African population size. *PLoS One* **9**: e97674.

Elsayed LEO, Drouet V, Usenko T, et al. (2016a). A novel nonsense mutation in DNAJC6 expands the phenotype of autosomal-recessive juvenile-onset Parkinson's disease. *Ann Neurol* **79**: 335–337.

Elsayed LEO, Mohammed IN, Hamed AAA, et al. (2016b). Hereditary spastic paraplegias: identification of a novel SPG57 variant affecting TFG oligomerization and description of HSP subtypes in Sudan. *Eur J Hum Genet* **25**: 100–110.

Hanein S, Martin E, Boukhris A, et al. (2008). Identification of the SPG15 gene, encoding spastizin, as a frequent cause of complicated autosomal-recessive spastic paraplegia, including Kjellin syndrome. *Am J Hum Genet* **82**: 992–1002.

Ibrahim M and Musa M (2016). Effective size and effectiveness: next generation sequencing and the practice of genomics in Africa. *Next Generat Sequenc Applic.* DOI: 10.4172/2469-9853.S1-008.

Klebe S, Stevanin G, and Depienne C (2015). Clinical and genetic heterogeneity in hereditary spastic paraplegias: from SPG1 to SPG72 and still counting. *Rev Neurol (Paris)* **171**: 505–530.

Martin E, Schule R, Smets K, et al. (2013). Loss of function of glucocerebrosidase GBA2 is responsible for motor neuron defects in hereditary spastic paraplegia. *Am J Hum Genet* **92**: 238–244.

Novarino G, Fenstermaker AG, Zaki MS, et al. (2014). Exome sequencing links corticospinal motor neuron disease to common neurodegenerative disorders. *Science* **343**: 506–511.

Romeo G and Bittles AH (2014). Consanguinity in the contemporary world. *Hum Hered* **77**: 6–9.

Ruano L, Melo C, Silva MC, and Coutinho P (2014). The global epidemiology of hereditary ataxia and spastic paraplegia: a systematic review of prevalence studies. *Neuroepidemiology* **42**: 174–183.

Salih MA (2010). Genetic disorders in Sudan. In Teebi AS, ed., *Genetic Disorders Among Arab Populations*. Springer, pp. 575–612.

Stevanin G, Santorelli FM, Azzedine H, et al. (2007). Mutations in *SPG11*, encoding spatacsin, are a major cause of spastic paraplegia with thin corpus callosum. *Nat Genet* **39**: 366–372.

14 Enabling Genomic Revolution in Africa

History and Impact of the African Society of Human Genetics and the Human Heredity and Health in Africa Initiative

Charles N. Rotimi

The African Society of Human Genetics (AfSHG) was established in 2003 after considerable consultation with African and international scientists interested in human genetics, genomics, and biomedical research on the African continent. Following these deliberations, including a satellite meeting held during the 2001 American Society of Human Genetics (ASHG) annual meeting in San Diego, California, it became clear that no other society, organization, or conference specifically addresses the critical needs for African researchers and institutions interested in genomics research and capacity development to enable full participation of African scientists and students in the ongoing global genomic revolution. The AfSHG was subsequently established with the aim of equipping the African scientific community and policy-makers with the information and practical knowledge they need to contribute to the field of genomics research and to attract global attention to the efforts of African scientists. It was decided that membership in the AfSHG will be open to any individual who is interested in issues of human genetics in Africa, including but not limited to issues of education, clinical care, research, public health, ethics, law, and policy. The AfSHG membership agreed that a major objective of the society will be to provide a forum for scientists interested in the broad enterprise of human genetics in Africa to meet, interact, network, and collaborate. The society has achieved, and continues to achieve, this objective through a series of annual workshops and meetings held in different African countries. To date, annual meetings have been held in Accra, Ghana (2003 and 2013), Washington, DC, USA (2004), Johannesburg, South Africa (2005), Addis Abba, Ethiopia (2006), Cairo, Egypt (2007 and 2017), Yaoundé, Cameroon (2009), Cape Town, South Africa (2011), Senegal (2016), and Kigali, Rwanda (2018); this year's meeting is planned to take place in Bamako, Mali (2019). The objectives of the society are also advanced through direct mailing to members of the society, specialized workshops, website,[1] and by its membership in and affiliation with international genetic organizations, including the International Federation of Human Genetics, the Human Variome Project Consortium, Human Genome Organization (HUGO), the Southern African Society of Human Genetics, and the ASHG. Consistent with its mission to enhance human genetics across the African continent, the AfSHG has facilitated the establishment of six national societies of human genetics in Mali, Cameroon, Senegal, the

[1] www.afshg.org.

Democratic Republic of the Congo, Rwanda, and Egypt. A complete list of member societies is available on the website.

Charles Rotimi was elected the AfSHG's Founding President and served as its president for over a decade until he stepped down in 2014, after providing scientific leadership for the successful establishment of the largest and most comprehensive ground-breaking genomic project on the African continent – the "Human Heredity and Health in Africa (H3Africa) Initiative" (see below for more details). Through its early years, the society was guided by a group of very dedicated African and international scientists who generously volunteered their time, efforts, and sometimes personal funds. Some of these individuals include Melanie Newport, Scott Williams, Adebowale Adeyemo, Branwen Hennig, Clement Adebamowo, Muntaser Ibrahim, Raj Ramesar, Giorgio Sirugo, Felix Okojie, Ambroise Wonkam, Georgia Dunston, Charmaine Royal, Himla Soodyall, Patricia Marshall, Sonia Abdelhak, Guida Landoure, Shirley Freeman, Sir Walter Bodmer, David Bentley, Myles Axton, Yehia Gad, and Amal M. Mohamed, to name a few. Professor Michele Ramsay, who served until 2019, was elected president following Rotimi stepping down in 2014. The president elect (2019) of the society is Professor Ambroise Wonkam. Other members of the executive committee include Raj Ramesar (Treasurer – South Africa), Adebowale Adeyemo (Communication and IT – Nigeria/USA), Rokhaya Ndiaye (Membership Secretary – Senegal), and Fasil Tekola-Ayele (Social Media – Ethiopia/USA). Current board members are Sonia Abdelhak (Tunisia), Clement Adebamowo (Nigeria/USA), Amadou Gaye (Senegal/USA), Melanie Newport (UK), Charles Rotimi (Nigeria/USA), and Scott Williams (USA). National society representatives in the AfSHG are Ambroise Wonkam (Cameroon), Prosper Lukusa (Democratic Republic of the Congo), Amal Mohamed (Egypt), Guida Landoure (Mali), Leon Mutesa (Rwanda), Rokhaya Ndiaye (Senegal), and Cheikh Anta Diop (South Africa).

Over the years, the society has been generously supported by several international not-for-profit and for-profit organizations, including the US National Institutes of Health (NIH), the UK-based Wellcome Trust Private Foundation, the International Centre for Genetic Engineering and Biotechnology, Affymetrix, Illumina, deCODe Genetics, Applied Biosystems, Sequenome, *Nature Genetics*, *PLoS Genetics*, and the Galton Institute, among others.

The inaugural conference of the AfSHG was held on December 8–9, 2003 in Accra, Ghana followed by a high-profile publication in *Nature Genetics* in June 2004 (Rotimi 2004). The grant that supported this inaugural meeting was written in the basement of Charles Rotimi's home, with Clement Adebamowo and Adebowale Adeyemo. The theme for the inaugural meeting was "Biomedical Research in Africa with Emphasis on Genetics." During this meeting, the importance of building a strong foundation for genetics and genomics in Africa was discussed by scientists from Africa and abroad. Genomic variation in the African diaspora was explored with attention to understanding the differential distribution of diseases such as malaria, tuberculosis, AIDS, diabetes, hypertension, and cancer among population groups. Challenges surrounding the design and implementation of biobanks, bioinformatics, and large-scale genetic epidemiology studies for communicable and non-communicable diseases in Africa and the African diaspora were critically assessed. Ethical issues surrounding voluntary participation and informed consent in genetic research conducted in African settings and the status of bioethics in medical education were discussed (Rotimi 2004). Given that the fundamental goals of the AfSHG are biomedical research capacity building, networking, and training, the society continued the exploration of these issues in future meetings and through advocacy across the African continent and internationally.

By successfully engaging the global genomic/genetic enterprise, the AfSHG brought continent-wide focus and excitement to human genetics and biomedical research in general. The society paid attention to the engagement of students, post-doctoral researchers, and early-career investigators in genomics and other aspects of biomedical research. For example, with the leadership of Branwen Hennig, special awards were created to recognize young scientists during the annual meeting. The AfSHG Young Researcher's Forum (YRF) was established, with the inaugural meeting held on March 5, 2011 at the Faculty of Health Sciences, University of Cape Town, South Africa. This YRF meeting included a full day of presentations and interactions between the young researchers themselves and between the young researchers and their national and international mentors. The full-day presentations provided the young researchers with a platform to showcase their research and to seek career development guidance. The YRF meeting was a resounding success, with 154 attendees from ten countries worldwide, including Cameroon, South Africa, Egypt, Zimbabwe, the UK, Italy, the USA, Kenya, Sudan, and Malawi. In addition to the young researchers' presentations, three distinguished speakers (Dr. Roger Glass, NIH; Dr. Myles Axton, *Nature Genetics*; and Dr. Charles Rotimi, NIH) gave keynote addresses with a focus on career development in biomedical research. The organizing committee of the YRF was co-chaired by Aisha Pandor and Kusha Kalideen. Although strongly supported by the entire membership of the AfSHG, Professor Raj Ramesar was the inspirational leader of the organizing committee of the YRF. The success story of the YRF continued, as shown during the 2016 and 2018 annual AfSHG meetings in Dakar, Senegal and Kigali, Rwanda respectively, where a day was devoted to presentation and mentoring of students and early-career scientists.

Another example of a very successful training workshop organized and supported by AfSHG members was one held in Khartoum, Sudan, on January 19–24, 2008. This training workshop, entitled "Molecular Medicine and Genomics in Africa," which was co-directed by Professor Muntaser Ibrahim of the University of Khartoum and Professor Francisco Baralle of the International Centre for Genetic Engineering and Biotechnology, Trieste, Italy, attracted over 30 course attendees and several national and international expert instructors (e.g., Professors Melanie Newport, Raj Ramesar, and Charles Rotimi), most of whom are members of the AfSHG.

The AfSHG has also impacted genomics on the African continent by helping to strengthen and bringing focus to existing genetic and non-genetic professional societies and activities. This goal has been achieved by organizing joint meetings with African and international professional organizations. Examples include annual meetings with the Ethiopian Medical Society, the African Society of Bioinformatics and Computational Biology, the International Society of Computational Biology, and the H3Africa Consortium. The society also organizes satellite meetings during its annual conferences and examples include genetic counseling, neurogenetics, and Affymetrix medical genomics workshops, which took place during the annual conference of the AfSHG that was jointly organized with the Southern Society of Human Genetics in 2011 in Cape Town, South Africa. Other supported meetings include the Keystone Symposium: Leveraging Genomic Diversity to Promote Animal and Human Health in November 2018 in Kampala, Uganda; and the Wellcome Genome Campus Advanced Courses and Scientific Conferences, in collaboration with the AfSHG NGS Analysis for Monogenic Diseases in African Populations in September 2018 in Kigali, Rwanda. Recently (April 2019), the AfSHG achieved a major milestone with the launch of a journal portal for the AfSHG in collaboration with F1000 on the African Academy of Sciences Open Research Platform.[2] This is a notable achievement of the Michele Ramsay presidency.

14.1 The Critical Role Played by Members of the AfSHG in the Establishment of the H3Africa Initiative

As the benefits of the Human Genome Project and the resulting genomic revolution accrued, members of the AfSHG became increasingly concerned that Africa and its people were not well positioned to contribute to this global effort and, more importantly, to take advantage of the many genomics-driven breakthroughs, including a fundamental understanding of the molecular basis of health and disease, and developing future biomedical technology and pharmaceuticals. AfSHG members believed strongly that, if properly deployed, genomics-driven advances could enable African people and their governments to leap-frog many years of health care and biomedical technological development while creating lasting opportunities to engage and train current and future generations of biomedical researchers. To achieve this goal, the AfSHG membership started to discuss, in its 2006 annual meeting in Addis Ababa, Ethiopia, how to fully engage African people, their scientists and health providers in the development and application of genomic tools to address Africa's problems, including health, food production, and economic development. Finally, due to the minimal participation of Africans in the genomic revolution as scientists and research participants, the AfSHG membership was particularly concerned that the current unacceptable inequality in health and economic development will be widened and that Africa will be left out of the genomic revolution (Newport and Rotimi 2009; Wonkam et al. 2010).

Because of these discussions, the AfSHG membership decided to lead an international effort to develop an ambitious new initiative referred to as the "African Genome Project (AGP)" during its annual meeting in 2007 in Cairo, Egypt. Consistent

[2] https://aasopenresearch.org.

with the AfSHG policy of engaging and enhancing existing African genetic and medical organizations in genomics, the Cairo meeting was a joint congress with the Egyptian National Society of Human Genetics and the Division of Human Genetics and Genome Research at the National Research Council (NRC). The society was privileged to have many eminent scientists from within and outside of Africa during the meeting, including Dr. Francis Collins who gave the keynote address entitled "Africa and the Genome Revolution." The society was particularly pleased to have Dr. Collins' input into the ongoing society-wide discussions on how to best design and implement the AGP, given Dr. Collins' leadership role in the design, implementation, and successful completion of the Human Genome Project and his desire to ensure that the benefits of genomics are shared by all human populations. Notably, Dr. Collins had to return to the USA immediately following his keynote address to receive the Presidential Medal of Freedom, the nation's highest civilian award. The award was presented by President George W. Bush in the East Room of the White House on November 5, 2007.

The Cairo meeting created excellent opportunities for African and international scientists to address the opportunities and challenges of developing the AGP with the goal of supporting large-scale, population-based application of genomics tools to offer novel insight into the causes of diseases that are important to African populations. It was discussed that the AGP will emphasize research, training, and infrastructure capacity development that will lead to improved health for all Africans. A critical point in these deliberations was how best to design the AGP to ensure it captures the vast genetic diversity that exists on the African continent, the birthplace of human populations. The initial agreed-upon design of the AGP project included four major components: population genetics, medical genetics, training, and infrastructure development. The design of the population genetics component of AGP called for a systematic sampling of a minimum of 100 ethnic groups across the continent from north to south and east to west; the medical genetics aspect of AGP will develop large-scale population-based biomedical resources to understand how genes interact with local African environments to increase susceptibility or resistance to diseases. Training and ensuring the full participation of African scientists in the global effort to use genomic tools to solve global health problems represented the third component of the AGP. Infrastructure development, the fourth component of the AGP, was deemed critical if the first components are going to be successful. For example, establishing fully functional laboratories was considered crucial for retention of well-trained scientists on the continent. Furthermore, there was considerable discussion about who to study and what Africa's health priorities are, with an emphasis on not ignoring the growing burden of non-communicable diseases (NCDs) such as hypertension, stroke, diabetes, cancer, and mental health disorders. Issues surrounding who the stakeholders are and how best to engage African populations and their political and economic leaders for full support for the project were also discussed in detail. Following the Cairo meeting and subsequent discussion between Dr. Collins, Dr. Mark Walport (then director of the Wellcome Trust), and Dr. Charles Rotimi (then President of the AfSHG), an international Frontier Meeting was convened to discuss a potential research agenda to study genetic diversity in health and disease in African populations. The two-day meeting (March 15–16, 2009), co-sponsored by the Wellcome Trust and the National Human Genome Research Institute (NHGRI), NIH, was planned to coincide with the sixth annual conference of the AfSHG in Yaoundé, Cameroon (March 13–15, 2009) (Sirugo et al. 2010). The Frontier Meeting brought together scientific experts from several different disciplines to identify the major scientific questions, and the ethical and logistic issues associated with pursuing a large-scale genomic research program on the African continent. The key aim of the meeting was to discuss how to efficiently engage African scientists and other stakeholders in the process of defining research priorities, appropriate study designs, and a careful assessment of existing biomedical infrastructure within Africa that will inform development of regional labs and clinical centers. It was envisioned that these sites will support sample collection, processing, and storage, and serve as training centers for African scientists and students in genomic research. The Frontier Meeting was a resounding success and as a result a high-level oversight working group of six members (Barry Bloom, Kay Davis, Pat Goodwin, Thomas Egwang, Eric Green, and Charles Rotimi) was constituted to provide leadership for the establishment of the project and to maintain the international momentum gained from the Frontier Meeting. The high-level oversight working group made two major decisions in their early deliberations. (1) The renaming of the original AfSHG concept title, "The African Genome Project," to reflect the broader scope of the new thinking behind the pan-African initiative, which included genomic infrastructure development and understanding disease burden and etiology by providing novel insights into the complex interactions between genes and lifestyle/cultural factors operating in the African context. With the initial suggestion from Charles Rotimi, the high-level oversight working group agreed to name the initiative "Human Heredity and Health in Africa" with the acronym H3Africa. (2) The establishment of two working groups, one focused on communicable diseases, led by Professor Sekou F. Traore from the University of Mali's Malaria Research and Training Center, in Bamako, Mali; and the other on NCDs, led by Professor Bongani M. Mayosi from the University of Cape Town, South Africa.

The two working groups were charged with developing a case for a pilot study for consideration by potential funders. The terms of reference for the working groups included (1) a review of relevant current genomic (biomedical) research activity in Africa; (2) identification and justification for studying specific disease(s); (3) identifying leaders in African institutions that could participate in the project; (4) agreeing on research programs and milestones; (5) describing how to facilitate interactions between African researchers and technology providers; (6) agreeing on a timetable and mechanism for developing guidelines for data and sample sharing; (7) considering how to address ethical, legal, and social issues; and (8) considering how to build infrastructure and human capacity in genomics research. The high-level working group charged Charles Rotimi with identifying and recruiting members for the two working groups. After considerable effort and consultation, Charles Rotimi circulated a potential list of African and international scientists who agreed to serve as members of the two working groups. After approval of the membership of the two working groups by the high-level oversight working group, a conference call to communicate the terms of reference for their deliberations was organized by Charles Rotimi and his program coordinator, Emmanuel Peprah. Subsequent activities of the two working groups – including arranging conference calls, travel for group meetings, and the development of the White Paper – were coordinated out of Charles Rotimi's office at the Center for Research on Genomics and Global Health, NHGRI at the NIH. As acknowledged by members of the working groups, Emmanuel Peprah's coordinating and Charles Rotimi's scientific leadership skills were instrumental in the successful completion of the task of the working groups, including the development and communication of the White Paper.[3] The White Paper discussed several aspects of the proposed initiative, including: the scientific background and justification for the H3Africa initiative, collaboration among African scientists, improvement of the infrastructure for genomic research in Africa, methods for facilitating interactions among African geneticists, clinicians, epidemiologists, and bioinformaticians, among others, and the legal, social, and data-sharing issues that are likely to arise during the implementation of this multinational and multi-institutional initiative.

Funding for the H3Africa initiative was announced in a press conference held at the Wellcome Trust Headquarters in London on June 22, 2010, with an initial commitment of about US$38 million from the US NIH and the UK-based Wellcome Trust.[4] Following this press release, which received worldwide attention, the two funding agencies announced separate requests for grant proposals with several stipulations, most notable of which was the requirement that the grant award be made directly to participating African institutions where the principal investigators are affiliated.[5] The H3Africa initiative was designed to enhance the capabilities and capacities of African scientists to use contemporary research approaches to study the genomics and non-genomic determinants of disease and to use this information to improve the health of African populations. To achieve these goals, H3Africa is to develop state-of-the-art genomics-based research infrastructure by leveraging the existing capacity and expertise of African scientists and institutions. The promise and potential of H3Africa were described by several members of the H3Africa leadership team in press releases following the launch of the initiative. According to Dr. Rotimi, "H3Africa will be fundamentally different than previous investments in medical research in Africa. In the past, many research projects took samples from Africa and conducted the studies back in Western labs. H3Africa will build the capacity for African researchers to study African populations to solve African problems." These sentiments were emphasized by Eric Green, the director of NHGRI, who was quoted as saying: "With support from these H3Africa grants, the African scientific community will be empowered to make key breakthroughs on a set of

[3] https://h3africa.org/index.php/about/white-paper.
[4] www.youtube.com/watch?v=G0H8KwmWaAY&t=559s.
[5] www.genome.gov/27540084/2010-release-nih-and-wellcome-trust-announce-partnership-to-support-populationbased-genome-studies-in-africa.

important diseases." Francis Collins, NIH director, said: "H3Africa aims to transform the way science is conducted in Africa, by creating a sustainable research infrastructure and catalyzing the use of advanced genomic technologies to improve our understanding of a variety of diseases... This is particularly relevant because Africa is the original cradle of all humanity, and in this era of expanded global travel and communication, it is becoming increasingly clear that we must think beyond our borders when it comes to understanding human biology and improving health." Sir Mark Walport, director of the Wellcome Trust, said: "If we are to help tackle the growing burden of disease in Africa, it is important that we build capacity within the continent for African researchers and their institutions in order to understand the genetic and environmental causes of illness. The geographical breadth of participating institutions shows that H3Africa is about doing just this, enabling the scientists themselves to drive forward the African research agenda."[6]

Following public request for proposal and review, a set of nine awards were made by the NIH and the Wellcome Trust in the summer of 2012 and the inaugural meeting of the first cohort of H3Africa investigators (the H3Africa Consortium) was held in Addis Ababa in October 2012. This inaugural meeting provided the funded investigators opportunities to interact with each other and members of the two funding agencies. It also provided opportunities to begin discussion on consortium-wide policies and several other issues important to the success of the funded projects. Charles Rotimi, whose vision laid the groundwork for H3Africa, opened the meeting with his presentation describing the history of the H3Africa Initiative. His presentation was followed by Pat Goodwin, who introduced the Wellcome Trust–H3Africa awardees and described how they will continue the legacy of the Wellcome Trust's historic international research portfolio. Following Goodwin was Eric Green, who introduced the NIH–H3Africa awardees and provided a brief description of the NIH, with an emphasis on the NIH Common Fund program that provided the initial funding for H3Africa. With a total budget of over US$74 million, the first phase (2012–2017) of H3Africa funded a total of 25 research projects in 27 countries, including biomedical and bioethical research projects, three pilot biorepositories in Nigeria, Uganda, and South Africa, a pan-African bioinformatics network (H3ABioNet), and the training of several students and investigators.[7]

The H3ABioNet[8] has been particularly successful in creating a pan-African network of bioinformaticians, with 32 different nodes in 15 African countries and 2 nodes in America (Mulder et al. 2016). Each node has a principal investigator who is part of the General Assembly, charged with the responsibility of developing high-level strategic planning for the network, maximizing the synergies between the project components, reviewing node status and progress, and fund-raising. The H3ABioNet has developed core bioinformatics infrastructure, including hardware and human resources, that is facilitating high-throughput genomic science across

[6] www.fic.nih.gov/News/GlobalHealthMatters/Aug2010; https://wellcome.ac.uk/press-release/first-recipients-research-grants-support-genomic-studies-africa-announced.

[7] https://commonfund.nih.gov/globalhealth/h3aresources. [8] www.h3abionet.org.

the African continent and beyond. Bioinformatics tools and solutions developed by the H3ABioNet are tailor-made for the exploitation and interpretation of biological information directly relevant to African people. These analytic resources are empowering the next generation of genomic scientists in Africa to perform state-of-the-art research, and perhaps more importantly are ensuring the retention of bioinformatics skills on the continent. The amazing success of the H3ABioNet was underscored by Dr. Kay Davies, deputy chair of the Wellcome Trust board of governors: "The genomics revolution in Africa is well under way... Networks are going particularly well in bioinformatics. Transformation is what we are producing." The overall success of the H3Africa first phase was described by NIH director Dr. Francis Collins as "Remarkable." In addition, he said of the developed biorepositories: "I do believe H3Africa is creating a critical mass of research capability in the institutions that are involved."

Following a very successful first phase, H3Africa was funded for another five years beginning in fiscal year 2017. This second phase, which will expand upon previous accomplishments, has two unique aspects, an Administrative Coordinating Center and a Bioinformatics Research Training Program. The NIH anticipate spending over US$76 million during the second phase.[9] It is anticipated that the three H3Africa biorepositories will house various types of biomedical specimens obtained from over 80,000 research participants across the African continent. These resources represent a "global good" that promises to facilitate biomedical research on health issues that are important to Africans and the global communities. The resources will also enhance African scientists' ability to compete for research grant funding at the national and international levels, especially for young investigators who would otherwise not have access to such resources. Over these years, the H3Africa research enterprise has been managed and supported by several dedicated individuals: NIH (Drs. Francis Collins, Eric Green, Mark Guyer, Jennifer Troyer, and the various institute directors and program managers); Wellcome Trust (Mark Walport, Jeremy Farrar, Pat Goodwin, Audrey Duncanson, Katherine Littler); and H3Africa Coordinating Centre, University of Cape Town, South Africa (Michelle Skelton), to name a few.

In addition to these wonderful success stories, H3Africa Consortium has created a new African-specific genome-wide association study DNA microarray chip.[10] This new chip, led by the H3ABioNet, contains 2.5 million different genetic markers that were selected for their African relevance; it was released in 2017 by Illumina, the chip manufacturer.[11] Given that humankind originated in Africa before migrating to populate different parts of the world about 100,000 years ago, the benefit of this H3Africa-led project is expected to extend to other global populations. Also, as this chip enables scientists to better tell the genetic story of Africans, it promises to enhance ongoing global efforts to use information coded in our DNA to tell the history of the world and to solve health problems afflicting humanity.

[9] https://commonfund.nih.gov/globalhealth.
[10] www.youtube.com/watch?v=HkiudBP5TgY.
[11] www.h3abionet.org/resources/h3africa-chip.

Perhaps more important, H3Africa is having a significant impact on the scientific, social, and political culture of biomedical research across Africa and internationally. For example, the impressive growth of research in Africa and with African populations spearheaded by H3Africa and similar large-scale projects led to the founding of a new funding entity (the Alliance for Accelerating Excellence in Science in Africa – AESA) by the African Academy of Sciences and the New Partnership for Africa's Development (NEPAD) Agency (Nordling 2015). AESA, which has been endorsed by the heads of states of the African Union, received its seed funding from the Wellcome Trust, the Gates Foundation, and the UK Department for International Development.[12] Notably, the Wellcome Trust will channel its funding for African research, including H3Africa, through AESA. The goal of AESA is to develop science strategies and to fund research in Africa by developing research leadership and promoting scientific excellence and innovation with the goal of overcoming some of Africa's most pressing developmental challenges. Also, AESA is partnering with the India Alliance to solve common health problems including infectious and non-infectious diseases, high rates of maternal and child mortality, and weak health-delivery systems (Jameel et al. 2016). This partnership, which met in September 2016, is discussing the best strategy to utilize the US$100 million Development Fund and $10 million Health Fund that the Indian government pledged to Africa in 2015 with the goal of increasing knowledge generation and innovation in Africa and India (Jameel et al. 2016).

As recently highlighted in a *Nature Genetics Editorial* (2018), there is much to be celebrated about genomic science in Africa because of the excitement and encouragement generated by the AfSHG and the H3Africa initiative. The editorial noted that human genetics is flourishing across the continent with examples of research excellence from Mali in neurogenetics, psychiatric genetics in Rwanda, cardiovascular genetics in Egypt, malaria genomics from Burkina Faso, hematology from Ghana and Tanzania, cancer genomics from Senegal and Nigeria, and bioinformatics from South Africa – as highlighted in the 2018 joint meetings of the AfSHG and H3Africa in Kigali, Rwanda. Another major achievement of the AfSHG is the successful bid to host the next International Congress of Human Genetics[13] in March 2021 in Cape Town, South Africa. Importantly, and with the goal of publishing an issue of the journal devoted to African genetics, this editorial invited pre-submission enquiries for articles, perspectives, and analyses on human, animal, plant, and microbial genetics and genomics from Africa with a stipulation that the corresponding authors live and work in African countries (*Nature Genetics* 2018).

In all, the AfSHG and H3Africa initiatives are a resounding biomedical (genomics) success story with transformative impact on the African continent and globally. With a current budget of over US$170 million, funding from the NIH and the Wellcome Trust, the H3Africa initiative has created pan-continental networks of labs that are applying leading-edge research to the study of genetic and environmental bases of disease and drug responses in Africans (H3Africa Consortium and Rotimi 2014). With

[12] https://aasciences.ac.ke/aesa. [13] www.ifhgs.org.

footprints in over 34 African countries, participation of over 500 African scientists in 48 funded projects and 197 publications, H3Africa has fully engaged the continent and the global scientific communities in genomic science and big data analysis. Notably, the biomedical infrastructure and human capacity being developed by the AfSHG, H3Africa, AESA, and other pan-African initiatives promise to position Africa to take full advantage of the ongoing genomics revolution, including precision medicine (Nordling 2017).

References

H3Africa Consortium and Rotimi C (2014). Research capacity: enabling the genomic revolution in Africa. *Science* 344(6190): 1346–1348.

Jameel S, Kariuki T, and Kay S (2016). Health research: Africa–India health-science partnerships. *Nature* 537(7621): 488.

Mulder NJ, Adebiyi E, Alami R, et al. (2016). H3ABioNet, a sustainable pan-African bioinformatics network for human heredity and health in Africa: H3ABioNet Consortium. *Genome Res* 26(2): 271–277.

Nature Genetics (2018). Editorial: African genetics for human society. *Nature Genetics* 50: 1495.

Newport MJ and Rotimi CN (2009). Reducing the global genomic inequity gap: development of an African genome project. *Public Health Genomics* 12(4): 251–252.

Nordling L (2015). African hub set up to boost research autonomy. *Nature News* April 8.

Nordling L (2017). How the genomics revolution could finally help Africa: new investments promise to get precision medicine and precision public health off the ground. But experts debate how much work needs to be done first. *Nature News* April 5.

Rotimi CN (2004). Inauguration of the African Society of Human Genetics. *Nature Genetics* 36(6): 544.

Sirugo G, Williams SM, Royal CD, et al. (2010). Report on the 6th African Society of Human Genetics (AfSHG) Meeting, March 12–15, 2009, Yaounde, Cameroon. *Am J Trop Med Hyg* 83(2): 226–229.

Wonkam A, Muna W, Ramesar R, Rotimi CN, and Newport MJ (2010). Capacity-building in human genetics for developing countries: initiatives and perspectives in sub-Saharan Africa. *Public Health Genomics* 13(7–8): 492–494.

Index